Communications in Computer and Information Science 2686

Series Editors

Rationale

The CCIS series is devoted to the publication of proceedings of computer science conferences. Its aim is to efficiently disseminate original research results in informatics in printed and electronic form. While the focus is on publication of peer-reviewed full papers presenting mature work, inclusion of reviewed short papers reporting on work in progress is welcome, too. Besides globally relevant meetings with internationally representative program committees guaranteeing a strict peer-reviewing and paper selection process, conferences run by societies or of high regional or national relevance are also considered for publication.

Topics

The topical scope of CCIS spans the entire spectrum of informatics ranging from foundational topics in the theory of computing to information and communications science and technology and a broad variety of interdisciplinary application fields.

Information for Volume Editors and Authors

Publication in CCIS is free of charge. No royalties are paid, however, we offer registered conference participants temporary free access to the online version of the conference proceedings on SpringerLink (http://link.springer.com) by means of an http referrer from the conference website and/or a number of complimentary printed copies, as specified in the official acceptance email of the event.

CCIS proceedings can be published in time for distribution at conferences or as post-proceedings, and delivered in the form of printed books and/or electronically as USBs and/or e-content licenses for accessing proceedings at SpringerLink. Furthermore, CCIS proceedings are included in the CCIS electronic book series hosted in the SpringerLink digital library at http://link.springer.com/bookseries/7899. Conferences publishing in CCIS are allowed to use our online conference service (Meteor) for managing the whole proceedings lifecycle (from submission and reviewing to preparing for publication) free of charge.

Publication process

The language of publication is exclusively English. Authors publishing in CCIS have to sign the Springer CCIS copyright transfer form, however, they are free to use their material published in CCIS for substantially changed, more elaborate subsequent publications elsewhere. For the preparation of the camera-ready papers/files, authors have to strictly adhere to the Springer CCIS Authors' Instructions and are strongly encouraged to use the CCIS LaTeX style files or templates.

Abstracting/Indexing

CCIS is abstracted/indexed in DBLP, Google Scholar, EI-Compendex, Mathematical Reviews, SCImago, Scopus. CCIS volumes are also submitted for the inclusion in ISI Proceedings.

How to start

To start the evaluation of your proposal for inclusion in the CCIS series, please send an e-mail to ccis@springer.com

Suparna Dhar · Sanjay Goswami ·
Dinesh Kumar Unni Krishnan · Indranil Bose ·
Rameshwar Dubey · Chandan Mazumdar ·
Xi Chen
Editors

Human-centered Data Analytics: Technology for Sustainable Development

Third Analytics Global Conference, AGC 2025
Kolkata, India, March 8–9, 2025
Revised Selected Papers

Springer

Editors
Suparna Dhar
NSHM Knowledge Campus Kolkata
Kolkata, West Bengal, India

Sanjay Goswami
NSHM Knowledge Campus Kolkata
Kolkata, West Bengal, India

Dinesh Kumar Unni Krishnan
Indian Institute of Management Bangalore
Bengaluru, Karnataka, India

Indranil Bose
NEOMA Business School
Reims, France

Rameshwar Dubey
Liverpool Business School
Liverpool, UK

Chandan Mazumdar
Jadavpur University
Kolkata, West Bengal, India

Xi Chen
Zhejiang University
Hangzhou, Zhejiang, China

ISSN 1865-0929 ISSN 1865-0937 (electronic)
Communications in Computer and Information Science
ISBN 978-3-032-09152-9 ISBN 978-3-032-09153-6 (eBook)
https://doi.org/10.1007/978-3-032-09153-6

This Springer imprint is published by the registered company Springer Nature Switzerland AG
The registered company address is: Gewerbestrasse 11, 6330 Cham, Switzerland

Preface

In continuation of AGC 2023 and AGC 2024, the Analytics Society of India Kolkata Chapter, conceived by NSHM Institute of Computing & Analytics, NSHM Knowledge Campus, Kolkata, organised the 3rd edition of the same conference, AGC 2025, on the 8th and 9th of March, 2025 at Dhono Dhanyo Auditorium, Kolkata. The conference theme "Human-centered Data Analytics: Technology for Sustainable Development" generated extensive interest from industry and academia. The conference was inaugurated by Swami Kamalasthananda, Principal, Ramakrishna Mission Vivekananda Centenary College, India, along with Raj Jain, CMD, RS Software Ltd. The inaugural ceremony was consecrated by Utpal Naresh Patel, NSHM Knowledge Campus Kolkata; Indranil Bose, Neoma Business School, France; U. Dinesh Kumar, IIM Bangalore; Ambarish Dasgupta, Intueri Consulting LLP, India; Deepankar Sinha, Indian Institute of Foreign Trade, Kolkata, India; and the Chief Convener of the conference Suparna Dhar. The objective of the conference was to create an industry-academia interface in the domains of artificial intelligence, machine learning, and data analytics, and their applications in solving societal problems, involving participants from industry, educational institutes in India and abroad, and enthusiastic students from institutions in Kolkata. Around 35 eminent speakers from industry and academia enlightened the audience over the two days of the conference. The speakers represented Ramakrishna Mission Vivekananda Centenary College; IIM Bangalore; Zhejiang University, China, Neoma Business School, France, Indian Statistical Institute Kolkata, AIIMS Kalyani, Jadavpur University, University of Calcutta, Indian Institute of Foreign Trade Kolkata, NIT Durgapur, CNMC Kolkata, KPC Medical College, Jagannath Gupta Institute of Medical Sciences & Hospital Kolkata, Amity University, Aliah University, RBI, IBM, IDRBT, TCS, RS Software, PwC, Intueri Consulting Kolkata, Kolkata Police, and the Department of Sundarban Affairs, Govt. of West Bengal, and other eminent institutions in India.

The discussions spanned different aspects of analytics, right from the fundamental statistical theories forming the base, to the technological aspects of data management, analysis, and their applications across various industrial and social portals, addressing topical problems. The areas covered in various panel discussions included transforming economy and alleviating poverty by drawing insights from data, geospatial data analytics for disaster risk reduction, responsible AI and data analytics for sustainable and inclusive growth of young minds, real-time data analytics for decision-making in public health, and data-driven insights for mental well-being. This conference carried forward the movement created in the previous year for nurturing analytics culture in the eastern region of India, attracting 400+ participants, including students, scholars, faculty members, and industry professionals from various relevant sectors.

The conference received a good response from the research community as well, consequent to the call for papers floated during the year. The topics were organized into three broad classes, namely, (1) Financial Security, (2) Social Well-Being, and (3) Health and Well-Being. A total of 46 papers were submitted through the Microsoft CMT system. The papers were subjected to a stringent peer review process, where each paper was subjected to three single-blind peer reviews. The Editorial Board of the Conference reviewed the revised papers again to check for compliance. A total of 33 papers were accepted for presentation at this stage in the conference. Scholars and practitioners of the accepted papers were invited to present their work to a highly accomplished jury appointed separately for each session. Finally, 16 (all full-length) papers were selected for publication in the proceedings after rigorous assessment by the juries. The final 16 papers include 4 papers on Financial Security, 6 papers on Health and Well-Being, and 6 papers on Social Well-Being.

The papers covered several aspects of analytics, machine learning, and artificial intelligence. Image and text analytics were widely studied, with papers spanning the identification of EEG signals, cancer cells, heat transfer patterns, vulgar language, CSR reports, and many more. The papers applied machine learning, deep learning, time series analysis, graph theory, and Gen AI approaches to analyze data collected from social media, scholarly databases, and other sources. Some scholars collected primary data for the analysis as well. Some papers focused on data preprocessing pipelines, such as multilayer data validation and feature selection. Deep neural networks and transfer learning were used for image and text analytics.

We express our gratitude to Springer for allowing us to publish the conference proceedings within the CCIS series and for their invaluable handholding throughout the publication process. Our appreciation extends to the authors for placing their trust in publishing with us, the reviewers for their punctual and thorough evaluations, and the Organizing Committee members, volunteers, and participants whose collective efforts ensured the conference's success. We thank our academic partners – the Operational Research Society of India, and the IEEE Computer Society Kolkata Section. Lastly, we extend our thanks to our sponsors and mentors for their trust and unwavering support in our conference organization endeavors. The conference series aims to motivate students and young scholars to engage in quality research and impactful projects in analytics that will help build a strong analytics ecosystem in the region. We hope that the rich experiences gained in making AGC 2023, AGC 2024, and AGC 2025 successful events will guide us in making future editions of the Analytics Global Conference even more relevant and impactful.

June 2025

Suparna Dhar
Sanjay Goswami
Dinesh Kumar Unni Krishnan
Indranil Bose
Rameshwar Dubey
Chandan Mazumdar
Xi Chen

Organization

Program Committee Chairs

Bose, Indranil	Neoma Business School, France
Chen, Xi	Zhejiang University, China
Dhar, Suparna	NSHM Knowledge Campus, Kolkata, India
Dubey, Rameshwar	Liverpool Business School, UK
Goswami, Sanjay	NSHM Knowledge Campus, Kolkata, India
Mazumdar, Chandan	Jadavpur University, India
Unni Krishnan Dinesh Kumar	IIM Bangalore, India

Reviewers

Aslam, Humaira	NSHM Knowledge Campus, Kolkata, India
Banerjee, Nisha	NSHM Knowledge Campus, Kolkata, India
Banerjee, Soumya	Amity University Kolkata, India
Basu, Moumita	NSHM Knowledge Campus, Kolkata, India
Bhoumick, Dwaipayan	NSHM Knowledge Campus, Kolkata, India
Chakraborty, Varsha	NSHM Knowledge Campus, Kolkata, India
Das, Ajanta	Amity University Kolkata, India
Das, Kousik	NSHM Knowledge Campus, Kolkata, India
Das, Swapan	NSHM Knowledge Campus, Kolkata, India
Datta Khan, Swapna	NSHM Knowledge Campus, Kolkata, India
Datta, Pratyay	NSHM Knowledge Campus, Kolkata, India
Debnath, Bikash	Brainware University, India
Dutta, Abhaya	Vidyasagar College for Women, India
Dutta, Aditya	IBM, India
Dutta, Ambar	Amity University Kolkata, India
Dutta, Arjama	NSHM Knowledge Campus, Kolkata, India
Ghosh, Debmitra	JIS University, India
Ghosh, Kaushik	NSHM Knowledge Campus, Kolkata, India
Ghosh, Malabika	NSHM Knowledge Campus, Kolkata, India
Goswami, Sanjay	NSHM Knowledge Campus, Kolkata, India
Gupta, Nishi	Pimpri Chinchwad College of Engineering, India
Kar, Ashutosh	NSHM Knowledge Campus, Kolkata, India
Mallik, Majid	NSHM Knowledge Campus, Kolkata, India
Masanta, Monalisa	IIT Guwahati, India

Mazumder, Sangita	NSHM Knowledge Campus, Kolkata, India
Mollah, Ayatullah Faruk	Aliah University, India
Mondal, Sayantani	NSHM Knowledge Campus, Kolkata, India
Mondal, Tuhin	M.S. Ramaiah University of Applied Sciences, India
Mukherjee, Dhritiman	Amity University Kolkata, India
Nag, Aniruddha	Techno India Group, India
Pal, Sanjay	NSHM Knowledge Campus, Kolkata, India
Paul, Abhijit	Swami Vivekananda University, India
Paul, Chayan	Swami Vivekananda University, India
Paul, Madhurima	NSHM Knowledge Campus, Kolkata, India
Roy, Moumita	NSHM Knowledge Campus, Kolkata, India
Sanyal, Atri	NSHM Knowledge Campus, Kolkata, India
Sarkar, Dhrubasish	Supreme Institute of Management and Technology, India
Sharib, Md	NSHM Knowledge Campus, Kolkata, India
Shinde, Rupali	Navrachana University, India
Sinha, Nilanjana	NSHM Knowledge Campus, Kolkata, India

NSHM Knowledge Campus, Kolkata, India

Federal Bank

91.9 Friends FM

The Telegraph Online Edugraph

Operational Research Society of India

IEEE Computer Society Kolkata Section Chapter

RGM

Euphoria GenX

Contents

Health and Well-Being

Deep Learning for Depression Detection Using EEG Signals: A RNN Based Approach

Agnish Paul[1], Maheak Dave[2], Mayukh Sinha[1], Debarshi Mondal[1], Dripta Patra[1], and Pawan Kumar Singh[1,3](✉)

[1] Department of Information Technology, Jadavpur University, Jadavpur University Salt Lake Campus, Plot No. 8, Salt Lake Bypass, LB Block, Sector III, Kolkata 700106, West Bengal, India

agnishpaul2002@gmail.com, mayukhs.it.ug@jadavpuruniversity.in, pawansingh.ju@gmail.com

[2] Department of Computer Science and Engineering, Techno India University, EM-4, EM-4/1, EM Block, Sector V, Kolkata 700091, West Bengal, India

[3] Shinawatra University, 99, Moo 10, Bang Toei, Sam Khok, Pathum Thani 12160, Thailand

Abstract. This paper presents a novel methodology for depression detection using Electroencephalogram (EEG) signals from the Multi-modal Open Dataset for Mental Disorder Analysis (MODMA). The proposed approach integrates advanced signal processing techniques with a Recurrent Neural Network (RNN) classification model to objectively identify depressive states. The methodology consists of three key stages: data preprocessing, feature extraction, and RNN-based classification. EEG data is preprocessed using bandpass filtering, reshaping, and splitting into training and testing sets, preparing it for deep learning model input. The RNN-based model is trained on these features, achieving an accuracy of 77%, outperforming previous studies that used traditional machine learning approaches, such as KNN (K-Nearest Neighbors). Results reveal potential challenges, including model overfitting and class imbalance, suggesting avenues for further optimization. This research highlights the capability of deep learning in mental health diagnostics, providing a data-driven approach that may offer earlier, more precise detection of depression. Future work will focus on enhancing diagnostic accuracy by combining EEG with neuroimaging techniques and exploring broader, more diverse patient populations.

Keywords: Depression detection · Electroencephalogram signals · Multi-modal Open Dataset for Mental Disorder Analysis (MODMA) · Recurrent Neural Network (RNN) · Deep learning

1 Introduction

Depression is a critical global mental health challenge that impacts more than 280 million individuals worldwide, with significant individual, societal, and economic consequences. Traditionally, diagnostics of depression have been based primarily on subjective methods, where clinical interviews and self-reported symptomatology have played a major

S. Dhar et al. (Eds.): AGC 2025, CCIS 2686, pp. 3–13, 2026.
https://doi.org/10.1007/978-3-032-09153-6_1

role. This results in methodological approaches often failing to capture the subtle neurophysiological markers that characterize depressive disorders, and interventions may be delayed or management strategies suboptimal [1]. The majority of the literature on depression detection has focused on feature extraction techniques from EEG signals prior to model training. This reflects the complexity of neurological signal processing and the challenges involved with identifying subtle neurophysiological markers of depressive conditions.

EEG signals emerge as a promising diagnostic tool that can provide a non-invasive window into brain electrical activity [2]. This neuroimaging technique captures synchronized neural electrical potentials using scalp-mounted electrodes and provides high-temporal-resolution insights into brain functioning. EEG signals are characterized by distinct frequency bands, such as delta (1–4 Hz), theta (4–8 Hz), alpha (8–13 Hz), beta (13–30 Hz), and gamma (30–100 Hz), each of which represents unique cognitive and neurological states.

Our study reveals a novel deep-learning strategy for treating depression. State classification. Using methodology involving bandpass filtering with strategic selection of channels during signal preprocessing in EEG and classification of a depressive state in the model by training on an RNN, an objective data-driven method towards depression classification is developed.

The contributions of our work are as follows:

1. We have developed a data-driven, objective approach to classify depressive states using EEG signals, addressing the limitations of subjective diagnostic methods commonly used in clinical settings.
2. We introduce a methodology incorporating bandpass filtering and strategic selection of EEG channels to enhance the extraction of relevant neurophysiological markers for accurate depression detection.
3. We have leveraged Recurrent Neural Networks (RNNs) for model training, capturing temporal dependencies in EEG data and improving the sensitivity to subtle neurophysiological patterns associated with depressive disorders.
4. We demonstrate the potential of EEG-based deep learning techniques as a non-invasive, high-temporal-resolution tool for early and effective detection of depression, paving the way for improved intervention strategies.

The research integrates neurophysiological signal analysis with computational intelligence, towards transforming mental health assessment into a more nuanced, objective diagnostic technique with potential for earlier and more precise disorder detection. Our paper is further divided into 4 sections:

- **Section 2 (Related Survey)** reviews existing literature on neuroimaging and machine-learning approaches for depression detection.
- **Section 3 (Proposed Methodology)** details the research design, data preprocessing, and RNN classification framework.
- **Section 4 (Result Analysis)** presents experimental outcomes and performance metrics.
- **Section 5 (Conclusion and Future Works)** highlights significant findings and gives recommendations for future research avenues.

The entire pipeline of our deep learning methodology for depression detection has been shown in Fig. 1.

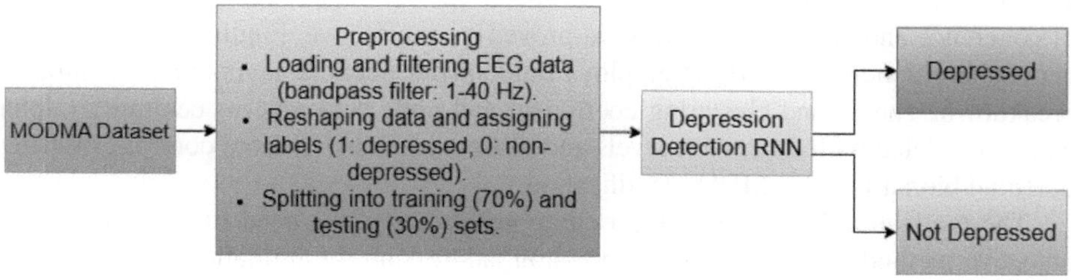

Fig. 1. The entire pipeline of our deep learning methodology for automatic depression detection.

2 Related Study

Neuroimaging and machine learning advancements have emerged significantly to develop objective methods for detecting depression. Many researchers have explored the possibility of using EEG signals and deep learning techniques in identifying neuropsychiatric disorders [3–5]. Preliminary work has shown that different architectures of deep learning can be used efficiently in the analysis of brain signals for mental health assessment [6, 7].

Several studies have explored the use of CNNs and RNNs in processing EEG signals for diagnosis of depression [8, 9]. These techniques have yielded promising results in extracting complex temporal and spatial patterns related to depressive conditions [10, 11]. Deep learning techniques have lately yielded much better performance in a binary classification task between depressed and non-depressed patients as compared to conventional diagnostic methods [12, 13].

The integration of advanced signal preprocessing techniques with deep learning algorithms has emerged as the most crucial approach in neurological disorder detection [14, 15]. Others have concentrated on strategies for feature extraction and selection to increase the accuracy of depression classification using EEG data [14, 15]. Such approaches may be promising in recognizing subtle neurophysiological markers not evident through standard diagnostic protocols.

Zhongyi Zhang et al. [16] investigate how deep neural networks can be used to improve the detection and classification of depression by EEG signals, filling in some of the deficiencies of the current models. In contrast with the traditional binary classification model, where a model could only identify whether a patient is depressed or not, the present study uses a multi-classification model based on GCNs to differentiate among several levels of depression. To address issues like class imbalance in multi-classification tasks, the authors have used a penalty coefficient for smaller classes. To reduce inter-subject variability in EEG signals, they have added a domain generalization module, thereby enhancing the robustness of the model to various individuals. The proposed method is more accurate and yields 75.47% on the MODMA dataset and 77.97% on the PRED + CT dataset, respectively, as compared to state-of-the-art models. Extensive ablation studies further support and validate the contributions of each module and demonstrate the model's overall effectiveness in advancing depression level identification.

Shuting Sun et al. [17] propose robust methods for constructing functional brain networks in MDD using high-density EEG data. The combination of the imaginary part of coherence and cluster-span threshold proved most effective. Findings revealed right hemisphere deficiencies, MDD employs randomized network layouts and symmetry breakdown. Theta band clustering coefficient and node betweenness centrality (alpha band) correlated with depressive levels and distinguished MDD from controls, offering potential biomarkers for MDD identification.

The motivation behind our proposed approach lies in the need for more accurate, objective methods for diagnosing depression, addressing the limitations of traditional subjective assessments. While EEG signals offer valuable insights into brain activity, existing deep learning models, such as CNNs and RNNs, have shown promise but face challenges like overfitting, class imbalance, and inconsistent performance. Our approach integrates advanced preprocessing techniques with a robust RNN classification model to capture subtle neurophysiological markers of depression, offering a more reliable and data-driven solution. By leveraging the MODMA dataset, which includes both EEG and speech data from clinically diagnosed patients, our goal is to enhance diagnostic accuracy, generalizability, and early detection, ultimately contributing to more personalized and effective mental health interventions.

3 Proposed Methodology

The proposed methodology for depression detection using EEG signals from the MODMA dataset comprises three primary stages: data preprocessing, feature extraction, and RNN-based classification.

3.1 Data Acquisition

The MODMA (Multi-modal Open Dataset for Mental Disorder Analysis) dataset [18] is a significant resource for advancing research in mental health and affective computing. It comprises data from clinically diagnosed depressed patients and healthy controls, offering both EEG and speech recordings for public use. The EEG data includes resting-state and stimulation recordings, collected through traditional 128-electrode caps for high-resolution studies and innovative 3-electrode wearable devices for mobile and pervasive applications. This dual approach allows researchers to analyze neural activity with varying levels of granularity, making it a valuable tool for identifying biomarkers associated with mental disorders, particularly depression.

Additionally, the dataset features a rich collection of speech data, recorded during interviews, reading sessions, and picture descriptions. These recordings provide opportunities to study various linguistic and acoustic features, such as prosody, emotional tone, and articulation, which are often impacted by mental disorders. The multi-modal nature of MODMA enables researchers to explore the interplay between neural activity and speech patterns, paving the way for cross-modal studies. With its comprehensive annotations and open-access availability, the dataset serves as a benchmark for developing predictive models and classification systems for mental health diagnostics. Its inclusion of wearable EEG devices also supports the exploration of real-world, non-invasive monitoring solutions, enhancing its relevance for clinical and practical applications in mental health care.

3.2 Pre-Processing

The preprocessing begins with loading and processing EEG data stored in.mat files from Google Drive. First, the script collects all.mat files and reads them into Python using the scipy.io.loadmat function. The dataset is filtered by selecting specific EEG channels based on predefined indices and cropping the data to 300.3 s. A bandpass filter (1–40 Hz) is applied to remove noise from the EEG signals. The data is then reshaped into a 4D array, where the dimensions correspond to the number of samples, channels, and time points. The dataset is labeled based on the file name, where "depressed" files are assigned a label of 1 and others a label of 0. After this, the dataset is further reshaped to prepare it for convolutional processing by adding a channel dimension, resulting in a 5D array suitable for feeding into deep learning models. Lastly, the dataset is split into 70–30 training and testing sets.

3.3 Our RNN-Based Model

The architecture of the model begins with an input layer that expects data with a shape of (274, 16 * 274), representing a sequence of 274-time steps, each with 16 * 274 features. This data is passed to a **SimpleRNN** layer with 128 units and ReLU activation, which processes the temporal dependencies within the sequence. To prevent overfitting, a **Dropout** layer with a rate of 0.2 follows the RNN, randomly setting 20% of the units to zero during training. The model is then subsequently regularized with a fully connected Dense layer with 64 units and ReLU activation, followed by a 0.2-rate Dropout layer.

Finally, the output is generated by a **Dense layer** with a single unit and **sigmoid activation**, which produces a probability value between 0 and 1, representing the likelihood of an individual being classified as "depressed" (positive class) in this binary classification task. The binary cross-entropy loss function and the Adam optimizer are used to train the model, and accuracy is used as the evaluation metric. This architecture effectively integrates **recurrent sequence learning**, **dropout regularization**, and **dense layers** to build a robust model capable of classifying EEG signal data for depression detection. The architecture of our RNN-based model is shown in Fig. 2.

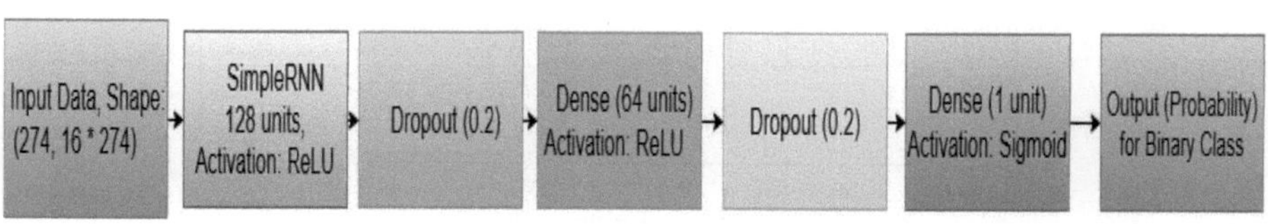

Fig. 2. Architecture of our RNN-based model.

4 Results Analysis

The plot in Fig. 3 illustrates the training and validation loss over the course of 50 epochs for a proposed model using the MODMA Dataset. The x-axis represents the number of epochs, while the y-axis shows the loss values. The blue line denotes the training loss, and the orange line represents the validation loss. Initially, both losses start at higher values and gradually decrease as training progresses. The training loss exhibits fluctuations, indicative of the model learning and adjusting its weights to minimize the loss. The validation loss follows a similar downward trend, although with its own fluctuations, reflecting the model's ability to generalize to unseen data. The plot ultimately provides valuable insights into the convergence of the model and helps in identifying overfitting or underfitting scenarios.

Figure 4 presents the training and validation accuracy across 50 epochs for the same model. The x-axis shows the number of epochs, while the y-axis indicates the accuracy, ranging from 0.1 to 1.0. The blue line represents the training accuracy, which starts lower and increases over time, showing the model's improved performance on the training data. The orange line, depicting the validation accuracy, follows a similar trend but with different fluctuations. This behavior highlights how well the model performs on unseen data and helps in assessing its generalization capabilities. The plot ultimately serves as a key indicator of the model's overall performance and effectiveness in achieving the desired accuracy.

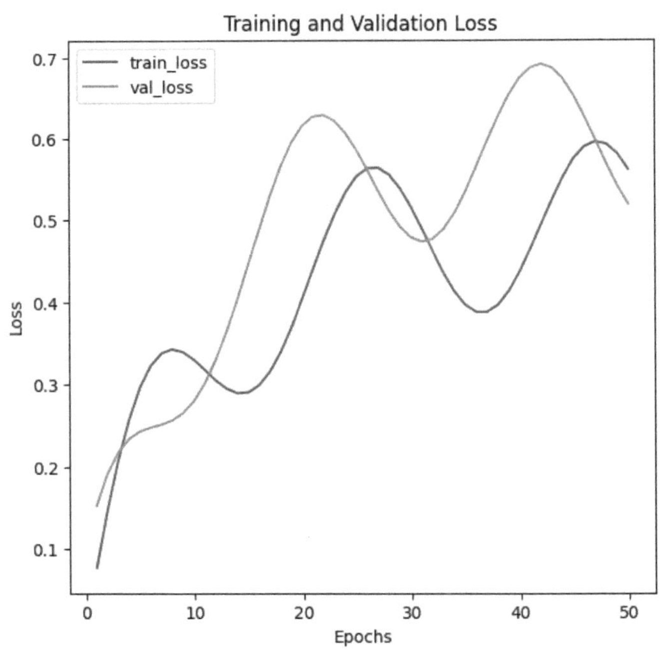

Fig. 3. Plot depicting loss over epoch of our RNN-based model over the MODMA dataset.

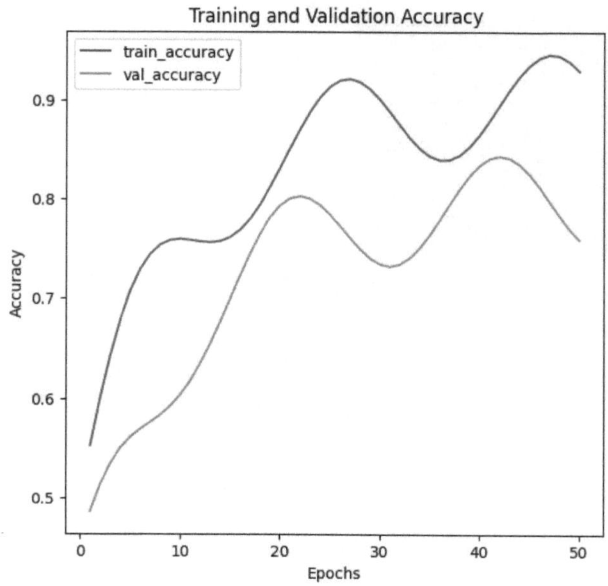

Fig. 4. Plot depicting accuracy over epoch of our RNN-based model over the MODMA dataset.

The confusion matrix in Fig. 5 for the binary classification problem of identifying depression, where 0 represents "not depressed" and 1 represents "depressed," provides a clear insight into the model's performance. The matrix shows the model's predictions against the actual classes. Here's the confusion matrix:

- True 0, Pred 0: 23 (True Negative)
- True 0, Pred 1: 6 (False Positive)
- True 1, Pred 0: 6 (False Negative)
- True 1, Pred 1: 18 (True Positive)

From the matrix, we can infer that the model correctly identified 23 out of 29 not depressed individuals and 18 out of 24 depressed individuals. However, it made some mistakes, falsely identifying 6 not depressed individuals as depressed (false positives) and 6 depressed individuals as not depressed (false negatives). This confusion matrix helps to understand the model's strengths and areas for improvement in classifying depression.

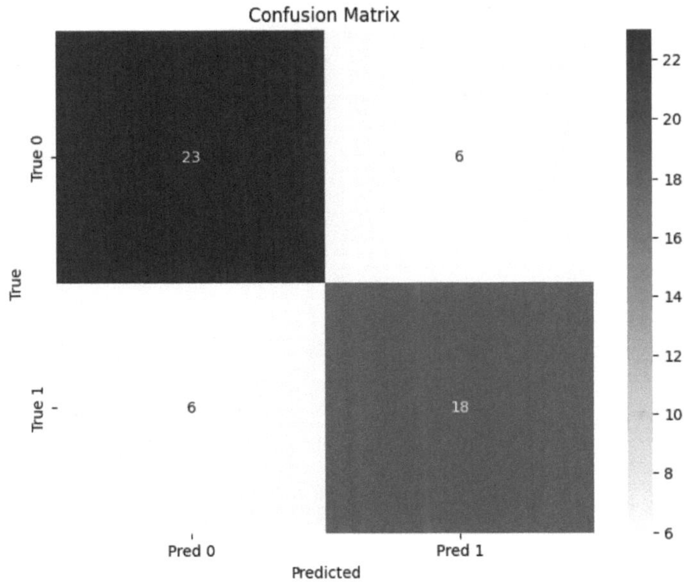

Fig. 5. Confusion matrix of our RNN-based model on the MODMA dataset. **Label '0'** indicates "non-depressed" individuals (healthy controls) whereas **Label '1'** indicates "depressed" individuals.

The bar graph in Fig. 6 shows precision, recall, and F1-score across each class (0 and 1) of our proposed RNN model on the MODMA dataset. The x-axis indicates classes: 0 means "not depressed", and 1 for "depressed". The Y-axis shows values going from 0.0 to 0.8. The bars are colored differently for each metric: blue for precision, green for recall, and red for F1-score. This graph shows the performance of the model with regards to precision, recall, and F1-score for each of the classes. For class 0 ("not depressed"), we have a value of about 0.79 for precision, recall, and F1-score, showing the capability of the model to identify non-depressed individuals. For class 1 ("depressed"), precision, recall, and F1-score sit at approximately 0.75, portraying the model's ability to determine the depressed individuals. From this immediate evidence, we have a clear sense of the classification competency of the model toward both classes.

The classification report in Fig. 7 provides a tabulation of model performance metrics across each class (namely: 0 and 1); in MODMA dataset are also informed by precision, recall, F1-score, and support for each of these classes. For class 0 (not depressed), the precision, recall, and F1-score are around 0.79 with a support of 29 instances; for class 1 (depressed), precision, recall, and F1-score are about 0.75 with a support of 24 instances. So the modules obtained an overall accuracy of 0.77, meaning that the prediction is correct in 77% of total instances. In addition, the report provides the macro and weighted averages of each of the metrics that the model performance on both classes represents. Thus, this classification report evaluates the applicability of the model to discriminate between depressed and non-depressed individuals.

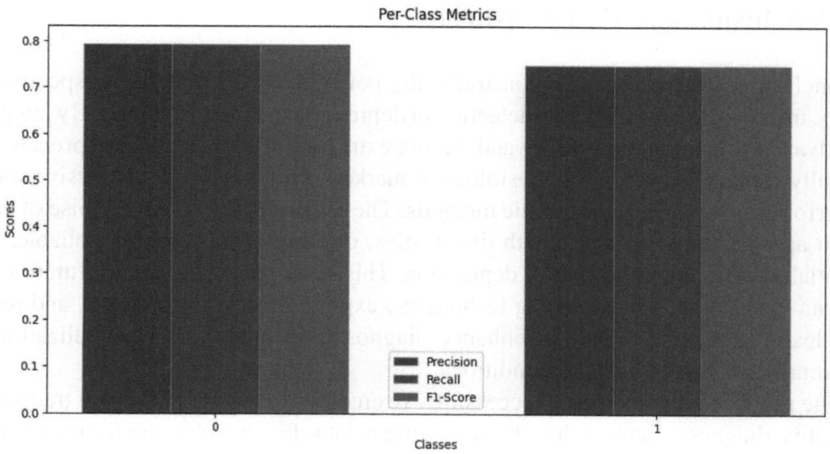

Fig. 6. Bar graph representing Precision, Recall, and F1-score of each class of RNN-based model on the MODMA dataset.

	precision	recall	f1-score	support
0.0	0.79	0.79	0.79	29
1.0	0.75	0.75	0.75	24
accuracy			0.77	53
macro avg	0.77	0.77	0.77	53
weighted avg	0.77	0.77	0.77	53

Fig. 7. Classification report of our RNN-based model on the MODMA dataset.

It is evident from Table 1, our approach outperforms other models and algorithms previously done on the MODMA dataset.

Table 1. Performance comparison of our RNN-based model with previous studies done on depression detection

Authors	Model	Accuracy
Zhongyi Zhang et al. [16]	Graph Convolutional Neural Network (GCN)	75.47%
Shon et al. [19]	KNN + GA	71.76%
Paul et al.	**RNN**	**77%**

5 Conclusion and Future Works

In conclusion, this research demonstrates the potential of deep learning, specifically RNNs, in improving the objective detection of depression using EEG signals. By integrating advanced signal preprocessing and feature extraction techniques, our approach successfully captures subtle neurophysiological markers associated with depressive states, outperforming traditional diagnostic methods. The results highlight the promise of data-driven approaches in mental health diagnostics, offering more accurate, reliable, and potentially earlier identification of depression. This study paves the way for future work that can incorporate neuroimaging techniques, explore diverse populations, and refine deep learning models to further enhance diagnostic accuracy and personalization of interventions for mental health conditions.

The proposed methodology successfully overcomes the shortcomings of traditional subjective diagnostic approaches, by providing a data-driven technique for the identification of depressive states. The key findings show the efficacy of RNNs in capturing subtle neurophysiological markers, thereby offering a more precise and potentially earlier method of depression detection.

Critical contributions include the development of a comprehensive preprocessing strategy, implementation of an innovative neural network architecture, and validation of machine learning techniques in neurological disorder identification. The research underlines the transformative potential of computational approaches in advancing mental health assessment methodologies.

Future studies will emphasize improving diagnostic accuracy by incorporating EEG with neuroimaging techniques and exploring more heterogeneous populations. Efforts will also focus on improving the detection of early-stage depression and personalizing interventions. Advanced deep learning architectures will be evaluated for their ability to capture complex neurophysiological markers, and the methodology's applicability across diverse neuropsychiatric conditions will be explored. These efforts aim to advance mental health diagnostics with more objective and comprehensive tools.

References

1. Li, R., et al.: A novel EEG decoding method for a facial-expression-based BCI system using the combined convolutional neural network and genetic algorithm. Front. Neurosci. **16**, 988535 (2022)
2. Bashivan, P., Rish, I., Yeasin, M., Codella, N.: Learning representations from EEG with deep recurrent-convolutional neural networks. arXiv preprint arXiv:1511.06448 (2015)
3. Liu, H., Zhang, Y., Li, Y., Kong, X.: Review on emotion recognition based on electroencephalography. Front. Comput. Neurosci. **15**, 758212 (2021)
4. Das, A.K., Naskar, R.: A deep learning model for depression detection based on MFCC and CNN generated spectrogram features. Biomed. Signal Process. Control **90**, 105898 (2024)
5. Wang, F., et al.: Emotion recognition with convolutional neural network and EEG-based EFDMs. Neuropsychologia **146**, 107506 (2020)
6. Suzuki, K., Laohakangvalvit, T., Sugaya, M.: Machine-Learning-Based Depression Detection Model from Electroencephalograph (EEG) Data Obtained by Consumer-Grade EEG Device. Brain Sci. **14**(11), 1107 (2024)

7. LeCun, Y., Bengio, Y., Hinton, G.: Deep learning. nature **521**(7553), 436–444 (2015)
8. Li, J.W., et al.: An innovative EEG-based emotion recognition using a single channel-specific feature from the brain rhythm code method. Front. Neurosci. **17**, 1221512 (2023)
9. Litjens, G., et al.: A survey on deep learning in medical image analysis. Med. Image Anal. **42**, 60–88 (2017)
10. Yan, B., et al.: Quantitative identification of major depression based on resting-state dynamic functional connectivity: a machine learning approach. Front. Neurosci. **14**, 191 (2020)
11. Rajkomar, A., et al.: Scalable and accurate deep learning for electronic health records. NPJ Digital Medicine **1**(1), 18 (2019)
12. Lotte, F., Congedo, M., Lécuyer, A., Lamarche, F., Arnaldi, B.: A review of classification algorithms for EEG-based brain–computer interfaces. J. Neural Eng. **4**(2), R1 (2007)
13. Luján, M.Á., Jimeno, M.V., Mateo Sotos, J., Ricarte, J.J., Borja, A.L.: A survey on eeg signal processing techniques and machine learning: Applications to the neurofeedback of autobiographical memory deficits in schizophrenia. Electronics **10**(23), 3037 (2021)
14. Craik, A., He, Y., Contreras-Vidal, J.L.: Deep learning for electroencephalogram (EEG) classification tasks: a review. J. Neural Eng. **16**(3), 031001 (2019)
15. Wu, H., & Liu, J.: A multi-stream deep learning model for EEG-based depression identification. In 2022 IEEE International Conference on Bioinformatics and Biomedicine (BIBM) (pp. 2029–2034). IEEE (2022, December)
16. Zhang, Z., Xu, C., Jin, L., Hou, H., & Meng, Q.: A Depression Level Classification Model Based on EEG and GCN Network with Domain Generalization. In *2024 43rd Chinese Control Conference (CCC)* (pp. 8582–8587). IEEE (2024, July)
17. Sun, S., Li, J., Chen, H., Gong, T., Li, X., & Hu, B.: A study of resting-state EEG biomarkers for depression recognition. *arXiv preprint* arXiv:2002.11039 (2020)
18. Cai, H., et al.: A multi-modal open dataset for mental-disorder analysis. Scientific Data **9**(1), 178 (2022)
19. Shon, D., Im, K., Park, J.H., Lim, D.S., Jang, B., Kim, J.M.: Emotional stress state detection using genetic algorithm-based feature selection on EEG signals. Int. J. Environ. Res. Public Health **15**(11), 2461 (2018)

Detection of Gynecological and Andrological Cancers Using AI: A Review

Abhaya Dutta[1] (✉), Priya Saha[2], and Aditya P. Dutta[3]

[1] Vidyasagar College for Women, 39, Sankar Ghose Lane, Kolkata 700006, West Bengal, India
abhayadutta8@gmail.com
[2] Hi Tech Animation, Tangra, Kolkata 700046, West Bengal, India
[3] IBM Consulting, DLF 1, Kolkata 700163, West Bengal, India

Abstract. Major challenges in finding solutions to cancer lie in lack of early symptoms and then rapid uncontrollable abnormal cell growth once it begins. Cancer continues to spread at an alarming rate, while progress in overcoming treatment barriers remains slow. Early detection holds the key to greatly improve outcomes by identifying cancer before it spreads. Stage zero diagnosis can prevent development of full-blown malignancies. Advancements in artificial intelligence are making early detection increasingly feasible, offering powerful tools for intervening in the disease's earliest stages. This review provides a systematic analysis to establish link between artificial intelligence techniques and early detection of gynecological and andrological cancers using a variety of imaging techniques and bio-molecular markers.

Keywords: Gynecological cancer · Prostate cancer · Molecular markers · Artificial intelligence · Imaging techniques · Machine learning

1 Introduction

Despite significant medical advancements, many diseases remain beyond our control, and cancer stands out as one of the most formidable challenges. According to estimates, one out of every five men and women were expected to develop cancer at some time in their lives. In 2022, there were over twenty million more cancer cases reported [1]. Among various cancer types, gynecological and andrological cancers in men and women are very common. Also, gynecological and andrological malignancies cases face additional challenges due to social taboo of talking about reproductive parts. These have high fatalities in various parts of the world, including a substantial number of cases in India, due to lack of awareness. Hence, we have selected five types of reproductive cancers: breast cancer, ovarian cancer, uterine cancer, cervical cancer, and prostate cancer for this review.

Our main objective is to perform a structured analysis to establish link between artificial intelligence techniques with early detection of these specific cancers with the aim to develop better diagnostic system for the most prevalent gynecological and andrological cancers, with the help of biomolecule markers and image detection techniques. With the application of AI, medical practitioners may use algorithms to identify individuals who do not exhibit any symptoms as potentially susceptible to developing cancer. [2].

S. Dhar et al. (Eds.): AGC 2025, CCIS 2686, pp. 14–38, 2026.
https://doi.org/10.1007/978-3-032-09153-6_2

2 Materials and Methods

A review of the literature revealed findings linking advanced artificial intelligence to gynecological and andrological cancer detection. The reviewers conducted systematic search in PubMed/Medline database with keywords = (Artificial intelligence OR machine learning) and (breast cancer or ovarian cancer or prostate cancer or endometrial cancer or cervix cancer) and (bio marker or molecule marker or imaging technique); where full text was available, and publication date was within 2019 and 2024 (both years inclusive). The search for papers was conducted on October 14th and 15th, 2024.

The reviewers used R Studio version 4.4.1 to assist in performing the bibliometric analysis presented in this paper.

3 AI in Detection of Gynecological and Andrological Cancers: A Bibliometric Analysis

3.1 Thematic Evolution Study

The shift in research focus from 2019–2022 to 2023–2024 was systematically studied. We used thematic evolution maps in form of Sankey diagrams to capture the evolution of research within each specific cancer type which is presented in the following subsections.

3.1.1 Breast Cancer Research Trends Evolution

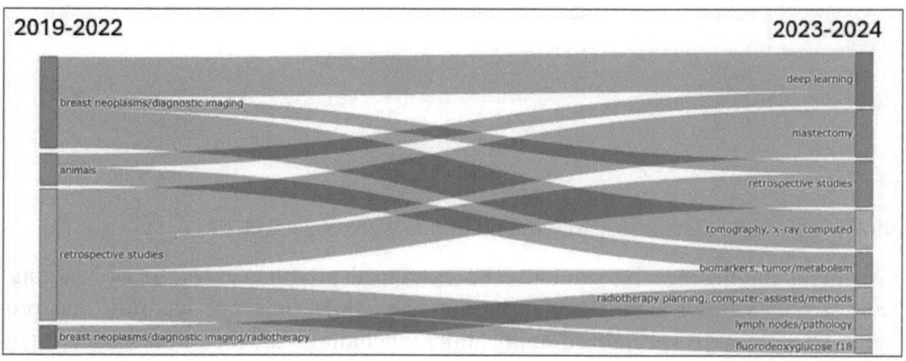

Breast cancer research theme trends based on the above schematic,

- Continuity in Retrospective Studies: The concept of retrospective studies remains a consistent research focus across both timeframes, indicating its sustained importance in medical research methodologies.
- Expansion from Breast Neoplasms & Imaging to broader Topics: The prior focus on "breast neoplasms/diagnostic imaging" has broadened to encompass several emerging areas, including "deep learning," "mastectomy," and "computed tomography." This suggests an increasing incorporation of artificial intelligence and advanced imaging techniques in oncology diagnostics and treatment approaches.

- Rise of Computational and Metabolic Research:
  - The transition to deep learning and computer-assisted methods in radiotherapy planning underscores the increasing significance of artificial intelligence and automation in medical diagnostics and treatment approaches.
  - The terms "biomarkers, tumor/metabolism" and "fluorodeoxyglucose F18" highlight the significance of metabolic imaging and biomolecular diagnostics, essential components of personalized medicine.

- Reduction in Animal Studies: The category "Animals" was significant from 2019 to 2022 but is nonexistent in the 2023–2024 timeframe. This indicates a shift from animal-based research, maybe due to ethical concerns or an increased focus on human-centered therapeutic investigations.
- The emergence of terms such as "lymph nodes/pathology" and "mastectomy" indicates an increasing emphasis on pathological evaluation and surgical approaches within cancer research.

3.1.2 Prostate Cancer Research Trends Evolution

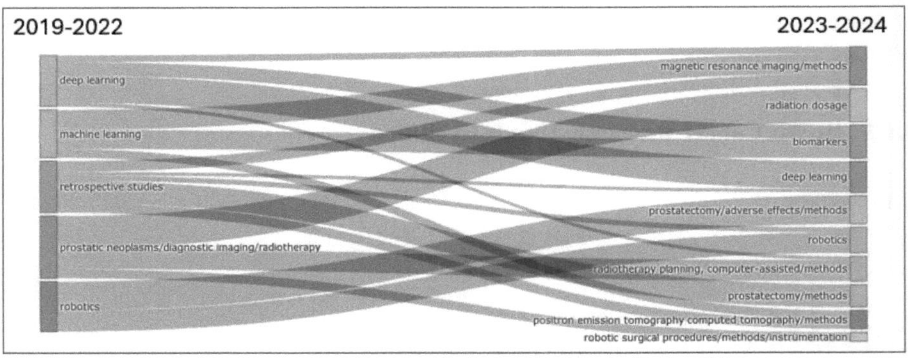

Following prostate cancer research theme trends were observed,

- Artificial Intelligence Integration: The continued prominence of "deep learning" across both periods signifies enduring interest, while "machine learning" has progressed into more specialized applications, including "magnetic resonance imaging/methods" and "radiation dosage" optimization. This indicates that AI is progressively utilized to improve imaging methodologies and treatment strategies.
- Transition in Prostatic Neoplasms Investigation: Previous research concentrated on "prostatic neoplasms/diagnostic imaging/radiotherapy," but has now expanded to include "prostatectomy/adverse effects/methods" and "prostatectomy/methods." This shift signifies a transition from a mostly imaging-centric methodology to an expanded emphasis on surgical treatments and their results.
- Progress in Imaging and Radiation Therapy: "Magnetic resonance imaging/methods" and "positron emission tomography computed tomography/methods" have surfaced,

reflecting an increasing dependence on sophisticated imaging modalities for the identification and monitoring of prostate cancer. The inclusion of "radiotherapy planning, computer-assisted/methods" indicates a shift towards AI-enhanced precision radiotherapy, improving treatment planning and reducing adverse effects.

- Expansion of Robotics in Surgical Applications: The initial emphasis on "robotics" has evolved into "robotic surgical procedures/methods/instrumentation," indicating progress in robotic-assisted surgeries, especially in prostatectomy. This indicates that robotic devices are assuming a greater role in precision surgery, diminishing complications and enhancing patient outcomes.
- Expansion of Biomarker Research: The rise of "biomarkers" as a pivotal research domain indicates a growing focus on molecular and genetic indicators for early detection, prognosis, and tailored treatment approaches.

3.1.3 Endometrial Cancer Research Trends Evolution

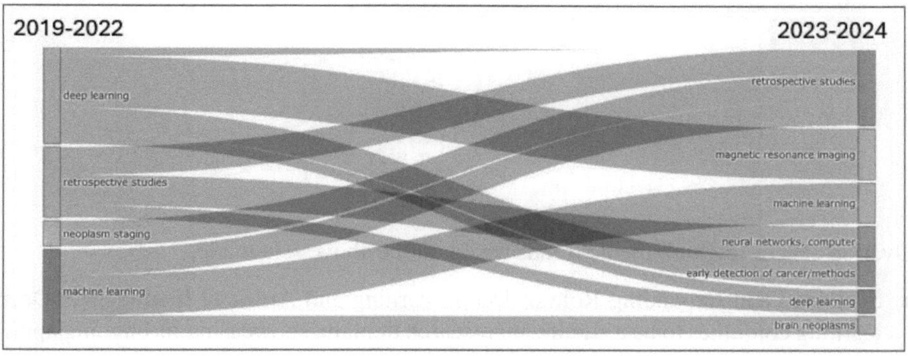

The trends observed in endometrial cancer research were,

- The Emergence of AI-Driven Methods in Endometrial Cancer Diagnosis: The ongoing development of deep learning and machine learning indicates further progress in AI applications for histopathological image analysis, radiomics, and predictive modeling related to endometrial cancer prognosis. The development of neural networks and computer-assisted techniques highlights an increasing dependence on automated feature extraction from imaging data to enhance diagnostic precision.
- Research emphasis has transitioned from neoplasm staging (2019–2022) to early detection of cancer and methods (2023–2024). This indicates a growing interest in non-invasive screening techniques, biomarker discovery, and AI-driven imaging interpretation for the detection of early-stage endometrial cancer.
- Expanded Use of Advanced Imaging Techniques: The shift from general "diagnostic imaging" to "magnetic resonance imaging (MRI)" underscores the growing significance of MRI-based radiomics and deep learning algorithms in characterizing endometrial tumors, evaluating myometrial invasion, and informing treatment decisions. AI-enhanced MRI interpretation may improve the precision of staging and differentiation of endometrial cancer subtypes, thereby decreasing the dependence on invasive biopsies.

- Integration of retrospective studies for the validation of artificial intelligence. The ongoing significance of retrospective studies underscores the need for extensive historical datasets to train, test, and validate AI algorithms in the diagnosis, treatment planning, and prognostic modeling of endometrial cancer.
- Analysis of Brain Neoplasms as a Parallel AI Application: The presence of brain neoplasms in 2023–2024 indicates that AI-driven methodologies initially designed for neurological tumor analysis could be adapted for gynecological cancers, especially in enhancing MRI segmentation and classification tasks for endometrial cancer.

3.1.4 Ovarian Cancer Research Trends Evolution

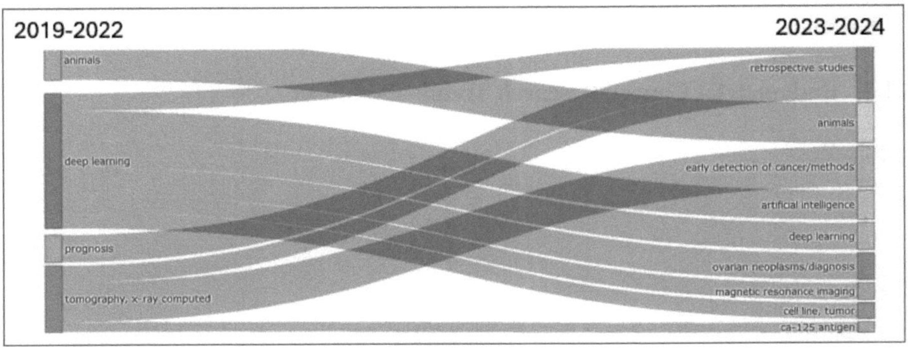

Ovarian cancer research theme trends based on the above schematic,

- Sustained and Expanding Role of Deep Learning and Artificial Intelligence: Deep learning continues to be a significant research domain, showcasing its ongoing application in medical diagnostics and prognostics. Artificial intelligence represents a distinct category, suggesting wider applications beyond deep learning, likely including machine learning, neural networks, and automation in medical imaging and diagnosis.
- Transition from Prognosis to Early Cancer Detection: The research focus has shifted from general "prognosis" to "early detection of cancer/methods," highlighting the increasing significance of early intervention strategies in oncology. This transition indicates progress in AI-assisted screening technologies and biomarker-driven methods for cancer detection.
- Expansion in Imaging Modalities: Research in "tomography, x-ray computed" has progressed to include "magnetic resonance imaging" (MRI), reflecting a growing dependence on MRI for improved diagnostic accuracy, particularly in oncology. This indicates progress in AI-driven imaging analysis and tumor characterization.
- Emerging Focus on Ovarian Cancer Diagnostics: The terms "ovarian neoplasms/diagnosis" and "CA-125 antigen" have emerged as significant research areas, indicating an increased emphasis on the detection of ovarian cancer. The mention of "cell line, tumor" suggests a focus on experimental and translational cancer research, likely utilizing in vitro tumor models.
- Continued Use of Retrospective Studies and Animal Models: Retrospective studies are pertinent, underscoring the significance of historical clinical data in the training of AI models for cancer diagnostics. The ongoing presence of "ani-mals" indicates

sustained preclinical research, which likely aids in the development and validation of new diagnostic techniques and therapies.

3.1.5 Cervix Cancer Research Trends Evolution

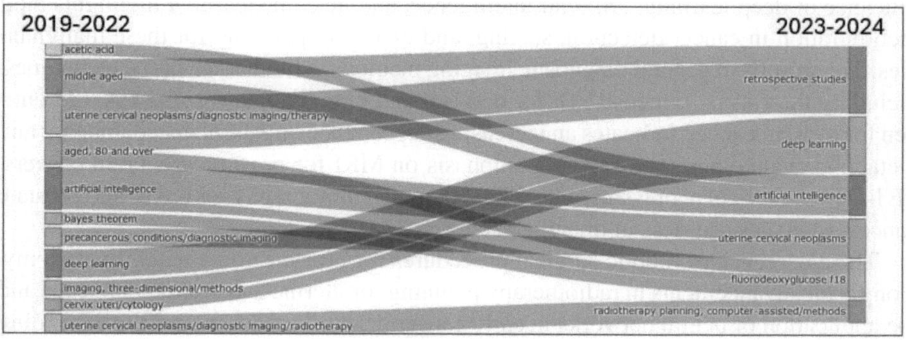

The research trends observed for Cervix cancer were,

- Enhanced Emphasis on AI and Deep Learning: "Artificial intelligence" and "deep learning" remain prominent research themes, highlighting their ongoing importance in medical imaging, diagnostics, and predictive analytics for cervical neoplasms. The association of these terms with "radiotherapy planning, computer-assisted methods" indicates the incorporation of AI in tailored cancer treatment planning.
- Transition from General Diagnostic Imaging to Specific Imaging Biomarkers: The terms "precancerous conditions/diagnostic imaging" and "imaging, three-dimensional/methods" have progressed to a concentrated emphasis on "fluorodeoxyglucose F18" (FDG-PET imaging), reflecting a transition toward metabolic imaging for the evaluation of cervical cancer. FDG-PET underscores the increasing significance of nuclear medicine in tumor metabolism detection and therapy response monitoring.
- Refinement in Cervical Neoplasm Research: The broad category "uterine cervical neoplasms/diagnostic imaging/therapy" has transitioned into "uterine cervical neoplasms," indicating a more focused approach to research in cervical cancer, encompassing early detection, treatment optimization, and prognosis. This indicates a shift in research towards targeted, AI-enhanced diagnostic and treatment strategies.
- Continued Importance of Retrospective Studies: Retrospective studies maintain a significant research focus, reflecting a dependence on historical clinical data for the development of AI models and the validation of diagnostic tools.
- Demographic and statistical methodologies have diminished. Research interest in specific demographic groups, such as "middle aged" and "aged, 80-and-over," along with statistical approaches like "Bayes theorem," seems to have decreased. This indicates a transition towards technology-driven, biomarker-based, and AI-assisted methodologies, moving away from traditional epidemiological or statistical modeling studies.

3.1.6 Research Trends Evolution: Summary Across Cancer Types

The studies and schematics presented in the previous section indicate a notable shift towards AI-driven diagnostics, advanced imaging, and precision oncology, especially concerning breast, prostate, cervical, ovarian, and prostate cancers. The ongoing significance of deep learning, artificial intelligence, and machine learning highlights their incorporation in cancer detection, staging, and treatment planning for these malignancies. The shift from general diagnostic imaging methods to biomarker-based techniques, including fluorodeoxyglucose F18 for cervical and brain neoplasms and CA-125 antigen for ovarian cancer, indicates an increasing dependence on AI-enhanced imaging and metabolic evaluations. The growing emphasis on MRI for prostate and brain cancers, PET-CT for cervical neoplasms, and X-ray computed tomography for breast and prostate cancers indicates progress in imaging accuracy.

The emergence of robotic surgical procedures for prostatectomy and hysterectomy, along with advancements in radiotherapy planning for uterine and breast neoplasms and the application of neural networks in early cancer detection, underscores the expanding role of AI in treatment optimization. Retrospective studies are crucial for the validation of AI models. The shift from traditional epidemiological methods to technology-driven oncology research signifies a progression toward personalized, AI-assisted cancer diagnostics and management. These trends underscore the increasing influence of computational intelligence in enhancing cancer detection, treatment strategies, and patient outcomes across various cancer types.

3.2 Co-Occurrence Network Study

A systematic study was performed for research topics within the search set of literature and presented below in form of co-occurrence network diagrams for each specific cancer type.

3.2.1 Breast Cancer Research Topics Co-Occurrence Study

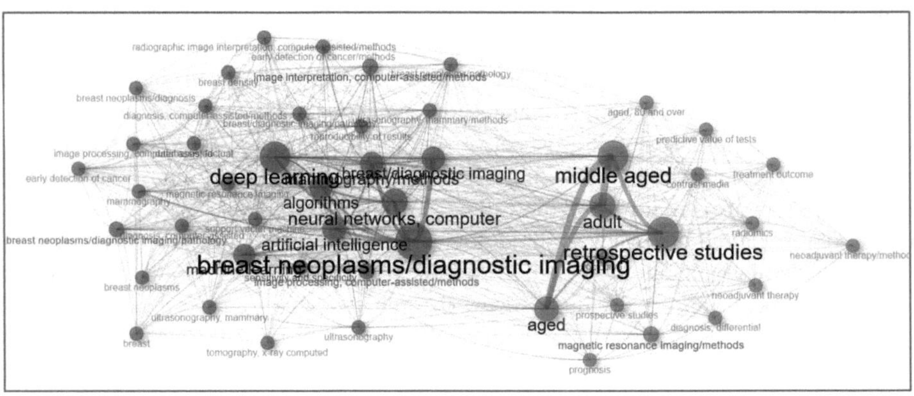

The diagram consists of two distinct clusters, each represented by different colors, indicating thematic groupings:

- Red Cluster – Artificial Intelligence and Diagnostic Imaging - The red cluster is characterized by a high degree of interconnectivity, focusing on the utilization of artificial intelligence (AI) in the diagnosis of breast cancer. Key concepts such as deep learning, neural networks, computer algorithms, and artificial intelligence underscore the growing significance of AI-driven diagnostic tools. These methods are primarily utilized in the diagnostic imaging of breast neoplasms, including mammography, ultrasonography, and magnetic resonance imaging (MRI). The dense interconnections indicate a dynamic research area aimed at improving cancer detection accuracy via computational methods.
- Blue Cluster – Clinical and Epidemiological Studies: The blue cluster denotes the clinical and epidemiological dimensions of breast cancer research. Terms like retrospective studies, middle-aged, adult, aged, and treatment outcome highlight a significant focus on patient demographics, treatment assessment, and results. Related terms encompass neoadjuvant therapy and radiomics, highlighting the increasing significance of personalized treatment strategies and predictive analytics in the management of breast cancer.

3.2.2 Prostate Cancer Research Topics Co-Occurrence Study

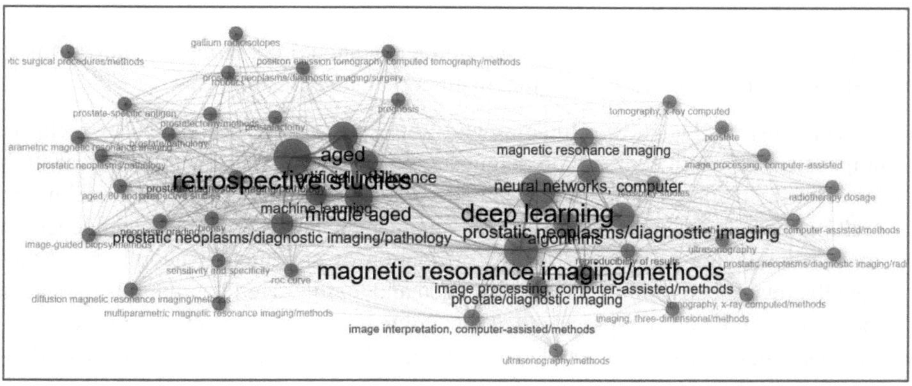

The network reveals two primary research domains, represented by distinct clusters:

- Red Cluster – Clinical and Pathological Studies: The red cluster emphasizes the clinical and pathological dimensions of prostate cancer. Terms like retrospective studies, aged, middle-aged, prostate neoplasms, diagnostic imaging, pathology, and prostatectomy highlight a significant focus on patient demographics, diagnostic pathology, and treatment outcomes. The inclusion of terms such as prostate-specific antigen and image-guided biopsy indicates active research in diagnostic biomarkers and minimally invasive diagnostic techniques. The incorporation of robotics and surgical methods underscores the integration of advanced techniques in the treatment of prostate cancer.
- Blue Cluster – Artificial Intelligence and Advanced Imaging: The blue cluster indicates the growing significance of artificial intelligence (AI) and imaging technologies in the diagnosis and management of prostate cancer. Important concepts encompass deep learning, neural networks, machine learning, algorithms, image processing,

and computer-assisted techniques. The significance of magnetic resonance imaging (MRI) and multiparametric magnetic resonance imaging indicates an increasing dependence on high-resolution imaging and AI-based analytical techniques for early detection and treatment planning. Terms such as radiomics and feasibility studies indicate a continuing transition towards AI-assisted precision diagnostics.

3.2.3 Endometrial Cancer Research Topics Co-Occurrence Study

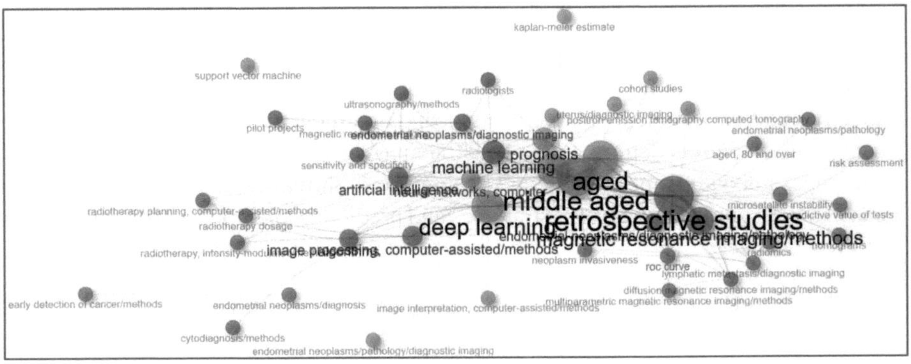

This network is constructed based on the frequency and co-occurrence of key terms in published scientific literature, revealing distinct thematic clusters.

- Brown Cluster – Diagnostic Imaging and Retrospective Studies: The brown cluster emphasizes research centered on diagnostic imaging and retrospective clinical studies. The inclusion of terms like magnetic resonance imaging/methods, radiomics, multiparametric imaging, and risk assessment indicates a growing dependence on advanced imaging techniques for the evaluation of endometrial neoplasms. Additionally, retrospective studies involving aged and middle-aged populations indicate a focus on patient demographics and clinical outcomes in retrospective analyses.
- Green Cluster – Prognostic and Cohort Studies: The green cluster denotes investigations into prognostic factors and cohort studies. Terms such as prognosis, cohort studies, and lymphatic metastasis reflect a significant focus on the investigation of disease progression and treatment outcomes. The association with microsatellite instability indicates an increasing emphasis on genetic and molecular markers for personalized medicine strategies.
- Red Cluster – Artificial Intelligence in Diagnostics: The red cluster emphasizes the utilization of artificial intelligence (AI) in the diagnostics of endometrial cancer. Terms like deep learning, machine learning, neural networks, and artificial intelligence reflect the incorporation of AI-driven computational techniques aimed at improving diagnostic precision. The presence of sensitivity and specificity indicates continuous efforts to validate AI-driven models within clinical practice.
- Purple Cluster – Radiotherapy and Treatment Planning: The purple cluster highlights research focused on radiotherapy and treatment planning. The terms radiotherapy planning, intensity-modulated radiotherapy, and radiotherapy dosage indicate progress in precision techniques for treating endometrial cancer.

- Blue Cluster – Early Detection and Cytodiagnosis: The blue cluster pertains to the early detection and cytodiagnosis of endometrial cancer. Terms such as early detection of cancer, cytodiagnosis, and endometrial neoplasms indicate ongoing research aimed at enhancing early screening and diagnostic techniques.

3.2.4 Ovarian Cancer Research Topics Co-Occurrence Study

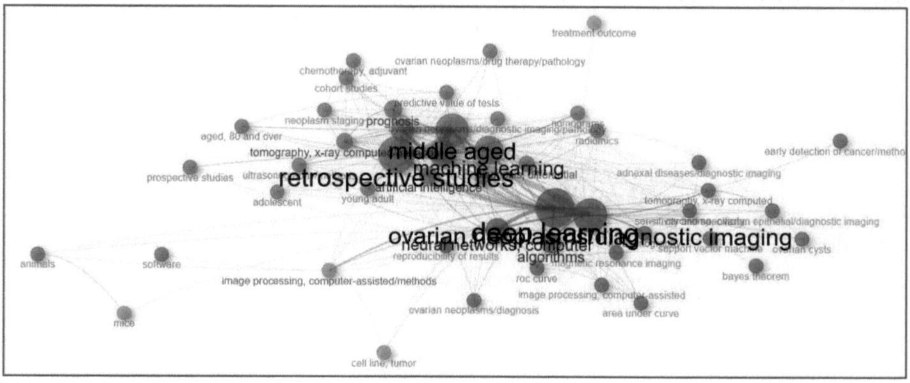

By analyzing the frequency and co-occurrence of key terms in scientific literature, this network reveals distinct research domains.

- Blue Cluster – Clinical and Epidemiological Studies: The blue cluster emphasizes clinical research, epidemiological studies, and prognostic factors related to ovarian cancer. Terms like retrospective studies, middle-aged, aged 80 and over, prospective studies, and cohort studies highlight a significant focus on patient demographics and long-term outcome evaluations. The presence of chemotherapy, adjuvant therapy, and treatment outcomes indicates the necessity for continuous assessment of therapeutic interventions and their efficacy. Ovarian neoplasms staging and prognosis underscore the significance of disease classification and outcome prediction.
- Red Cluster – Artificial Intelligence and Imaging Techniques: The red cluster indicates the growing significance of artificial intelligence (AI) and imaging technologies in the diagnosis of ovarian cancer. Key terms such as deep learning, machine learning, neural networks, computer algorithms, and image processing underscore the integration of AI-driven diagnostic tools. The significance of magnetic resonance imaging, computed tomography, and radiomics highlights the dependence on sophisticated imaging techniques for early detection and disease characterization. Furthermore, concepts such as support vector machine, area under curve, and Bayes theorem indicate the utilization of predictive models to improve diagnostic accuracy.
- Green Cluster – Experimental and Preclinical Research: The green cluster emphasizes preclinical and experimental research dimensions of ovarian cancer. Terms like cell line, tumor, and image processing, computer-assisted denote laboratory investigations centered on molecular and cellular mechanisms. This cluster indicates an increasing focus on translational research aimed at connecting experimental results with clinical applications.

- Purple and Orange Clusters – Animal Studies and Computational Tools: Smaller clusters, including the purple and orange clusters, highlight the importance of supporting research methodologies. Terminology such as mice, animals, and software indicate the continued utilization of animal models and computational tools to further ovarian cancer research.

3.2.5 Cervical Cancer Research Topics Co-Occurrence Study

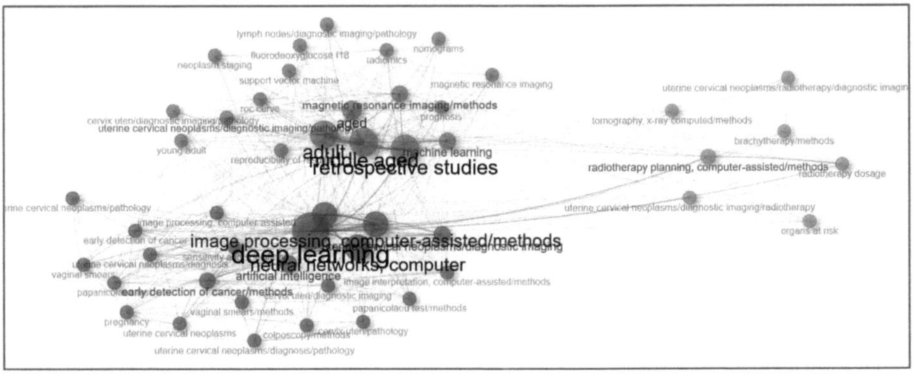

This network is derived from the frequency and co-occurrence of terms within published scientific literature.

- Red Cluster – Artificial Intelligence and Early Detection: The red cluster emphasizes the growing significance of artificial intelligence (AI) in the diagnostics and early detection of cervical cancer. Terminology including deep learning, neural networks, computer algorithms, artificial intelligence, and image processing reflects a significant research emphasis on AI-driven methodologies aimed at enhancing diagnostic precision. Furthermore, the early detection of cancer through methods such as the Papanicolaou test (Pap test), colposcopy, and vaginal smears illustrates the traditional screening approaches, highlighting the progressive integration of artificial intelligence with established cytological diagnostics.
- Blue Cluster – Clinical Studies and Imaging-Based Diagnostics: The blue cluster denotes clinical research and imaging diagnostics related to cervical cancer. Terms like retrospective studies, middle-aged, adult, aged, and neoplasm staging highlight an emphasis on patient demographics and disease progression. The utilization of magnetic resonance imaging (MRI), computed tomography (CT), radiomics, and fluorodeoxyglucose F18 indicates an increasing dependence on advanced imaging modalities for tumor characterization and treatment planning.
- Green Cluster – Radiotherapy and Treatment Planning: The green cluster emphasizes radiotherapy and treatment planning for cervical cancer. Terms like radiotherapy planning, computer-assisted methods, brachytherapy, radiotherapy dosage, and organs at risk reflect ongoing research aimed at optimizing radiotherapy techniques to enhance patient outcomes while reducing treatment-related side effects.

4 Detection of Gynaecological and Andrological Cancers with Help of AI

AI-driven biomarker identification and image analysis represent advanced methodologies transforming the detection of gynecological cancers. Biomarker identification employs artificial intelligence to examine genomic, proteomic, and metabolic data, thereby identifying molecular signatures specific to cancer. Machine learning (ML) and deep learning (DL) techniques facilitate the processing of extensive datasets derived from genomic sequencing, transcriptomics, and liquid biopsies. This approach facilitates early detection, occasionally prior to the visibility of tumors, and aids in personalized risk evaluation. Challenges include the necessity for extensive labeled datasets, the complexity of biomarker validation, and elevated costs.

Conversely, AI-driven image analysis improves the interpretation of medical imaging modalities including MRI, CT, ultrasound, and histopathology. Convolutional neural networks (CNNs) and deep learning models facilitate accurate tumor segmentation, feature extraction, and automated classification. This method is non-invasive and offers essential anatomical and morphological information, facilitating real-time clinical decision-making. Nonetheless, the accuracy is contingent upon the availability of high-quality imaging datasets, and there exists a potential for bias and misinterpretation.

AI-driven biomarker identification in gynecological cancers has facilitated the detection of CA-125 and HE4 in ovarian cancer, HPV DNA integration in cervical cancer, and MSI status or PTEN mutations in endometrial cancer. AI-assisted colposcopy enhances the diagnosis of cervical cancer, CNN-based MRI segmentation improves the detection of ovarian tumors, and digital pathology AI optimizes the grading of endometrial cancer. Biomarker identification serves important roles in screening, risk prediction, and assessing treatment response, whereas image analysis is essential for tumor classification, staging, and treatment planning.

Future integration of AI with multi-omics data and real-time biomarker analysis via liquid biopsy or wearables has the potential to enhance early detection strategies. AI-driven radiomics and histomics will improve precision oncology by providing real-time image analysis during surgical procedures. Each approach presents unique advantages; however, the integration of AI-powered biomarkers with imaging may yield a highly accurate and early detection strategy for gynecological cancers.

4.1 Detection of Gynaecological and Andrological Cancers with Help of AI Through Bio Markers

Due to a combination of an inherently complex physiological system, lack of visible symptoms and societal taboos around open discussions around them, abnormalities in and surrounding the human gynecological and andrological systems are commonly undetected till late stages. Yet, if we look for markers there are detection possibilities although quite difficult. The uncontrolled division of cell changes the basic body homeostasis. As a result, different types of inflammatory or molecular markers start to secrete or alter their quantity and quality. For example, some studies suggest that Aptamer is a synthetic antibody with a significant affinity for binding for a particular biomarker which are used to target cancers in the reproductive organs, the detection efficiency was superior to that of other probes [3].

Breast Cancer More than one out of ten new cancer diagnoses in women are breast cancer, which is also the second leading cause of cancer-related deaths among women internationally [4]. Just 5% to 10% of patients have a detectable genetic change, and 90% to 95% of cases of breast cancer are sporadic and the most prevalent linked to genes are BRCA 1 and 2 [5]. As a result, we can now see how important it is to detect breast cancer in order to prevent and save many lives. However, AI is now actively involved in this task. Deep learning, a branch of artificial intelligence, seeks to improve diagnostic results by reducing errors made by humans and bettering precision while working together with specialists [6].

There has been application of AI to forecast precise measurements of biomarkers such as a patient's HER2, Ki-67, progesterone receptor (PR), and estrogen receptor (ER) in cases of breast cancer has been investigated [7]. Pathologists generally agreed with the suggested AI result and validate it in their ultimate diagnosis, according to this analysis. In this study, the results of the suggested AI assistance were accepted by the majority of pathologists in 95.8% of Ki-67 and 93.2% of ER/PR cases [7]. A study examined CD3, CD8, and FOXP3 markers in different tumor sites using automated deep learning assessment for tumor-infiltrating lymphocytes evaluation [8].

Early detection biomarkers exhibit variable expression among patients and may fall below detectable levels in the initial stages. Therefore, peripheral blood mononuclear cells (PBMCs) exhibiting changes in gene expression due to interaction with tumor antigens may function as improved early detection biomarkers. Moreover, alterations in the immune gene profile in PBMCs are more readily detectable, despite variability among different breast cancer mutants. Research indicates that AI techniques, specifically XGBoost machine learning models combined with SHAP for explainability, can identify significant genes such as SVIP, BEND3, and MDGA2, which may serve as potential early, non-invasive diagnostic and prognostic biomarkers for breast cancer patients [9].

Ovarian Cancer Over 95% of ovarian cancers are epithelial, while the remaining 5% are most likely non-epithelial [10]. The stage of illness of a patient at detection also affects survival and recurrence rates. Initial stage of carcinoma of the ovary has a five-year survival rate of 93.1 per cent, while later-stage cancer has a five-year survival rate of 30.8% and 90 per cent of patient at stage IV experience recurrence, whereas the chance for recurrence in stage I cancer is under 10 percent [11].

A study was conducted with 202 ovarian tumor patients where the XGBoost algorithm had the best AI accuracy of 0.80, then came random forest (0.78), logistic regression (0.67), and support vector machine (0.62) and the correlation coefficient, regression coefficient, and random forest feature importance were computed in order to evaluate the clinical outcomes. Based on correlation coefficients, ascites, CRP and albumin were found to be the best predictors. On the other hand, platelet, albumin, and CA125 were the most significant in the random forest feature importance, and CA125, CA19–9, and CRP displayed the best regression coefficient [12].

In a study elsewhere, through support vector machines, serum CA125 generated by cancer has a unique and distinctive postprandial pattern that sets it apart from the typical CA125 rising in a benign disease condition [13]. With the aid of AI, this analytical study assists in cancer condition prediction.

Another study with 302 patients employed Gradient Boosting Machine (GBM), Support Vector Machine, Random Forest (RF), Conditional RF (CRF), Naïve Bayes, Neural Network, and Elastic Net to forecast evaluation with an accuracy of 0.92 using 32 clinical indicators from a blood test [14].

Endometrial Cancer The most prevalent tumor in female reproductive organs is endometrial cancer, which primarily affects postmenopausal women [15]. The overall incidence of endometrial cancer has risen by 132% in recent decades, and it is now a major global health concern [16].

In a very recent study, using a deep learning RF model, a five-biomarker group of cervico-vaginal fluid proteins (APOE, GGCT, CFAB, LY6D, and CEAM5) estimating late-stage endometrial cancer having an AUC of 0.96 (0.92–1.00) and a six-biomarker group of plasma proteins (SERPINA4, APOA2, TTR, CRP, CLEC3B, and APOD) estimated developed stage endometrial cancer in an AUC of 0.93 (0.85–0.99) [17]. The study involved 118 symptomatic post-menopausal women.

Numerous studies indicate that the SHAP value is useful for making an accurate prediction in cases of endometrial cancer. Clinicians may be prompted to think about customized treatments that target the pathways impacted by the EWRS1 protein if a SHAP analysis shows that biomarkers such as this protein have a significant impact on prognosis [18].

Cervical Cancer One of the most common gynaecologic cancers in the world is still cervical cancer. Based on the available data, it is the fourth leading cause of cancer in women globally and ranks 14th overall [19]. Over 99 percent of cervical cancers are caused by a persistent HPV infection. All over the globe, there are roughly 250,000 cervical cancer-related deaths and over 500,000 new cases of the disease each year [20]. AI is primarily used in the imaging process for cervical cancer; there is not much research comparing AI to cervical cancer biomarkers.

Prostate Cancer Prostate cancer is the fifth most common cause of cancer-related mortality among men worldwide and the most frequently diagnosed cancer in men [21]. After initial contradictory trial outcomes, in the last few 10–15 years, prostate specific antigen (PSA) tests combined with other risk calculators as well as advanced detection techniques like MRI has increased detection/classification of prostate cancer cases. [22].

Using a cascade approach and Gleason grading, a machine learning method has been built to separate prostate cancer biopsy samples into multiple classes [23]. An artificial algorithm which had high sensitivity and accuracy was designed for automated Gleason grading [24].

4.2 Detection of Gynaecological and Andrological Cancers with Help of AI Through Imaging

Imaging technology is a powerful tool for oncologists for detection, precise location, treatment planning and treatment effectiveness tracking. Numerous machine learning and deep learning algorithms have emerged to help oncologists and radiologists by automating various tasks including image augmentation, detection/segmentation/classification, biological characterization of the cancer such as staging and subtyping, prediction of clinical outcomes. [25]. The advent of digital workflows and artificial intelligence (AI) in radiology and pathology specialties presents encouraging opportunities to overcome some of the constraints [26].

From machine learning (ML) to deep learning (DL) and, more recently, transformer models which employ multimodal data as inputs, artificial intelligence (AI) systems have evolved rapidly during the past two decades. Since they can extract particular and spatial data from images utilizing many layers of data and convolutional neural network operations, convolutional neural networks (CNNs) are among the most often used deep learning architectures for imaging analysis [27].

Breast Cancer When used as a supplement to a single human reader in a double-reading evaluation workflow, the conventional computer-aided detection (CAD) system showed encouraging results and was thought to be effective [28]. This double reading detection is difficult to do manually. The first step in developing AI-CAD is gathering a sizable dataset that reflects the imaging device and target population. In order to detect breast cancer, human readers then work together to recognize and identify lesions in mammograms using verified pathological reports [29]. These improving techniques increase the accuracy to detect the abnormality and reduce the limitation of a radiologist.

Retrospective studies permit the testing of AI algorithms through a variety of simulations and make use of pre-existing data that represents target populations. For comparison, the radiologists' choices and the histopathological information were required. In order to identify screen-detected cancers, it uses a standard procedure to establish the actual situation based on at least two successive screening episodes [30]. While retaining sensitivity, the application of AI resulted in a notable decrease in biopsy requests by 27.8% and false-positive rates by 37.3%. Additionally, an independently. AI system demonstrated an AUROC improvement of 0.038 (95% CI, 0.028–0.052; $p < 0.001$), that AI performed more accurate than any manual or human hand [31].

Using input of 256 × 256-pixel patches from whole-slide images of breast sentinel lymph nodes, Wang et al. trained classification models including FaceNet, GoogLeNet Patch, VGG16 and AlexNet. The patch classification accuracy was reported as 96.8, 98.4, 97.9 and 92.1% separately. Among the algorithms, GoogLeNet had the best performance with AUC of 0.925 for whole-slide images classification. With the assistance of deep-learning system, the accuracy of pathologist's diagnoses improved significantly as the AUC increased from 0.966 to 0.995, representing ~ 85% reduction of human error rate [32].

Ovarian Cancer Al changes the traditional images and tries to reduce the errors with false interpretation. AI basically simplifies the images which were having many variations. As a result, more accuracy are given by the test reports. MRI and CT images are already giving images but with the collaboration with machine learning it gives enhance correctness.

In comparison to other algorithms, research revealed that collaborative learning methods such as RandomForest, AdaBoost, XGBoost, and Gradient Boosting Machine performed better in terms of AUC values. The AUC results of logistic regression and eight machine learning algorithms on the internal validation set were 0.957 (0.927–0.988) for XGBoost, 0.957 (0.928–0.986) for RF, 0.950 (0.918–0.982) for AdaBoost, and 0.949 (0.911–0.987) for GBM, 0.922 (0.882–0.962) for NNET, 0.916 (0.872–0.959) for LR, 0.914 (0.863–0.966) for DT, 0.882 (0.827–0.936) for SVM, and 0.860 (0.7891–0.9317) for NB [33].

Deep learning helps to identify the gene mutation through stained images, which reveal the genetical cause as well as cancer detection [34].

A study classified malignant and benign ovarian cysts based on their fluid characteristics utilizing a texture-based MRI radiomics system. Using filtration-histogram-based textural research, they discovered the fact that the intensity of pixel standard deviation was an important indicator of ovarian cyst malignancy [35].

Endometrial Cancer Just like breast cancer and ovarian cancer, AI reduces the radiomics error also in endometrial cancer. Here multiomics create its own world to detect and treat the problem. In addition to portals that facilitate the examination and presentation of multiomics data, such as Oncomine, UALCAN, LinkedOmics, and miRDB, there are a number of multiomics dataset repositories, including ones for endometrial cancer data [36].

Excellent resolution images and AI-based U/S positioning will lower the possibility of mistakes being made in the case of U/S, which is dependent on the practitioner's competence. In a comparable vein even if AI-driven states are routine, radiologists must ignore them [37]. To identify the possibility of endometrial cancer, some studies employed hysteroscopic influenced visuals of endometrial lesions. 1851 photos from 454 individuals were used in this study. After preliminary processing the photos, an initial set of 6478 photos was fed into a tuned VGGNet-16 model, and 250 photos were used to evaluate the strategy performance. Following the model's assessment, they contrasted it with the gynaecologists' diagnosis. This study demonstrates that endometrial lesion visuals can be evaluated via deep learning [38]. A retrospective study conducted to assess the efficacy of CT-based AI waist skeletal muscle volume; the results indicated that waist skeletal muscle may be a good indicator for EC and aid in disease prognosis prediction [39].

Cervical Cancer Cervical cancer detection highly depends on the imaging techniques. Many possible ways of artificial intelligence are used to detection of this neoplasm. The merged classifier was developed using three algorithms: random forest (RF), multilayer perceptron (MLP), and least square support vector machine (LSSVM). The models used had an accuracy of 99.01% at the cell level and 98.11% at the sample level (53). Efficacy is increased and observer bias reduced by cutting down on personal observation time [40].

In order to precisely identify and categorize cancer of the cervical cavity in diffuse-weighted imaging (DWI), a study created a U-Net CNN. With a dice coefficient of 0.8, a threshold of 0.89, and a favourable predictive value of 0.92, they demonstrated the most capacity for learning during image conditioning [41].

Using the new artificial pharmacokinetic images for transfer learning (APITL), some researchers created a CNN model, which they discovered performed more effectively than an SVM classification algorithm [42].

Prostate Cancer Due to its widespread incidence, it is the only andrological cancer that we have selected for our review. On non-invasive visualization, a precise distinction between aggressive and passive prostate cancer can aid in setting priorities and facilitating prompt treatment planning for patients with aggressive prostate cancer. Therapy options, such as whether to perform constant monitoring, specific counselling, or radical prostatectomy, can also be influenced by the position and size of aggressive cancer on non-invasive imaging [43].

The study assessed the prostate entire slide image structures in order to avoid spending time and money on pixel-by-pixel annotations that are manually performed. The prostate dataset (N = 24,859) produced the most satisfactory results, with an AUC of 0.989 at twenty times the transparent clarity and no discernible loss after 75% of the slides were autonomously eliminated [44].

In order to more accurately recognize, defines, identify, and estimate prognoses for prostate cancer, fresh MRI procedures are being developed to improve the resolution and acquisition duration of graphic sequences, both quantitative imaging, computer-aided diagnosis, and artificial intelligence (MR fingerprinting (MRF), segmented or multi-shot DWI, hybrid multi-dimensional MRI (HM-MRI), and illumination water imaging (LWI) [45].

4.3 Limitations

AI excels in visual image evaluation and data processing. It is becoming more prominent in clinical studies because of these features, which also lessen the burden of physicians and reduce the incidence of erroneous detection in visual checks. Yet it still carries many challenges which are opportunities for future. Even though AI algorithms have shown impressive performance in many studies, medical verification using huge data sets must be performed before implementing these algorithms in the real world.

A majority of AI algorithms are often regarded as "black boxes" due to the fact that it can be difficult to interpret and explains their behaviour. In this sense, methods for creating explainable AI algorithms may increase clinician trust, make decision-making more transparent, and reduce biases [46].

Table 1. Methodology, Datasets, AI Techniques observed

Study (year)	Title	Objectives & Methodology	Dataset	Key AI techniques included	Key Findings
Ovarian Cancer					
HASHIMOTO, 2020	Artificial Intelligence in Ovarian Cancer Diagnosis	Predict the pathological diagnosis of ovarian tumors using patient information and data from preoperative examinations	202 patients with ovarian tumors were enrolled, including 53 with ovarian cancer, 23 with borderline malignant tumors, and 126 with benign ovarian tumors	5 machine learning classifiers, including support vector machine, random forest, naive Bayes, logistic regression, and XGBoost	Highest accuracy was 0.80 in the machine learning algorithm of XGBoost

(continued)

Table 1. *(continued)*

Study (year)	Title	Objectives & Methodology	Dataset	Key AI techniques included	Key Findings
Wu, 2022	Artificial intelligence-based preoperative prediction system for diagnosis and prognosis in epithelial ovarian cancer	develop a specific interpretable machine learning (ML) prediction model for the diagnosis and prognosis of epithelial ovarian cancer (EOC) based on a variety of biomarkers	A total of 521 patients with EOC and 144 patients with benign gynecological diseases were enrolled including derivation datasets and an external validation cohort	9 supervised ML methods, through 34 parameters. Behind predicted reasons for the best ML were improved by using the SHapley Additive exPlanations (SHAP) algorithm. In addition, the prognosis of EOC was analyzed by unsupervised clustering and Kaplan-Meier (KM) survival analysis	ML technology was superior to conventional logistic regression in predicting EOC diagnosis and XGBoost performed best in the external validation datasets. The AUC values of distinguishing EOC and benign disease patients, determining pathological type, grade and clinical stage were 0.958 (0.926–0.989), 0.792 (0.701–0.8834), 0.819 (0.687–0.950) and 0.68 (0.573–0.788) respectively. For negative CA-125 EOC patients, the AUC performance of XGBoost model was 0.835(0.763–0.907). Also, SHAP analysis can improve the interpretability of ML models and contribute to precision medicine
Ho, 2022	Deep Interactive Learning-based ovarian cancer segmentation of H&E-stained whole slide images to study morphological patterns of BRCA mutation	Automated cancer segmentation using deep learning can be used to diagnose malignancy and to find novel morphological patterns to predict molecular subtypes. To train pixel-wise cancer segmentation models, manual annotation from pathologists is generally a bottleneck due to its time-consuming nature. In this paper, we propose Deep Interactive Learning with a pretrained segmentation model from a different cancer type to reduce manual annotation time	609 high-grade serous ovarian cancer cases at Memorial Sloan Kettering Cancer Center	Deep Multi-Magnification Network (DMMN) with multi-encoder, multi-decoder, and multi-concatenation for ovarian cancer segmentation. Deep Interactive Learning (DIaL) to reduce time for manual annotation	Accurately segmented cancer regions presented in H&E-stained whole slide images with intersection-over-union of 0.74, recall of 0.86, and precision of 0.84. Deep Interactive Learning (DIaL) were found useful to efficiently label multiple tissue subtypes in whole slide image-level to reduce time for manual annotation but to have accurate segmentation
Lupean, 2020	Radiomic Analysis of MRI Images is Instrumental to the Stratification of Ovarian Cysts	The imaging diagnosis of malignant ovarian cysts relies on their morphological features, which are not always specific to malignancy. The histological analysis of these cysts shows specific fluid characteristics, which cannot be assessed by conventional imaging techniques. This study investigates whether the texture-based radiomics analysis (TA) of magnetic resonance (MRI) images of the fluid content within ovarian cysts can function as a noninvasive tool in differentiating between benign and malignant lesions	Twenty-eight patients with benign ($n = 15$) and malignant ($n = 13$) ovarian cysts who underwent MRI examinations were retrospectively included. TA of the fluid component was undertaken on an axial T2-weighted sequence	Univariate, multivariate, multiple regression, and receiver operating characteristics analyses	The prediction model was able to identify malignant lesions with 84.62% sensitivity and 80% specificity (AUC = 0.841). TA of the fluid contained within the ovarian cysts can differentiate between malignant and benign lesions and potentially act as a noninvasive tool augmenting the imaging diagnosis of ovarian cystic lesions
Breast Cancer					
Wang, 2016	Deep learning for identifying metastatic breast cancer	Train classification models	Whole-slide images of breast sentinel lymph nodes	Deep Learning Networks. Classification Models. Random Forest classification	Achieved near human level classification performance. Combining deep learning with the pathologist produced a major 87% reduction in pathologist error rate

(continued)

Table 1. *(continued)*

Study (year)	Title	Objectives & Methodology	Dataset	Key AI techniques included	Key Findings
Abele, 2023	Noninferiority of Artificial Intelligence–Assisted Analysis of Ki-67 and Estrogen/Progesterone Receptor in Breast Cancer Routine Diagnostics	Image analysis of cropped sections of scanned core biopsy specimens of 204 female patients with invasive breast cancer	Cropped sections of scanned core biopsy specimens of 204 female patients with invasive breast cancer	Neural network architecture consisting of 21 convolutional layer blocks with a rectified linear unit activation function and batch normalization and separate branches for cell detection and classification	AI–human agreement was on the same level as human–human agreement, indicating that the AI tool without human intervention achieves the same level of accuracy as pathologists
Kumar S, 2023	Peripheral blood mononuclear cell derived biomarker detection using eXplainable Artificial Intelligence (XAI) provides better diagnosis of breast cancer	Identification of blood-based biomarkers for breast tumor diagnostics	PBMCs of breast cancer patients and healthy individuals	XAI applied to XGBoost model	Predicted key genes i.e. SVIP, BEND3, MDGA2, LEF1-AS1, PRM1, TEX14, MZB1, TMIGD2, KIT, and FKBP7. The predicted biomarkers could help in early detection even in case of triple-negative breast cancer. Classification with 96.7% accuracy
Sun, 2019	Multi-View Convolutional Neural Networks for Mammographic Image Clas-sification	Most of the existing applications of deep learning to ammographic image classification use single mammography view. To improve the mammographic image classification performance, this study uses a novel multi-view convolutional neural network (CNN) based on multiple mammography views	Mediolateral oblique (MLO) and craniocaudal (CC) views of the mammographic images. 1,445 samples selected from Digital Database for Screening Mammography (DDSM) database	Multi-view convolutional neural network (CNN). Utilize two subnetworks to extract features from the MLO and CC views of the breast mass images. Embed the dilated convolution layers with different factors. Add penalty term to the cross entropy loss function to reduce misclassification rate	Accuracy improvements of the study methods were 8.76%-10.60% over the baseline LeNet model and 2.76%-4.60% over the LeNet-BN. This aspect illustrates that the proposed multi-view based methods can extract more powerful features from two different views of the breast mass images simultaneously compared to those obtained by the LeNet and LeNet-BN
Shen, 2021	Artificial intelligence system reduces false-positive findings in the interpre-tation of breast ultrasound exams	Though consistently shown to detect mammographically occult cancers, breast ultrasound has been noted to have high false-positive rates. This study presents an AI system that achieves radiologist-level accuracy in identifying breast cancer in ultrasound images	44,755 Ultrasound (US) exams (25,003 patients, 79,156 breasts)	Deep learning model (DLM) with CNN and Gated Attention Mechanism	The ten radiologists achieved an average AUROC of 0.924 (SD: 0.020, 95% CI: 0.905, 0.944) and an average AUPRC of 0.565 (SD: 0.072, 95% CI: 0.465, 0.625). Compared to the average radiologist in this study, the AI system achieved a higher AUROC of 0.962 (95% CI: 0.943, 0.979) with an AUROC improvement of 0.038 (95% CI: 0.028, 0.052, $P < 0.001$) and a higher AUPRC of 0.752 (95% CI: 0.675, 0.849) with an AUPRC improvement of 0.187 (95% CI: 0.140, 0.256, $P < 0.001$)
Endometrial Cancer					
Zhang, 2021	Deep learning model for classifying endometrial lesions	All histopathologically confirmed endometrial lesion images were obtained from the Shengjing Hospital of China Medical University, including endometrial hyperplasia without atypia, atypical hyperplasia, endometrial cancer, endometrial polyps, and submucous myomas	The study included 1851 images from 454 patients	After the images were preprocessed (histogram equalization, addition of noise, rotations, and flips), a training set of 6478 images was input into a tuned VGGNet-16 model	The overall accuracy of the VGGNet-16 model in classifying endometrial lesions is 80.8%

(continued)

Table 1. (*continued*)

Study (year)	Title	Objectives & Methodology	Dataset	Key AI techniques included	Key Findings
Kim, 2021	Prognostic role of computed tomography-based, artificial intelligence-driven waist skeletal muscle volume in uterine endometrial carcinoma	To investigate the impact of computed tomography (CT)-based, artificial intelligence-driven waist skeletal muscle volume on survival outcomes in patients with endometrial cancer. The retrospectively identified endometrial cancer patients who received primary surgical treatment	pre-treatment CT scans were available (n = 385)	Using an artificial intelligence-based tool, the skeletal muscle area (cm2) at the third lumbar vertebra (L3) and the skeletal muscle volume (cm3) at the waist level were measured. These values were converted to the L3 skeletal muscle index (SMI) and volumetric SMI by normalisation with body height. The relationships between L3, volumetric SMIs, and survival outcomes were evaluated	Setting 39.0 cm2/m2 of L3 SMI as cut-off value for sarcopenia, sarcopenia ($<$ 39.0 cm2/m2, n = 177) and non-sarcopenia ($\geq$ 39.0 cm2/m2, n = 208) groups showed similar progression-free survival (PFS; p = 0.335) and overall survival (OS; p = 0.241)
Cervical Cancer					
Bora, 2017	Automated classification of Pap smear images to detect cervical dysplasia	The main objective of this database generation is to categorize the images according to the Bethesda system of classification both of which require lots of training and expertise	The system is evaluated on two generated databases obtained from two diagnostic centers, one containing 1610 single cervical cells and the other 1320 complete smear level images	Ripplet Type I transform, Histogram first order statistics and Gray Level Co-occurrence Matrix	Pap smear images performing significantly better
Lin, 2020	Deep learning for fully automated tumor segmentation and extraction of magnetic resonance radiomics features in cervical cancer	Evaluate the performance of U-Net for fully automated localization and segmentation of cervical tumors in magnetic resonance (MR) images	images from 169 patients with cervical cancer stage IB-IVA, diffusion-weighted (DW) images from 144 patients, another 25 patients were recruited for testing	U-Net convolutional network	Combining b0, b1000, and ADC images as a triple-channel input exhibited the highest learning efficacy in the training phase and had the highest accuracy in the testing dataset, with a dice coefficient of 0.82, sensitivity 0.89, and a positive predicted value 0.92
Shao, 2020	DCE-MRI pharmacokinetic parameter maps for cervical carcinoma prediction	study aims to develop machine learning approaches for cervical carcinoma prediction based on pharmacokinetic parameters	In total 26 parameters were derived from the 5 pharmacokinetic models	support vector machines (SVMs), APITL (artificial pharmacokinetic images for transfer learning)	parameter possesses very significant power to differentiate cancerous tissue from normal tissue (with a p-value $< 10e-8$)
Postate Cancer					
Campanella, 2019	Clinical-grade computational pathology using weakly supervised deep learning on whole slide images	Uses only the reported diagnoses as labels for training, thereby avoiding expensive and time-consuming pixel-wise manual annotations	Evaluated this framework at scale on a dataset of 44,732 whole slide images from 15,187 patients	Multiple instance learning-based deep learning system	This system has the ability to train accurate classification models at unprecedented scale
Chatterjee, 2022	New prostate MRI techniques and sequences	Improve prostate cancer (PCa) diagnosis which are reviewed in this article	Did not know exact number	Hybrid multi-dimensional MRI; Hyperpolarized 13C MRI; Luminal water imaging; MR fingerprinting	HM-MRI measures volumes of prostate tissue components, where PCa has reduced lumen and stromal and increased epithelium volume similar to results shown in histology
Postate Cancer					

(*continued*)

Table 1. *(continued)*

Study (year)	Title	Objectives & Methodology	Dataset	Key AI techniques included	Key Findings
Doyle, 2012	Cascaded discrimination of normal, abnormal, and confounder classes in histopathology: Gleason grading of prostate cancer	A cascaded (CAS) approach to classifying prostate biopsy tissue samples, where images from different classes are grouped to maximize intra-group homogeneity while maximizing inter-group heterogeneity	2000 tissue samples taken from 214 patient studies	Seven classes: epithelium, stroma, atrophy, prostatic intraepithelial neoplasia (PIN), and prostate cancer Gleason grades 3, 4, and 5. A series of increasingly granular binary classifiers are used to split the different tissue classes until the images have been categorized into a single unique class. Automatically-extract image feature set includes architectural features based on location of the nuclei within the tissue sample as well as texture features extracted on a per-pixel level	The CAS strategy yields a positive predictive value (PPV) of 0.86 in classifying the 2000 tissue images into one of 7 classes, compared with the OVA (0.77 PPV) and OSC approaches (0.76 PPV). In classification problems such as histopathology, where multiple class groups exist with varying degrees of heterogeneity, the CAS system can intelligently assign class labels to objects by performing multiple binary classifications according to domain knowledge
Marginean, 2021	An Artificial Intelligence-based Support Tool for Automation and Stand-ardisation of Gleason Grading in Prostate Biopsies	Develop an artificial intelligence (AI) algorithm, based on machine learning and convolutional neural networks, as a tool for improved standardisation in Gleason grading in prostate cancer biopsies	A total of 698 prostate biopsy sections from 174 patients were used for training. The training sections were annotated by two senior consultant pathologists. The final algorithm was tested on 37 biopsy sections from 21 patients, with digitised slide images from two different scanners	CNN based algorithm trained on annotated scans	The algorithms used show high accuracy in detecting cancer areas (sensitivity: 100%, specificity: 68%). Compared with the pathologists, the algorithm also performed well in detecting cancer areas (intraclass correlation coefficient [ICC]: 0.99) and assigning the Gleason patterns correctly: Gleason patterns 3 and 4 (ICC: 0.96 and 0.94, respectively), and to a lesser extent, Gleason pattern 5 (ICC: 0.82)

5 Conclusions and Authors' Perspective

Our goal in this review was to study the current state of application of artificial intelligence in detection of andrological and gynaecological cancers. 'We discussed about significant research that has already been done and their accuracy levels to identify a particular type of cancer. With results that have already been found, we also expressed our concern about correlation and forecasting with high quality efficacy. We looked at two separate areas of detection methods where AI and ML could be applicable – biomarkers and images. We found evidence of successful usage of AI, machine learning and imaging techniques to assist oncologists and radiologists. Beyond accuracy, we inferred that explainability and interpretability would be critical factors in putting AI and machine learning to use in clinical usage for detection and treatment of andrological and gynaecological cancers.

Disclosure of Interests. The authors have no competing interests.

Abbreviations

AI Artificial Intelligence

DL	Deep learning
ML	Machine learning
BRCA	Breast cancer gene
HER2	Human epidermal growth factor receptor 2
Ki-67	Antigen Kiel 67
PR	Progesterone receptor
ER	Oestrogen Receptor
CD3	Cluster of differentiation 3
FOXP3	Forkhead box protein P3
CRP	C-reactive protein
CA-125	Cancer antigen 125
GBM	Used Gradient Boosting Machine
SVM	Support Vector Machine
RF	Random Forest
CRF	Conditional RF
APOE	Apolipoprotein E
HPV	Human papillomavirus
PSA	Prostate-specific antigen
CNN	Convolutional neural network
(CAD)	Computer-aided detection
MLP	Multilayer perceptron
LSSVM	Least square support vector machine
MRF	MR fingerprinting
HM-MRI	Hybrid multi-dimensional MRI
LWI	Illumination water imaging

References

1. Bray, F. L.: Global cancer statistics 2022: GLOBOCAN estimates of incidence and mortality worldwide for 36 cancers in 185 countries. CA: a cancer journal for clinicians, 74(3), 229–263. https://doi.org/10.3322/caac.21834 (2024)
2. Hunter, B.H.: The role of artificial intelligence in early cancer diagnosis. Cancers **14**(6), 1524 (2022). https://doi.org/10.3390/cancers14061524
3. Xiaoqun Ma, T.L.: Recent advances in identifying biomarkers and high-affinity aptamers for gynecologic cancers diagnosis and therapy. J. Anal. Methods Chem. (2019). https://doi.org/10.1155/2019/5426974
4. Watkins EJ.: Overview of breast cancer. JAAPA, (10):13–17. https://doi.org/10.1097/01.JAA.0000580524.95733.3d (2019)
5. Cancer, C. G. (n.d.).: Collaborative Group on Hormonal Factors in Breast Cancer Familial breast cancer: collaborative reanalysis of individual data from 52 epidemiological studies including 58,209 women with breast cancer and 101,986 women without the disease. Lancet (London, England), **358**(9291), 1389–1399. https://doi.org/10.1016/S0140-6736(01)06524-2 (2001)
6. Cui, M. &.: Artificial intelligence and computational pathology. Laboratory investigation; a journal of technical methods and pathology, **101**(4), 412–422. https://doi.org/10.1038/s41374-020-00514-0 (2021)

7. Abele, N. T.: Noninferiority of Artificial Intelligence-Assisted Analysis of Ki-67 and Estrogen/Progesterone Receptor in Breast Cancer Routine Diagnostics. an official journal of the United States and Canadian Academy of Pathology, Inc, **36**(3), 100033. https://doi.org/10.1016/j.modpat.2022.100033 (2023)
8. Soliman, A.L.: Artificial intelligence's impact on breast cancer pathology: a literature review. Diagn. Pathol. **19**(38), 2024 (2024). https://doi.org/10.1186/s13000-024-01453-w
9. Kumar S.: Das A Peripheral blood mononuclear cell derived biomarker detection using eXplainable Artificial Intelligence (XAI) provides better diagnosis of breast cancer. Computational Biology and Chemistry, https://doi.org/10.1016/j.compbiolchem.2023.107867 Volume 104 (2023)
10. Lheureux, S., Braunstein, M., & Oza, A. M.: Epithelial ovarian cancer: Evolution of management in the era of precision medicine. CA: a cancer journal for clinicians, **69**(4), 280–304. https://doi.org/10.3322/caac.21559 (2019)
11. Burke, W.B.-K., et al.: Executive summary of the ovarian cancer evidence review conference. Obstet. Gynecol. **142**(1), 179–195 (2023). https://doi.org/10.1097/AOG.0000000000005211
12. HASHIMOTO, M. A. Artificial Intelligence in Ovarian Cancer Diagnosis. Anticancer Research August 2020, **40**(8) 4795–4800; https://doi.org/10.21873/anticanres.14482 (2020)
13. Gu, Z.H.: Postprandial increase in serum CA125 as a surrogate biomarker for early diagnosis of ovarian cancer. J. Transl. Med. **16**(1), 114 (2018). https://doi.org/10.1186/s12967-018-1489-4
14. Kawakami, E.T.: Application of artificial intelligence for preoperative diagnostic and prognostic prediction in epithelial ovarian cancer based on blood biomarkers. Clinical cancer research: an official journal of the American Association for Cancer Research **25**(10), 3006–3015 (2019). https://doi.org/10.1158/1078-0432.CCR-18-3378
15. Paleari, L.P.: New insights into endometrial cancer. Cancers **13**(7), 1496 (2021). https://doi.org/10.3390/cancers13071496
16. Crosbie, E.J.: Endometrial cancer. Lancet (London, England) **399**(10333), 1412–1428 (2022). https://doi.org/10.1016/S0140-6736(22)00323-3
17. Njoku K., Pierce A., Chiasserini D., Geary B., Campbell A.E., Kelsall J., Reed R., Geifman N., Whetton A.D., Crosbie E.J.: Detection of endometrial cancer in cervico-vaginal fluid and blood plasma: leveraging proteomics and machine learning for biomarker discovery. EBioMedicine **102**, 105064. https://doi.org/10.1016/j.ebiom.2024.105064 (2024)
18. Yasar, S.Y.: Integrating proteomics and explainable artificial intelligence: a comprehensive analysis of protein biomarkers for endometrial cancer diagnosis and prognosis. Front. Mol. Biosci. **11**, 1389325 (2024). https://doi.org/10.3389/fmolb.2024.1389325
19. Brisson, M.: Global elimination of cervical cancer as a public health problem. . The Lancet. Oncology, **20**(3), 319–321 (2019). https://doi.org/10.1016/S1470-2045(19)30072-5
20. Habtemariam, L. W.: Cervix Type and Cervical Cancer Classification System Using Deep Learning Techniques. Medical devices (Auckland, N.Z.), 15, 163–176 (2022). https://doi.org/10.2147/MDER.S366303
21. Mattiuzzi, C. &. Current Cancer Epidemiology. . Journal of epidemiology and global health, **9**(4), 217–222 (2019). https://doi.org/10.2991/jegh.k.191008.001
22. Van Poppel, H., Albreht, T., Basu, P. et al. Serum PSA-based early detection of prostate cancer in Europe and globally: past, present and future. Nat Rev Urol 19, 562–572 (2022). https://doi.org/10.1038/s41585-022-00638-6
23. Doyle, S.F.: Cascaded discrimination of normal, abnormal, and confounder classes in histopathology: Gleason grading of prostate cancer. BMC Bioinformatics **13**, 282 (2012). https://doi.org/10.1186/1471-2105-13-282
24. Marginean, F.A.: An artificial intelligence-based support tool for automation and standardisation of gleason grading in prostate biopsies. Eur. Urol. Focus **7**(5), 995–1001 (2021). https://doi.org/10.1016/j.euf.2020.11.001

25. Zhang J., Wu J., Zhou X.S., Shi F., Shen D.: Recent advancements in artificial intelligence for breast cancer: Image augmentation, segmentation, diagnosis, and prognosis approaches. Seminars in Cancer Biology. https://doi.org/10.1016/j.semcancer.2023.09.001 Pages 11–25, Volume 96 (2023)
26. Ahn, J.S.: Artificial intelligence in breast cancer diagnosis and personalized medicine. J. Breast Cancer **26**(5), 405–435 (2023). https://doi.org/10.4048/jbc.2023.26.e45
27. LeCun, Y.B.: Deep learning. Nature **521**(7553), 436–444 (2015). https://doi.org/10.1038/nature14539
28. Gilbert, F. J. Single reading with computer-aided detection for screening mammography. . The New England journal of medicine, **359**(16), 1675–1684. https://doi.org/10.1056/NEJMoa0803545(2008)
29. Sun, L., Multi-View, J. W.: Convolutional Neural Networks for Mammographic Image Classification. IEEE Access, vol. 7, pp. 126273–126282 (2019). https://doi.org/10.1109/ACCESS.2019.2939167
30. McKinney, S.M.-V.-B.: International evaluation of an AI system for breast cancer screening. Nature **577**(7788), 89–94 (2020). https://doi.org/10.1038/s41586-019-1799-6
31. Shen, Y.S.-S.: Artificial intelligence system reduces false-positive findings in the interpretation of breast ultrasound exams. Nat. Commun. **12**(1), 5645 (2021). https://doi.org/10.1038/s41467-021-26023-2
32. Wang, D., Khosla, A., Gargeya, R., Irshad, H. and Beck, A.H.: Deep learning for identifying metastatic breast cancer. arXiv preprint (2016) arXiv:1606.05718
33. Wu M, Z. Y.: Artificial intelligence-based preoperative prediction system for diagnosis and prognosis in epithelial ovarian cancer: A multicenter study. Front. Oncol. 12:975703. (2022) https://doi.org/10.3389/fonc.2022.975703
34. Ho, D.J.: Deep Interactive Learning-based ovarian cancer segmentation of H&E-stained whole slide images to study morphological patterns of BRCA mutation. Journal of pathology informatics **14**, 100160 (2022). https://doi.org/10.1016/j.jpi.2022.100160
35. Lupean, R.A.: Radiomic analysis of MRI images is instrumental to the stratification of ovarian cysts. Journal of personalized medicine **10**(3), 127 (2020). https://doi.org/10.3390/jpm10030127
36. Boroń, D.Z.-S.: Recent multiomics approaches in endometrial cancer. Int. J. Mol. Sci. **23**(3), 1237 (2022). https://doi.org/10.3390/ijms23031237
37. Hardy, M.: Artificial intelligence in diagnostic imaging: impact on the radiography profession. The British journal of radiology, **93**(1108), 20190840. (2020) https://doi.org/10.1259/bjr.20190840
38. Zhang, Y.W.: Deep learning model for classifying endometrial lesions. J. Transl. Med. **19**(1), 10 (2021). https://doi.org/10.1186/s12967-020-02660-x
39. Kim, S.I.: Prognostic role of computed tomography-based, artificial intelligence-driven waist skeletal muscle volume in uterine endometrial carcinoma. Insights Imaging **12**(1), 192 (2021). https://doi.org/10.1186/s13244-021-01134-y
40. Bora, K.C.: Automated classification of Pap smear images to detect cervical dysplasia. Comput. Methods Programs Biomed. **138**, 31–47 (2017). https://doi.org/10.1016/j.cmpb.2016.10.001
41. Lin, Y.C.: Deep learning for fully automated tumor segmentation and extraction of magnetic resonance radiomics features in cervical cancer. Eur. Radiol. **30**(3), 1297–1305 (2020). https://doi.org/10.1007/s00330-019-06467-3
42. Shao, J.Z.: DCE-MRI pharmacokinetic parameter maps for cervical carcinoma prediction. Comput. Biol. Med. **118**, 103634 (2020). https://doi.org/10.1016/j.compbiomed.2020.103634
43. Ahmed, H. U.-S.-M.: Diagnostic accuracy of multi-parametric MRI and TRUS biopsy in prostate cancer (PROMIS): a paired validating confirmatory study. Lancet (London, England). (2017) https://doi.org/10.1016/S0140-6736(16)32401-1

44. Campanella, G.H.: Clinical-grade computational pathology using weakly supervised deep learning on whole slide images. Nat. Med. **25**(8), 1301–1309 (2019). https://doi.org/10.1038/s41591-019-0508-1
45. Chatterjee, A.H.: New prostate MRI techniques and sequences. Abdominal radiology (New York) **45**(12), 4052–4062 (2020). https://doi.org/10.1007/s00261-020-02504-8
46. Gastounioti, A., Kontos, D.: Is it time to get rid of black boxes and cultivate trust in AI? Radiology. Artificial intelligence **2**(3), e200088 (2020). https://doi.org/10.1148/ryai.2020200088

Developing a Framework Driven by Analytics to Forecast the Binding Free Energy Level and, Consequently, the New Inhibitor Nature of Medicinal Chemical Compounds

Ashutosh Kar[1], Dipayan Chakraborty[2(✉)], Sangita Mazumder[1], and Pratyay Ranjan Datta[1]

[1] NSHM Institute of Computing and Analytics, Kolkata, India
mazumdersangi93@gmail.com, pratyaydata@gmail.com
[2] TCS, Kolkata, India
dipayanchakrabortymail@gmail.com

Abstract. Pharmaceuticals play a critical role in both the prevention and treatment of diseases. However, the drug development process is often lengthy, complex, and unpredictable, despite the growing demand for novel and effective treatments. To streamline this process, there is a pressing need for virtual in-silico screening frameworks that can reduce computation time, minimize the selection of non-viable compounds, and improve the accuracy of potential drug candidates.

This framework is based on the hypothesis that low binding free energy is a key indicator of a compound's ability to function as an effective inhibitor for a specific target protein. By focusing on this metric, we aim to efficiently classify compounds with high therapeutic potential while eliminating those with less favorable profiles. Our study focuses on the Sirtuin6 protein, a member of the class-III histone deacetylase enzyme family. Sirtuin6 is critical for regulating a wide range of biological processes, including transcription, cell survival, and the de-acylation of proteins such as histones and transcription factors. Given its role in longevity and metabolic regulation, Sirtuin6 is an attractive target for drug development.

To enhance the efficiency of the virtual screening process, we employ a machine-learning-driven approach utilizing supervised learning algorithms, including Decision Trees, Random Forests, Logistic Regression, and Support Vector Machines (SVM). These models help optimize resource usage, accelerate computational tasks, and improve the selection of novel inhibitors with low binding free energy.

Keywords: Binding Free Energy · Novel Inhibitors · Machine Learning · Support Vector Machine (SVM) · Virtual Screening · Drug Discovery

1 Introduction

Despite the constant demand for novel and efficient medications, the processes of drug development and design are often slow and unpredictable. This research was driven by the need for a streamlined, data-driven framework based on supervised learning that

S. Dhar et al. (Eds.): AGC 2025, CCIS 2686, pp. 39–64, 2026.
https://doi.org/10.1007/978-3-032-09153-6_3

could quickly classify or label drug molecules into High BFE and Low BFE categories (binding free energy). The aim is to ensure that the most promising drug molecules are identified for molecular docking and can potentially become novel inhibitors.

The study focuses on employing various classification methods to categorize drug molecules according to their binding free energy and to recommend the most suitable model for accurately predicting these classifications. By dividing the binding free energy (BFE) into two classes—High-level and Low-level—the research seeks to develop a classification model that uses continuous input variables and produces categorical outputs, enabling efficient identification of drug candidates based on their binding properties.

1.1 Objectives of the Research

i. To filter out unsuitable compounds from initial small molecule libraries, using a classification system to screen for potential therapeutic candidates against specific target proteins.
ii. To predict the binding free energy (BFE) and binding constant for a drug candidate's activity at the target site.
iii. To identify novel pharmacologically inhibitory compounds (Low BFE) that could serve as potential inhibitors

2 Literature Review

A significant amount of research has been conducted to predict the binding free energy (BFE) and binding constant of drug candidates at target sites. This section reviews relevant studies and methodologies for identifying novel inhibitory (low-BFE) compounds.

2.1 Methodology

An extensive review of the literature was undertaken to identify synthetic compounds with potential as novel inhibitors (low-BFE). Relevant research articles were analyzed through a systematic process of identification, synthesis, scoping, and planning.

2.2 Scoping

Various machine learning techniques, such as decision trees, logistic regression, random forests, and support vector machines (SVM), were examined to develop predictive models. These models were aimed at forecasting the binding free energy (BFE) and binding constant of drug candidates at target sites. The primary objective was to improve predictive accuracy in identifying low-BFE compounds, which are more likely to serve as effective inhibitors.

2.3 Planning

Relevant literature was identified and reviewed during the planning phase. Articles that, although not directly related, provided significant insights were included to ensure a balance between sensitivity and specificity. Journal databases such as Web of Science, Science Direct, Emerald, and JSTOR were searched using key terms such as "binding free energy," "small molecules," "novel inhibitors," and "classification algorithms."

The inclusion and exclusion criteria were established as follows:

2.3.1 Inclusion Criteria

Articles applying classification algorithms to filter small molecule libraries, identifying compounds unsuitable for screening against specific target proteins. Also, the Publications forecasting binding free energy (BFE) levels. Besides, the Studies addressing a drug candidate's binding constant at the target site. And lastly, Researches focused on novel inhibitory pharmaceutical substances with low-BFE

2.3.2 Exclusion Criteria

Articles emphasizing small molecules with strong binding affinities that were not relevant to the study's classification objectives were excluded from further analysis.

2.4 Identification

Published and unpublished articles were identified through a snowball review approach (Wohlin, 2014). References from original papers were reviewed, and relevant contributions from citing articles were incorporated. This iterative method ensured comprehensive coverage of relevant research, while limitations from previous studies were noted to outline future research directions.

2.4.1 Synthesis of Literature Review

Once the relevant literature was identified, it was imported into Zotero (reference manager) for final inclusion. This process ensured that the most pertinent studies were synthesized, providing a balanced and thorough review of the existing research (Tables 1 and 2) .

Table 1. Classification of Themes

Serial No.	Theme
A	The papers concern the application of a classification algorithm for a drug candidate to filter the first small molecule libraries that contain compounds that might be considered inappropriate for screening against a certain target protein.
B	Publications that forecast the binding free energy level (BFE).
C	The publications having information about a drug candidate's binding constant for its action at the target location.
D	Researches that concentrate on the new inhibitor (low-BFE) pharmaceutical substances.

Table 2. Synthesis of Literature

Serial No	Citation	Reason of Inclusion	Themes
1	Armutlu, P., Ozdemir, M.E., Uney-Yuksektepe, F., Kavakli, I.H., & Turkay, M. (2008)	Discusses classification of drug molecules using computational approaches.	A, C
2	Armutlu, P., Ozdemir, M.E., Ozdas, S., Kavakli, I.H., & Turkay, M. (2009).	Focuses on discovery of novel inhibitors for prostate cancer.	D
3	Bento, A.P., Gaulton, A., Hersey, A., Bellis, L.J., Chambers, J., Davies, M., Kruger, F.A., Light, Y., Mak, L., McGlinchey, S., Nowotka, M., Papadatos, G., Santos, R., & Overington, J.P. (2014).	Provides a comprehensive database for bioactivity data, useful for screening and predictive modeling.	A, B
4	Bolton, E.E., Wang, Y., Thiessen, P.A., & Bryant, S.H. (2008).	Offers a platform for small molecules and biological activity data, aiding in drug discovery.	A, B
5	Cakir, B., Dagliyan, O., Dagyildiz, E., Baris, I., Kavakli, I.H., Kizilel, S., & Turkay, M. (2012).	Investigates structure-based drug design for regulating enzyme activity.	D
6	Cleary, J.G., & Trigg, L.E. (1995).	Describes a classification method applicable in drug discovery.	A
7	Dagliyan, O., Kavakli, I.H., & Turkay, M. (2009).	Discusses classification of inhibitors based on binding free energy and molecular descriptors.	A, B
8	Bolton, E.E., Wang, Y., Thiessen, P.A., & Bryant, S.H. (2008).	Applies optimization techniques for tumor classification.	A
9	Etchegaray, J.-P., Zhong, L., & Mostoslavsky, R. (2013).	Reviews the role of SIRT6 in disease and metabolism.	D
10	Finkel, T., Deng, C.-X., & Mostoslavsky, R. (2009)	Discusses the biological role of sirtuins, relevant to drug discovery.	D
11	Frank, E., Hall, M., Trigg, L., Holmes, G., & Witten, I.H. (2004).	Provides insights on using machine learning for bioinformatics, applicable to drug discovery.	A
12	Friedman, J., Hastie, T., & Tibshirani, R. (2000).	Introduces boosting techniques for model optimization, useful in drug discovery	A
13	Frye, R.A. (2000).	Explores the classification of Sir2 proteins, relevant for drug targeting.	A

(*continued*)

Table 2. (*continued*)

Serial No	Citation	Reason of Inclusion	Themes
14	Heckerman, D. (1998).	Discusses Bayesian networks, applicable in drug discovery for predictive modeling.	A
15	IBM ILOG. (2010).	Provides information on using CPLEX for optimization problems in drug discovery	A
16	Irwin, J.J., & Shoichet, B.K. (2005).	Offers a database for virtual screening in drug discovery	A, B
17	Jorgensen, W.L. (2004).	Reviews computational approaches in drug discovery.	A, B, D
18	Kahraman, P., & Turkay, M. (2007).	Discusses classification of calcium channel antagonists.	A
19	Kaidi, A., Weinert, B.T., Choudhary, C., & Jackson, S.P. (2010)	Studies the role of SIRT6 in DNA repair.	D
20	Kanfi, Y., Naiman, S., Amir, G., Peshti, V., Zinman, G., Nahum, L., Bar-Joseph, Z., & Cohen, H.Y. (2008)	Investigates SIRT6 regulation, relevant to drug discovery.	D
21	Kawahara, T.L.A., Michishita, E., Adler, A.S., Damian, M., Berber, E., Lin, M., McCord, R.A., Ongaigui, K.C.L., Boxer, L.D., Chang, H.Y., & Chua, K.F. (2009).	Explores the role of SIRT6 in gene expression and lifespan.	D
22	Kahraman, P., & Turkay, M. (2007).	Reviews docking and scoring techniques for virtual screening.	A, B
23	Kaidi, A., Weinert, B.T., Choudhary, C., & Jackson, S.P. (2010)	Studies the role of SIRT6 in DNA repair.	D
24	Lipinski, C.A., Lombardo, F., Dominy, B.W., & Feeney, P.J. (2012).	Reviews solubility and permeability estimation in drug discovery.	B
25	Liszt, G., Ford, E., Kurtev, M., & Guarente, L. (2005)	Studies SIRT6 enzyme activity, relevant to drug targeting.	D
26	MacKerell, A.D., Bashford, D., Bellott, M., Dunbrack, R.L., Evanseck, J.D., Field, M.J., Fischer, S., Gao, J., Guo, H., Ha, S., Joseph-McCarthy, D., Kuchnir, L., Kuczera, K., Lau, F.T.K., Mattos, C., Michnick, S., Ngo, T., Nguyen, D.T., Prodhom, B., Reiher, W.E., Roux, B., Schlenkrich, M., Smith, J.C., Stote, R., Straub, J., Watanabe, M., Wiorkiewicz-Kuczera, J., Yin, D., & Karplus, M. (1998).	Provides computational tools for protein modeling.	B
27	Mao, Z., Hine, C., Tian, X., Van Meter, M., Au, M., Vaidya, A., Seluanov, A., & Gorbunova, V. (2011).	Investigates the role of SIRT6 in DNA repair mechanisms	D
28	Michishita, E., Park, J.Y., Burneskis, J.M., Barrett, J.C., & Horikawa, I. (2005).	Reviews the functions of sirtuin proteins, relevant for drug discovery.	D
29	Mitchell, T., & Showell, G.A. (2001).	Discusses strategies for designing drug-like compounds.	D

(*continued*)

Table 2. (*continued*)

Serial No	Citation	Reason of Inclusion	Themes
30	Mostoslavsky, R., Chua, K.F., Lombard, D.B., Pang, W.W., Fischer, M.R., Gellon, L., Liu, P., Mostoslavsky, G., Franco, S., Murphy, M.M., Mills, K.D., Patel, P., Hsu, J.T., Hong, A.L., Ford, E., Cheng, H.L., Kennedy, C., Dausman, J., Bronson, R., & Alt, F.W. (2006). G315–329.	Studies the role of SIRT6 in genomic stability and aging.	D
31	Pan, P.W., Feldman, J.L., Devries, M.K., Dong, A., Edwards, A.M., & Denu, J.M. (2011).	Explores the structure and function of SIRT6, relevant for drug discovery.	D
32	Phillips, J.C., Braun, R., Wang, W., Gumbart, J., Tajkhorshid, E., Villa, E., Chipot, C., Skeel, R.D., Kale, L., & Schulten, K. (2005).	Provides computational tools for molecular dynamics simulations.	B
33	Platt, J. (1999)	Discusses optimization techniques for SVMs, applicable in drug discovery.	A
34	Qu, Y., Adam, B.-L., Yasui, Y., Ward, M.D., Cazares, L.H., Schellhammer, P.F., Feng, Z., Semmes, O.J., & Wright, G.L. (2002)	Applies machine learning for cancer detection.	A
35	Rosenthal, R.E. (2015)	Provides guidance on using GAMS for optimization in drug discovery.	A
36	Szachniuk, M., De Cola, M.C., Felicia, G., & Blazewicz, J. (2014).	Discusses algorithms applicable in drug discovery.	A
37	Trott, O., & Olson, A.J. (2010).	Introduces tools for virtual screening in drug discovery.	A, B
38	Uney, F., & Turkay, M. (2006)	Discusses classification techniques in drug discovery.	A
39	Villar, H.O., & Hansen, M.R. (2009).	Reviews strategies for chemical library design.	A
40	Whitley, D.C., Ford, M.G., & Livingstone, D.J. (2000).	Introduces methods for feature selection in predictive modeling.	A
41	Witten, I.H., & Frank, E. (2005).	Discusses machine learning tools, relevant for drug discovery.	A
42	Wold, S., Sjöström, M., & Eriksson, L. (2001).	Reviews PLS regression for chemometrics, applicable in drug discovery.	B
43	Yap, C.W. (2011)	Provides tools for calculating molecular descriptors, relevant for drug discovery.	A
44	Zhong, L., D'Urso, A., Toiber, D., Sebastian, C., Henry, R.E., Vadysirisack, D.D., Guimaraes, A., Marinelli, B., Wikstrom, J.D., Nir, T., Clish, C.B., Vaitheesvaran, B., Iliopoulos, O., Kurland, I.J., & Mostoslavsky, R. (2010).	Investigates the role of SIRT6 in glucose metabolism.	D
45	Djokovic, N., Rahnasto-Rilla, M., Lougiakis, N., Lahtela-Kakkonen, M., & Nikolic, K. (2023).	Introduces a predictive model for SIRT2 inhibitors.	A, D
46	Sakkiah, S., Baek, A., & Lee, K.W. (2012).	Applies molecular modeling for SIRT2 inhibitor discovery.	D
47	Karaman, B., & Sippl, W. (2015).	Discusses docking and binding free energy for sirtuin inhibitors.	B, D
48	Kokkonen, P., Kokkola, T., Suuronen, T., Poso, A., Jarho, E., & Lahtela-Kakkonen, M. (2015).	Applies virtual screening for sirtuin inhibitors.	A, D

(*continued*)

Table 2. (*continued*)

Serial No	Citation	Reason of Inclusion	Themes
49	Worachartcheewan, A., Toropova, A.P., Toropov, A.A., Pratiwi, R., Prachayasittikul, V., & Nantasenamat, C. (2021).	Uses QSAR modeling for sirtuin inhibitors.	D
50	Khanfar, M.A., & Taha, M.O. (2017).	Combines pharmacophore modeling and QSAR for SIRT2 inhibitors.	D
51	Sakkiah, S., Thangapandian, S., Park, C., Son, M., & Lee, K.W. (2012).	Uses docking and molecular dynamics for SIRT2 inhibitor discovery.	D
52	Kokkonen, P., Mellini, P., Nyrhilä, O., Rahnasto-Rilla, M., Suuronen, T., Kiviranta, P., Lahtela-Kakkonen, M. (2014).	Provides insights for designing SIRT1 inhibitors.	D
53	Sakkiah, S., Arooj, M., Lee, K.W., & Torres, J.Z. (2014)	Uses theoretical methods for sirtuin activator discovery.	D
54	Aldisa, O., Azminah, A., Erlina, L., Hayun, H., & Yanuar, A. (2017).	Applies virtual screening to find sirtuin activators.	A, D
55	Tervo, A.J., Kyrylenko, S., Niskanen, P., Salminen, A., Leppänen, J., Nyrönen, T.H., & Poso, A. (2004).	Uses in silico methods for discovering sirtuin inhibitors.	D
56	Sun, Y., Zhou, H., Zhu, H., & Leung, S.W. (2016).	Applies virtual screening and learning for SIRT1 inhibitors.	A, D
57	Xu, J., & Stevenson, J. (2000)	Introduces an index for measuring drug-likeness.	D
58	Fang, X., Bagui, S., & Bagui, S. (2017).	Uses machine learning to improve virtual screening accuracy.	A, B
59	Byvatov, E., & Schneider, G. (2004).	Discusses SVM-based feature selection in drug discovery.	A
60	Armutlu, P., Ozdemir, M.E., Uney-Yuksektepe, F., Kavakli, I.H., & Turkay, M. (2008).	To binding constant for the activity of a drug candidate on the target site.	C
61	Le-Thi-Thu, H., Casanola-Martín, M., Marrero-Ponce, Y., Rescigno, A., Abad, C., & Khan, M.T.H. (2014).	Discusses virtual screening workflow for tyrosinase inhibitors.	A, B
62	Ibrahim, M.A., Abdeljawaad, K.A., Roshdy, E., Mohamed, D.E., Ali, T.F., Gabr, G.A., & Abdelrahman, A.H. (2023)	Uses in silico methods for discovering sirtuin inhibitors as anticancer agents.	D
63	Langer, T., & Bryant, S.D. (2008).	Reviews in silico screening techniques for drug discovery.	A, B
64	Rodrigues, T., & Schneider, G. (2015)	To binding constant for the activity of a drug candidate on the target site.	C
65	Chiba, S., Ohue, M., Gryniukova, A., Borysko, P., Zozulya, S., Yasuo, N., Koyanagi, Y., Saito, Y., & Sekijima, M. (2019).	Applies compound screening to find sirtuin inhibitors.	D
66	Roche, O., Schneider, P., Zuegge, J., Guba, W., Kansy, M., Alanine, A., Bleicher, K., Danel, F., Fischli, W., & Gibbs, J. (2002).	Develops a method to identify frequent hitters in compound libraries.	A, B
67	Goodarzi, A., Valipour, M., & Irannejad, H. (2024).	Predicts affinities of Shikonins for sirtuins using computational methods.	D

(*continued*)

Table 2. (*continued*)

Serial No	Citation	Reason of Inclusion	Themes
68	Sakkiah, S., Krishnamoorthy, N., Gajendrarao, P., Thangapandian, S., Lee, Y., Kim, S., & Lee, K.W. (2009).	Applies pharmacophore mapping for SIRT1 activator discovery.	A, D
69	Fang, J., Huang, D., Zhao, W., Ge, H., Luo, H.B., & Xu, J. (2011).	Introduces a protocol for predicting GSK-3β inhibitors.	D
70	Yilmaz, H., Sizochenko, N., Rasulev, B., Toropov, A., Guzel, Y., Kuz'min, V., & Leszczynski, J. (2015).	Discusses SAR approaches for KDR inhibitors.	D
71	Kahraman, P., & Turkay, M. (2007).	Classification of 1,4-dihydropyridine calcium channel antagonists using the hyperbox approach.	A
72	Tardu, M., Rahim, F., Kavakli, I.H., & Turkay, M. (2016)	Applies MILP for drug design in SIRT6 inhibitors.	A, D
73	Wohlin, C., Kalinowski, M., Felizardo, K. R., & Mendes, E. (2022).	Successful combination of database search and snowballing for identification of primary studies in systematic literature studies.	A

3 Research Methodology

3.1 Biological Importance of Sirtuin6

Sirtuins, a family of class III histone deacetylase enzymes, play a pivotal role in regulating key biological processes such as transcription, cell survival, and protein deacetylation. These enzymes depend on NAD+ as a cofactor for their activity, which is crucial for maintaining cellular function and longevity. Among the seven mammalian sirtuins, each is distinct in its intracellular localization and the specific proteins it deacetylates. Sirtuin6 (SIRT6), a nuclear enzyme, stands out due to its involvement in several critical molecular pathways, including telomere maintenance, DNA repair, aging, and cancer prevention.

SIRT6's biological importance extends to its role in metabolic regulation. Recent research has revealed that a deficiency in SIRT6 leads to enhanced glycolysis and upregulation of glycolytic gene expression, such as GLUT1, in a hypoxia-inducible factor 1α (HIF1α)-dependent manner. This highlights its influence on cellular energy metabolism and its potential impact on metabolic disorders

Given its involvement in critical cellular processes, SIRT6 has emerged as a promising therapeutic target for diseases related to metabolism, aging, and cancer. The potential for SIRT6 modulation, through small-molecule inhibitors, offers an exciting avenue for therapeutic intervention. The crystal structure of SIRT6 (PDB ID: 3K35) provides valuable insights into its molecular architecture, enabling structure-based in-silico screening approaches. This facilitates a deeper understanding of SIRT6's interactions with chemical modulators, aiding in the development of novel inhibitors that can modulate its activity for therapeutic purposes.

3.2 Biological Importance of Sirtuin6

Molecular descriptors are quantitative representations of a molecule's properties, derived through statistical and mathematical algorithms. These descriptors capture key chemical and physical attributes of the molecule, offering valuable insights into its behavior, structure, and potential biological activity. By translating complex molecular features into numerical values, molecular descriptors serve as critical tools in computational chemistry and drug discovery, helping to predict molecular interactions, reactivity, and binding efficiency.

In this framework, the binding free energy of medicinal compounds is predicted using six widely recognized chemical descriptors:

SC-5 - A shape coefficient that reflects the overall geometric structure of the mole-cule, impacting how it fits into target sites.
SP-6 - A spatial descriptor that provides information about the molecular surface area, which influences solubility and interaction with biological environments.
SHBd - This descriptor represents the hydrogen bond donor count, which is essential for assessing the molecule's ability to form hydrogen bonds, a key interaction in mo-lecular binding.
minHaaCH - The minimum hydrogen-to-carbon ratio in the molecule, providing insights into molecular rigidity and structural complexity.
maxwHBa - This descriptor captures the maximum weighted hydrogen bond accep-tor count, important for evaluating the molecule's potential to engage in hydrogen bonding interactions.
FMF (Fractional Molecular Framework) - A descriptor that quantifies the fraction of the molecule's framework that contributes to its structural core, aiding in the as-sessment of molecular stability and interaction with target proteins.

These descriptors collectively provide a comprehensive view of the molecular properties relevant to predicting binding free energy levels. By leveraging these descriptors, the model can effectively classify and prioritize compounds with favorable binding characteristics, accelerating the virtual screening process and guiding the identification of novel drug candidates.

3.3 Research Design

To ensure accurate model training and evaluation, the dataset was randomized to prevent skewed results. Inconsistent distribution of target variables, such as an uneven proportion of loan defaults between the training and test datasets, could lead to biased model performance. For example, training a model on a dataset with a disproportionately high incidence of defaults and testing it on a dataset with fewer defaults, or vice versa, would result in inaccurate predictions and unfavorable outcomes.

To address this, every observation in the dataset was randomized and stored in a separate object. This object contains a vector of randomly selected rows, representing a random permutation of the entire dataset. This approach ensures that each combination of data points is equally likely to be chosen during training and testing. A training sample is then drawn from the randomized dataset to train the statistical model, while a separate test sample is used to evaluate the model's effectiveness.

By using this randomized approach, we mitigate the risk of overfitting and ensure the generalizability of the machine learning algorithm. The randomization process maintains the integrity of the data distribution across both training and test sets, leading to more reliable and robust predictions.

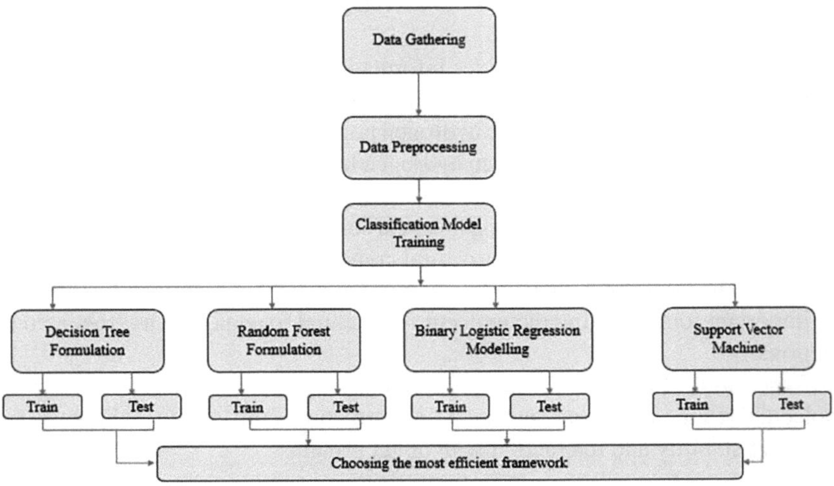

Fig. 1. Research Diagram

3.4 Sources of Data Collection

The data utilized in this study was obtained from a widely recognized machine learning database repository, commonly used by the machine learning community for empirical analysis and testing of various algorithms. Specifically, the dataset has been sourced from the following repository:

UCI Machine Learning Repository - **Sirtuin6 Small Molecules Dataset** (https://archive.ics.uci.edu/dataset/748/sirtuin6+small+molecules-1)

This dataset provides secondary data necessary for building a robust data-driven framework aimed at predicting the binding free energy (BFE) and binding constant for the activity of potential drug candidates on the Sirtuin6 target site. The availability of such curated datasets enables the application of machine learning models in virtual drug screening and helps accelerate the discovery of novel inhibitors with favorable binding characteristics.

4 Data Analysis and Interpretation

4.1 Formulation of a Decision Tree Model for Classification

A binary classification Decision Tree model was developed to categorize the binding free energy (BFE) of chemical compounds into two distinct classes: High-level and Low-level. The six molecular descriptors (SC-5, SP-6, SHBd, minHaaCH, maxwHBa, and FMF), all of which are metric in nature, were used as predictor variables for this classification task.

4.1.1 Pre-processing of Data

Since the dataset used in this analysis was free of missing values, null entries, or other erroneous data points, the only pre-processing step required was the randomization of the dataset. This randomization was carried out to ensure that potential biases were eliminated prior to model training. The dataset was randomized and stored as an object containing a vector of randomly selected rows, representing all possible combinations of data points. This process ensured that each observation had an equal probability of being selected during the training and testing phases.

The Decision Tree model was trained using 30% of the randomized dataset. The remaining 70% of the data was reserved for testing and validating the model's performance. This approach allowed the model to learn the relationships between the molecular descriptors and the BFE classes from the training data, while the validation sample was used to assess the model's effectiveness. Performance metrics such as accuracy, precision, recall, and F1-score were calculated to evaluate the efficacy of the classification (see Fig. 1 for detailed results).

By categorizing BFE into binary classes and employing a Decision Tree model, the process of distinguishing between high and low binding energy molecules was simplified, contributing to the identification of potential inhibitors.

```
Evaluation on training data (30 cases):

            Decision Tree
          ----------------
          Size      Errors

             3    1( 3.3%)   <<


          (a)   (b)    <-classified as
          ----  ----
            12     1   (a): class High_BFE
                  17   (b): class Low_BFE


        Attribute usage:

        100.00% FMF
         46.67% SC-5
```

Fig. 2. Decision Tree Output

4.1.2 Training Data Evaluation

Since the objective is to determine the level of binding free energy (BFE), a true positive in this context refers to a compound with actual Low BFE correctly classified as Low BFE, while a true negative corresponds to a compound with actual High BFE accurately categorized as High BFE. False positives are instances where molecules with High BFE are mistakenly classified as Low BFE, and false negatives occur when compounds with Low BFE are incorrectly classified as High BFE.

According to the confusion matrix, the model achieved an overall accuracy of 96.67%, indicating its high effectiveness in classifying BFE levels (see Fig. 2). This high level of predictability suggests that the model performs well in distinguishing between compounds with high and low binding free energy.

dt_pred1	High_BFE	Low_BFE
High_BFE	12	0
Low_BFE	1	17

Fig. 3. Confusion matrix for the training data (decision tree)

4.1.3 Test Data Evaluation

In this analysis, the level of binding free energy (BFE) was predicted. A true positive was defined as a compound with actual Low BFE that was correctly classified as Low BFE, while a true negative referred to a compound with actual High BFE that was correctly classified as High BFE. False negatives occurred when compounds with High BFE were incorrectly classified as Low BFE, and false positives were instances where compounds with Low BFE were mistakenly classified as High BFE.

The predictive accuracy of the model in identifying BFE levels was measured at 75.71%, as indicated by the confusion matrix (see Fig. 3). This accuracy demonstrates that the model performed reasonably well in distinguishing between compounds with high and low binding free energy.

dt_pred	High_BFE	Low_BFE
High_BFE	24	4
Low_BFE	13	29

Fig. 4. Confusion matrix for the test data (decision tree)

4.2 Formulation of a Random Forest Model for Classification

A Random Forest model was developed to classify the binding free energy (BFE) into two categories: High-level and Low-level, utilizing all the six provided molecular descriptors—SC-5, SP-6, SHBd, minHaaCH, maxwHBa, and FMF—all of which are metric in nature. The Random Forest algorithm was chosen due to its ability to handle multiple predictor variables and improve classification accuracy through the aggregation of decision trees.

```
Call:
 randomForest(formula = Class ~ ., data = train, proximity = TRUE)
               Type of random forest: classification
                     Number of trees: 500
No. of variables tried at each split: 2

        OOB estimate of  error rate: 6.06%
Confusion matrix:
         High_BFE Low_BFE class.error
High_BFE       17       1  0.05555556
Low_BFE         1      14  0.06666667
```

Fig. 5. Random Forest Output

4.2.1 Training Data Evaluation

Since the binding free energy (BFE) level was the focus of this analysis, a true positive was defined as actual Low BFE being correctly categorized as Low BFE, while a true negative was actual High BFE correctly classified as High BFE. False positives were instances where compounds were mistakenly classified as Low BFE despite having High BFE, and false negatives occurred when compounds with Low BFE were incorrectly classified as High BFE. According to the provided confusion matrix, the prediction accuracy was recorded at 100%, suggesting that the binding free energy (BFE) levels were accurately identified by the model (see Fig 4).

```
predTrain  High_BFE Low_BFE
  High_BFE       18       0
  Low_BFE         0      15
```

Fig. 6. Confusion matrix for the training data (Random Forest)

4.2.2 Test Data Evaluation

The objective was to determine the level of binding free energy (BFE). A true positive was defined as actual Low BFE being correctly classified as Low BFE, while a true negative was actual High BFE correctly classified as High BFE. False negatives occurred when compounds with High BFE were mistakenly classified as Low BFE, and false positives were instances where compounds with Low BFE were incorrectly classified as High BFE. According to the provided confusion matrix, the accuracy of this prediction was 80.6%, indicating that the model performed well in identifying the binding free energy (BFE) levels (see Figs. 5, 6, and 7).

predValid	High_BFE	Low_BFE
High_BFE	31	12
Low_BFE	1	23

Fig. 7. Confusion matrix for the test data (Random Forest)

4.3 Formulation of a Binary Logistic Model for Classification

In light of this, all six molecular descriptors—SC-5, SP-6, SHBd, minHaaCH, maxwHBa, and FMF—were used as input variables, while the output variable, the label, was treated as a binary categorical variable classified into two categories: High-level and Low-level. To classify the binding free energy (BFE) levels, a binomial logistic regression model or binomial logit model could be employed (Fig. 8).

```
Call:
glm(formula = Class ~ SC5 + SP6 + SHBd + minHaaCH + maxwHBa +
    FMF, family = binomial(link = "logit"), data = train)

Deviance Residuals:
       Min          1Q      Median          3Q         Max
-5.091e-05  -2.100e-08   2.100e-08   2.100e-08   5.953e-05

Coefficients:
             Estimate Std. Error z value Pr(>|z|)
(Intercept)   -893.84  598178.67  -0.001    0.999
SC5           -280.45  362749.39  -0.001    0.999
SP6             95.37   77617.39   0.001    0.999
SHBd          -220.80  436356.91  -0.001    1.000
minHaaCH       533.76  953748.94   0.001    1.000
maxwHBa         22.55  219335.88   0.000    1.000
FMF           1430.15  989971.79   0.001    0.999

(Dispersion parameter for binomial family taken to be 1)

    Null deviance: 4.5475e+01  on 32  degrees of freedom
Residual deviance: 9.2488e-09  on 26  degrees of freedom
AIC: 14

Number of Fisher Scoring iterations: 25
```

(a)

```
$Pseudo.R.squared.for.model.vs.null
                             Pseudo.R.squared
McFadden                             1.000000
Cox and Snell (ML)                   0.747922
Nagelkerke (Cragg and Uhler)         1.000000

$Likelihood.ratio.test
 Df.diff LogLik.diff  Chisq    p.value
      -6     -22.737 45.475 3.7663e-08

$Number.of.observations

Model: 33
Null:  33
```

(b)

```
        Hosmer and Lemeshow test (binary model)

data:  train$Class, fitted(logitMod)
X-squared = 5.8956e-10, df = 2, p-value = 1
```

(c)

Fig. 8. (a), (b), (c). Binary Logistic Regression Output

4.3.1 Training Data Evaluation

In this case, the misclassification error was found to be zero, indicating a 100% accuracy in classifying the training dataset. A predictive algorithm is generally considered reliable if the concordance exceeds 0.90. In the confusion matrix, High BFE was represented by the number 1 and Low BFE by the number 0. Since the objective was to identify Low-BFE compound molecules, true positives were defined as actual Low BFE classified as Low BFE, while true negatives were actual High BFE correctly identified as High BFE.

It can be concluded that the model performed well, as both the sensitivity (true positive rate) and specificity (true negative rate) were found to be 1, which exceeds the ideal cut-off probability value of 0.01 for this model when using the training dataset (see Fig. 9. (a), (b), (c)).

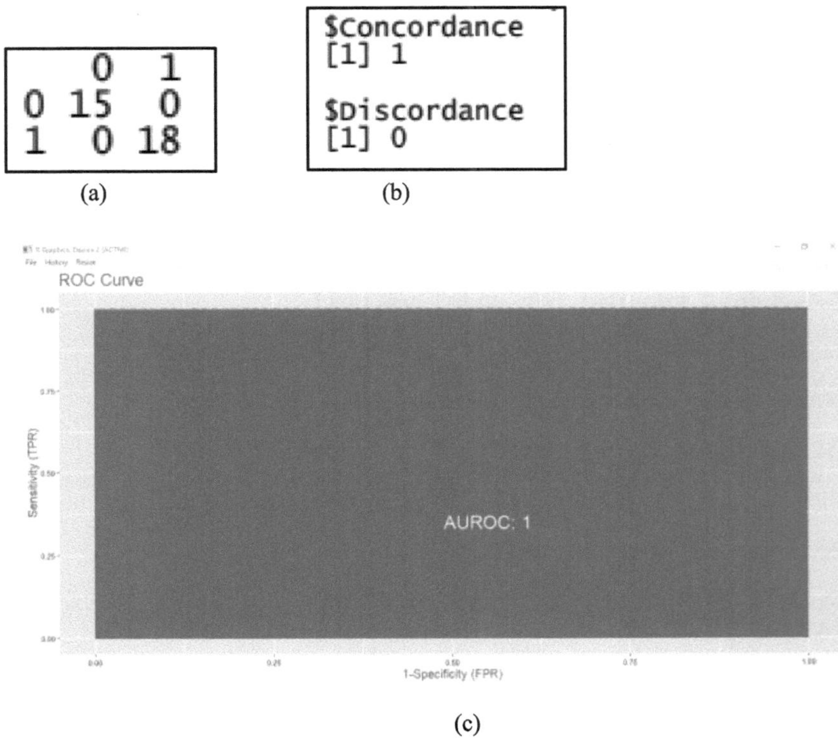

Fig. 9. **(a), (b), (c)** Confusion Matrix, Concordance value & ROC curve for the training data (Binary Logistic Regression)

4.3.2 Test Data Evaluation

The misclassification error was found to be 32.84%, indicating a 67.16% probability that the model correctly classified the test dataset. The algorithm demonstrated reasonable predictive performance, with a concordance of approximately 0.41. In the confusion matrix, High BFE was represented by the number 1 and Low BFE by the number 0. Since the primary objective was to identify Low-BFE compound molecules, true positives were defined as actual Low BFE classified as Low BFE, and true negatives as actual High BFE correctly classified as High BFE.

The real positive rate (0.9375) and true negative rate (0.4285) were observed, with the sensitivity found to be close to the optimal cut-off probability value (1 for this model with the test dataset). Therefore, it can be concluded that the model demonstrated moderate performance (see Fig. 10. (a), (b), (c)).

	0	1
0	15	2
1	20	30

(a)

```
$Concordance
[1] 0.4142857

$Discordance
[1] 0.5857143
```

(b)

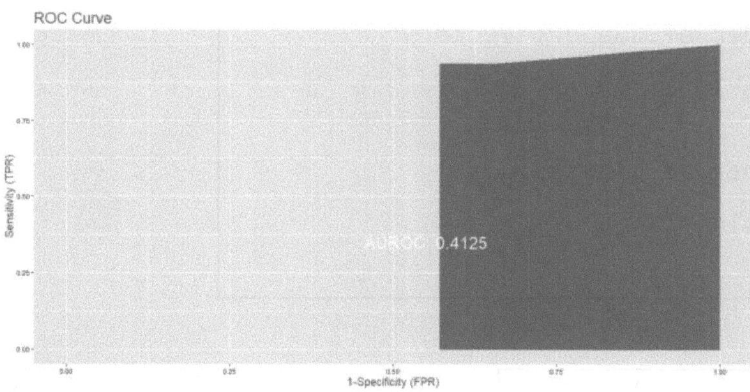

(c)

Fig. 10. **(a), (b), (c)** Confusion Matrix, Concordance value & ROC curve for the test data (Binary Logistic Regression)

4.4 Formulation of a Support Vector Machine Model for Classification

As all predictor variables—SC-5, SP-6, SHBd, minHaaCH, maxwHBa, and FMF—are continuous, a Support Vector Machine (SVM) model was formulated to classify the binding free energy (BFE) into two categories: High-level and Low-level.

In the confusion matrix, High BFE is represented by the number 1 and Low BFE by the number 0. Since the primary objective was to identify Low BFE compound molecules, true positives were defined as actual Low BFE categorized as Low BFE, while true negatives referred to actual High BFE correctly classified as High BFE.

To determine the most suitable SVM model, four different SVM specifications—Linear, Radial, Polynomial, and Sigmoid—were tested and compared to evaluate which model best fit the data.

4.4.1 Linear SVM

(See Fig. 11)

```
Parameters:
   SVM-Type:  C-classification
 SVM-Kernel:  linear
       cost:  1

Number of Support Vectors:  19

 ( 8 11 )


Number of Classes:  2

Levels:
 0 1
```

(a)

```
          Actual
Predicted  0  1
        0 12  4
        1  3 11

[1] 0.7666667
```

(b)

```
          Actual
Predicted  0  1
        0 29 10
        1  6 25

[1] 0.7714286
```

(c)

Fig. 11. **(a), (b), (c)** Linear SVM summary, Confusion Matrix & Correctness of Prediction for the training and test data

4.4.2 Radial SVM

(See Fig. 12)

```
Parameters:
   SVM-Type:  C-classification
 SVM-Kernel:  radial
       cost:  1

Number of Support Vectors:  26

 ( 12 14 )


Number of Classes:  2

Levels:
 0 1
```

(a)

```
          Actual
Predicted  0  1
        0 12  2
        1  3 13
[1] 0.8333333
```

(b)

```
          Actual
Predicted  0  1
        0 31  6
        1  4 29
[1] 0.8571429
```

(c)

Fig. 12. **(a)**, **(b)**, **(c)** Radial SVM summary, Confusion Matrix & Correctness of Prediction for the training and test data

4.4.3 Polynomial SVM

(See Fig. 13)

```
Parameters:
   SVM-Type:  C-classification
 SVM-Kernel:  polynomial
       cost:  1
     degree:  3
     coef.0:  0

Number of Support Vectors:  23

 ( 10 13 )


Number of Classes:  2

Levels:
 0 1
```

(a)

```
         Actual
Predicted  0  1
        0 10  0
        1  5 15
[1] 0.8333333
```

(b)

```
         Actual
Predicted  0  1
        0 17  2
        1 18 33
[1] 0.7142857
```

(c)

Fig. 13. **(a), (b), (c)** Polynomial SVM summary, Confusion Matrix & Correctness of Prediction for the training and test data

4.4.4 Sigmoid SVM

(See Fig. 14)

```
Parameters:
   SVM-Type:  C-classification
 SVM-Kernel:  sigmoid
       cost:  1
     coef.0:  0

Number of Support Vectors:  20

 ( 10 10 )


Number of Classes:  2

Levels:
 0 1
```

(a)

```
         Actual
Predicted  0  1
        0 13  4
        1  2 11
[1] 0.8
```

(b)

```
         Actual
Predicted  0  1
        0 30 10
        1  5 25
[1] 0.7857143
```

(c)

Fig. 14. **(a), (b), (c)** Sigmoid SVM summary, Confusion Matrix & Correctness of Prediction for the training and test data

In comparison to the other Support Vector Machine model subcategories, this comparative study demonstrated that the radial SVM classification provided a superior model fit, achieving over 80% prediction accuracy.

5 Conclusion

A novel data analytics-based approach was proposed in this research to determine the binding constant and binding free energy (BFE) for a drug candidate's activity at the target site. It was experimentally demonstrated that the binding free energy of drug candidates can be accurately predicted using advanced machine learning algorithms, leveraging molecular descriptor features and a supervised learning-based technique. As a result, it was shown that classification algorithms are effective for forecasting the binding free energy (BFE) of drug candidates at the target site.

6 Suggestions & Recommendations

- The binding free energy of drug candidates can be categorized into High BFE and Low BFE labels using Decision Tree, Random Forest, and Binary Logistic Regression models.
- The Random Forest method is recommended over the Decision Tree approach due to its superior accuracy in classifying drug candidates based on their binding free energy.
- It is recommended to use the largest feasible test dataset to ensure that the training sample accurately represents the entire dataset. This practice will enhance the model's performance, leading to more reliable and realistic outcomes.

7 Limitations of the Study

A few restrictions that have been encountered are:

- The analysis was constrained by system capacity and, allowing only a limited number of records to be processed. It is agreed with the reviewer comments and expected as well, that increasing the size of the training dataset would have led to further improvements in model accuracy and robustness, which is hindered by limitation of data availability.
- Due to lack of accessibility to latest real-time data set, the research was unable to cite a real-life case study.
- While the primary objective was to develop classification models to accurately predict the binding free energy of drug candidates, opportunities for enhancing the model's overall performance remain. Future work could explore more advanced techniques or optimize model parameters for better predictive accuracy.

8 Scopes for Future Research

Future planning of the theme of the current research includes experiment with some recent real-life dataset and test the existing framework's efficiency.

References

Aldisa, O., Azminah, A., Erlina, L., Hayun, H., Yanuar, A.: Virtual screening of Indonesian herbal database to find sirtuin 1 activators using the docking method. Asian J. Pharm. Clin. Res. **10**, 158–162 (2017)

Armutlu, P., Ozdemir, M.E., Ozdas, S., Kavakli, İH., Turkay, M.: Discovery of novel CYP17 inhibitors for the treatment of prostate cancer with structure-based drug design. Lett. Drug Des. Discovery **6**(5), 337–344 (2009)

Armutlu, P., Ozdemir, M.E., Uney-Yuksektepe, F., Kavakli, I.H., Turkay, M.: Classification of drug molecules considering their IC 50 values using mixed-integer linear programming based hyper-boxes method. BMC Bioinformatics **9**, 1–14 (2008)

Bento, A.P., Gaulton, A., Hersey, A., Bellis, L.J., Chambers, J., Davies, M., ... & Overington, J. P.: The ChEMBL bioactivity database: an update. Nucleic Acids Res. **42**(D1), D1083–D1090 (2014)

Bolton, E.E., Wang, Y., Thiessen, P.A., Bryant, S.H.: PubChem: integrated platform of small molecules and biological activities. In Annual Reports in Computational Chemistry, Vol. 4, pp. 217–241. Elsevier (2008)

Byvatov, E., Schneider, G.: SVM-based feature selection for characterization of focused compound collections. J. Chem. Inf. Comput. Sci. **44**(3), 993–999 (2004)

Cakir, B., et al.: Structure based discovery of small molecules to regulate the activity of human insulin degrading enzyme. PLoS ONE **7**(2), e31787 (2012)

Chakraborty, D., Mazumder, S., Kar, A.: a machine learning framework developed by leveraging the topological pattern analysis of bitcoin addresses in blockchain network for ransomware identification. In Analytics Global Conference, pp. 23–39. Cham: Springer Nature Switzerland (2023, April).

Chiba, S., Ohue, M., Gryniukova, A., Borysko, P., Zozulya, S., Yasuo, N., ... & Sekijima, M. : A prospective compound screening contest identified broader inhibitors for Sirtuin 1. Sci. Rep. **9**(1), 19585 (2019)

Cleary, J.G., Trigg, L.E.: K*: an instance-based learner using an entropic distance measure. In Machine Learning Proceedings 1995, pp. 108–114. Morgan Kaufmann (1995)

Dagliyan, O., Kavakli, I.H., Turkay, M.: Classification of cytochrome P450 inhibitors with respect to binding free energy and pIC50 using common molecular descriptors. J. Chem. Inf. Model. **49**(10), 2403–2411 (2009)

Dagliyan, O., Uney-Yuksektepe, F., Kavakli, I.H., Turkay, M.: Optimization based tumor classification from microarray gene expression data. PLoS ONE **6**(2), e14579 (2011)

Djokovic, N., Rahnasto-Rilla, M., Lougiakis, N., Lahtela-Kakkonen, M., Nikolic, K.: SIRT2i_Predictor: A machine learning-based tool to facilitate the discovery of novel SIRT2 inhibitors. Pharmaceuticals **16**(1), 127 (2023)

Etchegaray, J.P., Zhong, L., Mostoslavsky, R.: The histone deacetylase SIRT6: at the crossroads between epigenetics, metabolism and disease. Curr. Top. Med. Chem. **13**(23), 2991–3000 (2013)

Fang, J., Huang, D., Zhao, W., Ge, H., Luo, H.B., Xu, J.: A new protocol for predicting novel GSK-3β ATP competitive inhibitors. J. Chem. Inf. Model. **51**(6), 1431–1438 (2011)

Fang, X., Bagui, S., Bagui, S.: Improving virtual screening predictive accuracy of Human kallikrein 5 inhibitors using machine learning models. Comput. Biol. Chem. **69**, 110–119 (2017)

Finkel, T., Deng, C.X., Mostoslavsky, R.: Recent progress in the biology and physiology of sirtuins. Nature **460**(7255), 587–591 (2009)

Frank, E., Hall, M., Trigg, L., Holmes, G., Witten, I.H.: Data mining in bioinformatics using Weka. Bioinformatics **20**(15), 2479–2481 (2004)

Friedman, J.: Additive logistic regression: a statistical view of boosting (with discussion and a rejoinder by the authors). Ann. Statist. **28**, 337–407 (2000)

Frye, R.A.: Phylogenetic classification of prokaryotic and eukaryotic Sir2-like proteins. Biochem. Biophys. Res. Commun. **273**(2), 793–798 (2000)

Goodarzi, A., Valipour, M., Irannejad, H.: Affinity Prediction of Shikonins towards Sirtuins and the Requisite Structural Motifs for the Selective Inhibition of SIRT2 and SIRT3. Lett. Drug Des. Discovery **21**(4), 724–737 (2024)

Heckerman, D.: A tutorial on learning with Bayesian networks. In: Learning in Graphical Models, pp. 301–354. (1998)

IBM ILOG: Cplex user's manual 12.2 (2010)

Ibrahim, M.A., Abdeljawaad, K.A., Roshdy, E., Mohamed, D.E., Ali, T.F., Gabr, G.A., ... & Abdelrahman, A.H.: In silico drug discovery of SIRT2 inhibitors from natural source as anticancer agents. Sci. Rep. **13**(1), 2146 (2023)

Irwin, J.J., Shoichet, B.K.: ZINC– a free database of commercially available compounds for virtual screening. J. Chem. Inf. Model. **45**(1), 177–182 (2005)

Jorgensen, W.L.: The many roles of computation in drug discovery. Science **303**(5665), 1813–1818 (2004)

Kahraman, P., Turkay, M.: Classification of 1, 4-dihydropyridine calcium channel antagonists using the hyperbox approach. Ind. Eng. Chem. Res. **46**(14), 4921–4929 (2007)

Kaidi, A., Weinert, B.T., Choudhary, C., Jackson, S.P.: Retracted: human Sirt6 promotes DNA end resection through ctip deacetylation. Science **329**(5997), 1348–1353 (2010)

Kanfi, Y., Shalman, R., Peshti, V., Pilosof, S.N., Gozlan, Y.M., Pearson, K.J., ... & Cohen, H.Y.: Regulation of SIRT6 protein levels by nutrient availability. FEBS Lett. **582**(5) 543–548 (2008)

Karaman, B., Sippl, W.: Docking and binding free energy calculations of sirtuin inhibitors. Eur. J. Med. Chem. **93**, 584–598 (2015)

Kawahara, T.L., Michishita, E., Adler, A.S., Damian, M., Berber, E., Lin, M., ... & Chua, K.F.: SIRT6 links histone H3 lysine 9 deacetylation to NF-κB-dependent gene expression and organismal life span. Cell **136**(1), 62–74 (2009)

Khanfar, M.A., Taha, M.O.: Unsupervised pharmacophore modeling combined with QSAR analyses revealed novel low micromolar SIRT2 inhibitors. J. Mol. Recognit. **30**(9), e2623 (2017)

Kitchen, D.B., Decornez, H., Furr, J.R., Bajorath, J.: Docking and scoring in virtual screening for drug discovery: methods and applications. Nat. Rev. Drug Discovery **3**(11), 935–949 (2004)

Kokkonen, P., Kokkola, T., Suuronen, T., Poso, A., Jarho, E., Lahtela-Kakkonen, M.: Virtual screening approach of sirtuin inhibitors results in two new scaffolds. Eur. J. Pharm. Sci. **76**, 27–32 (2015)

Kokkonen, P., et al.: Quantitative insights for the design of substrate-based SIRT1 inhibitors. Eur. J. Pharm. Sci. **59**, 12–19 (2014)

Kubinyi, H.: Similarity and dissimilarity: a medicinal chemist's view. Perspect. Drug Discovery Des. **9**, 225–252 (1998)

Langer, T., Bryant, S.D. In silico screening: Hit finding from database mining. In The Practice of Medicinal Chemistry, pp. 210–227. Academic Press (2008)

Le-Thi-Thu, H., M Casanola-Martín, G., Marrero-Ponce, Y., Rescigno, A., Abad, C., Tareq Hassan Khan, M.: A rational workflow for sequential virtual screening of chemical libraries on searching for new tyrosinase inhibitors. Current Top. Med. Chem. **14**(12), 1473–1485 (2014)

Lipinski, C.A., Lombardo, F., Dominy, B.W., Feeney, P.J.: Experimental and computational approaches to estimate solubility and permeability in drug discovery and development settings. Adv. Drug Deliv. Rev. **23**(1–3), 3–25 (1997)

Liszt, G., Ford, E., Kurtev, M., Guarente, L.: Mouse Sir2 homolog SIRT6 is a nuclear ADP-ribosyltransferase. J. Biol. Chem. **280**(22), 21313–21320 (2005)

MacKerell Jr, A.D., Bashford, D., Bellott, M.L.D.R., Dunbrack Jr, R.L., Evanseck, J.D., Field, M.J., ... & Karplus, M.: All-atom empirical potential for molecular modeling and dynamics studies of proteins. J. Phys. Chem. B **102**(18), 3586–3616

Mao, Z., Hine, C., Tian, X., Van Meter, M., Au, M., Vaidya, A., ... Gorbunova, V.: SIRT6 promotes DNA repair under stress by activating PARP1. Science **332**(6036), 1443–1446 (2011)

Michishita, E., Park, J.Y., Burneskis, J.M., Barrett, J.C., Horikawa, I.: Evolutionarily conserved and nonconserved cellular localizations and functions of human SIRT proteins. Mol. Biol. Cell **16**(10), 4623–4635 (2005)

Mitchell, T., Showell, G.A.: Design strategies for building drug-like chemical libraries. Curr. Opin. Drug Discov. Devel. **4**(3), 314–318 (2001)

Mostoslavsky, R., Chua, K.F., Lombard, D.B., Pang, W.W., Fischer, M.R., Gellon, L., ... & Alt, F.W. Genomic instability and aging-like phenotype in the absence of mammalian SIRT6. Cell **124**(2), 315—329 (2006)

Pan, P.W., Feldman, J.L., Devries, M.K., Dong, A., Edwards, A.M., Denu, J.M.: Structure and biochemical functions of SIRT6. J. Biol. Chem. **286**(16), 14575–14587 (2011)

Phillips, J.C., Braun, R., Wang, W., Gumbart, J., Tajkhorshid, E., Villa, E., ... & Schulten, K.: Scalable molecular dynamics with NAMD. J. Comput. Chem. **26**(16), 1781–1802 (2005)

Platt, J.C.: Fast training of support vector machines using sequential minimal optimization (1998)

Qu, Y., Adam, B.L., Yasui, Y., Ward, M.D., Cazares, L.H., Schellhammer, P.F., ... & Wright Jr, G.L.: Boosted decision tree analysis of surface-enhanced laser desorption/ionization mass spectral serum profiles discriminates prostate cancer from noncancer patients. Clin. Chem. **48**(10), 1835–1843 (2002)

Roche, O., et al.: Development of a virtual screening method for identification of "frequent hitters" in compound libraries. J. Med. Chem. **45**(1), 137–142 (2002)

Rodrigues, T., Schneider, G.: In Silico screening: hit finding from database mining. In The Practice of Medicinal Chemistry, pp. 141–160. Academic Press (2015)

Rosenthal, R.E. Gams – a user's guide (2015)

Sakkiah, S., Arooj, M., Lee, K.W., Torres, J.Z.: Theoretical approaches to identify the potent scaffold for human sirtuin1 activator: Bayesian modeling and density functional theory. Med. Chem. Res. **23**, 3998–4010 (2014)

Sakkiah, S., Baek, A., Lee, K.W.: Pharmacophore modeling and molecular dynamics simulation to identify the critical chemical features against human sirtuin 2 inhibitors. J. Mol. Struct. **1011**, 66–75 (2012)

Sakkiah, S., Krishnamoorthy, N., Gajendrarao, P., Thangapandian, S., Lee, Y., Kim, S., ... & Lee, K.W.: Pharmacophore mapping and virtual screening for SIRT1 activators. Bull. Korean Chem. Soc. **30**(5), 1153

Sakkiah, S., Thangapandian, S., Park, C., Son, M., Lee, K.W.: Molecular docking and dynamics simulation, receptor-based hypothesis: application to identify novel sirtuin 2 inhibitors. Chem. Biol. Drug Des. **80**(2), 315–327 (2012)

Sun, Y., Zhou, H., Zhu, H., Leung, S.W.: Ligand-based virtual screening and inductive learning for identification of SIRT1 inhibitors in natural products. Sci. Rep. **6**(1), 19312 (2016)

Szachniuk, M., Cristina De Cola, M., Felici, G., Blazewicz, J.: The orderly colored longest path problem-a survey of applications and new algorithms. RAIRO-Operations Research-Recherche Opérationnelle **48**(1), 25–51 (2014)

Tardu, M., Rahim, F., Halil Kavakli, I., Turkay, M.: Milp-hyperbox classification for structure-based drug design in the discovery of small molecule inhibitors of Sirtuin6. RAIRO-Operations Research-Recherche Opérationnelle **50**(2), 387–400 (2016)

Tervo, A.J., Kyrylenko, S., Niskanen, P., Salminen, A., Leppänen, J., Nyrönen, T.H., ... & Poso, A.: An in silico approach to discovering novel inhibitors of human sirtuin type 2. J. Med. Chem. **47**(25), 6292–6298 (2004)

Trott, O., Olson, A.J.: AutoDock Vina: improving the speed and accuracy of docking with a new scoring function, efficient optimization, and multithreading. J. Comput. Chem. **31**(2), 455–461 (2010)

Üney, F., Türkay, M.: A mixed-integer programming approach to multi-class data classification problem. Eur. J. Oper. Res. **173**(3), 910–920 (2006)

Villar, H.O., Hansen, M.R.: Design of chemical libraries for screening. Expert Opin. Drug Discov. **4**(12), 1215–1220 (2009)

Whitley, D.C., Ford, M.G., Livingstone, D.J.: Unsupervised forward selection: a method for eliminating redundant variables. J. Chem. Inf. Comput. Sci. **40**(5), 1160–1168 (2000)

Witten, I.H., Frank, E., Hall, M.A., Pal, C.J., Data, M.: Practical machine learning tools and techniques. In Data Mining, Vol. 2, No. 4, pp. 403–413. Elsevier, Amsterdam, The Netherlands (2005, June)

Wohlin, C., Kalinowski, M., Felizardo, K.R., Mendes, E.: Successful combination of database search and snowballing for identification of primary studies in systematic literature studies. Inf. Softw. Technol. **147**, 106908 (2022)

Wold, S., Sjöström, M., Eriksson, L.: PLS-regression: a basic tool of chemometrics. Chemom. Intell. Lab. Syst. **58**(2), 109–130 (2001)

Worachartcheewan, A., Toropova, A.P., Toropov, A.A., Pratiwi, R., Prachayasittikul, V., Nantasenamat, C.: Interpretable SMILES-based QSAR model of inhibitory activity of sirtuins 1 and 2. Comb. Chem. High Throughput Screening **24**(8), 1217–1228 (2021)

Xu, J., Stevenson, J.: Drug-like index: a new approach to measure drug-like compounds and their diversity. J. Chem. Inf. Comput. Sci. **40**(5), 1177–1187 (2000)

Yap, C.W.: PaDEL-descriptor: An open source software to calculate molecular descriptors and fingerprints. J. Comput. Chem. **32**(7), 1466–1474 (2011)

Yilmaz, H., Sizochenko, N., Rasulev, B., Toropov, A., Guzel, Y., Kuz'min, V., ... & Leszczynski, J.: Amino substituted nitrogen heterocycle ureas as kinase insert domain containing receptor (KDR) inhibitors: Performance of structure–activity relationship approaches. J. Food Drug Analysis **23**(2), 168–175 (2015)

Zhong, L., et al.: The histone deacetylase Sirt6 regulates glucose homeostasis via Hif1α. Cell **140**(2), 280–293 (2010)

Dynamic Weighted Deep Learning Ensemble Model for Accurate Cervical Cancer Classification from Pap Smear Images

Agnish Paul[1], Chintan Jana[1], and Pawan Kumar Singh[1,2](✉)

[1] Department of Information Technology, Jadavpur University, Jadavpur University Salt Lake Campus, Plot No. 8, Salt Lake Bypass, LB Block, Sector III, Kolkata 700106, West Bengal, India
pawansingh.ju@gmail.com

[2] Shinawatra University, 99, Moo 10, Bang Toei, Sam Khok, Pathum Thani 12160, Thailand

Abstract. Cervical cancer, a disease that can be prevented with early detection, remains a global health concern, mainly in the remote and underrepresented regions of the world. Our study introduces an ensemble method combining Swin Transformer, Vision Transformer (ViT), and ResNet50 to classify cancerous and precancerous cells in Pap smear images. Each model has almost unique characteristics—hierarchical vision processing, global attention mechanisms, and robust feature extraction—the ensemble uses Softmax-weighted averaging to combine the classification power of each model and reduce false positives. Our ensemble model is evaluated on the SipakMed dataset of 4,049 images and 966 manually cropped cell clusters, achieving an impressive accuracy of 97.87% outperforming many state-of-the-art models for cervical cancer classification. These results also strengthen the fact that deep learning models and ensembles have the required capability to improve diagnostic accuracy and reliability in healthcare, building the path for more efficient and trustworthy screening tools.

Keywords: Cervical Cancer Classification · Pap Smear Image Analysis · Deep Learning Ensemble · Softmax-Weighted Averaging · Medical Image Classification · Diagnostic Accuracy · Healthcare AI · Vision Transformers · Computer-Aided Diagnosis

1 Introduction

Cervical cancer is a malignant tumor of the cervix, the portion of the uterus that joins the vagina at the bottom. It is mostly brought on by ongoing exposure to high-risk strains of the sexually transmitted virus known as the Human Papillomavirus (HPV) [1]. While cervical cancer is preventable through vaccination and early screening, it remains a significant public health concern, particularly in low and middle-income countries.

Cervical cancer remains a preventable disease through vaccination and regular screening. Global efforts focused on increasing access to these preventive measures,

S. Dhar et al. (Eds.): AGC 2025, CCIS 2686, pp. 65–77, 2026.
https://doi.org/10.1007/978-3-032-09153-6_4

particularly in underserved regions, are crucial to reducing the global burden of the disease. With continued advances in vaccination, screening, and treatment, the incidence and mortality of cervical cancer are expected to decline significantly in the coming decades.

In this study, we have proposed an ensemble method that combines the outputs of three advanced models—Swin Transformer, Vision Transformer (ViT), and ResNet50—to classify cancerous and precancerous cells in Pap smear images. Each model has been selected for its unique strengths: the Swin Transformer has hierarchical vision processing capability for detailed image understanding [2], ViT makes use of a global attention mechanism to identify important image regions [3], and ResNet50 has robust feature extraction capabilities [4]. By using softmax-weighted averaging, our ensemble automatically balances the contributions of these models, increasing classification accuracy and reducing false positives. It minimizes reliance on expert pathologists by automating accurate cell classification, making it suitable for low-resource settings. Its low computational demands enable deployment on portable or cloud-based platforms, ensuring accessibility in remote areas. By reducing false positives and lowering diagnostic costs, the model supports scalable, affordable mass screenings, aligning with government and NGO health programs. The entire pipeline of the methodology that we have employed in our paper is shown in **Fig.** 1.

The following are the contributions that our work has made:

1. Using images of isolated cells that have been manually cropped from 966 cluster images of Pap smear slides we have conducted a thorough analysis of our ensemble model, of how well it can classify the images as normal, abnormal, and benign.
2. A standard dataset which is an extensive collection of 4049 images has been used to assess our model, which demonstrates its flexibility across many data sources and guarantees a strong assessment of its performance.
3. Our analysis shows that our ensemble model has reached impressive accuracy rates on the SipakMed dataset. These findings provide important guidance for future implementations and improvements by proposing a model for medical image recognition and classification.
4. Since cervical cancer detection involves identifying cancerous and precancerous cells it has important applications in healthcare, especially in women's health. We propose an efficient model for the classification of cytological imaging, enhancing diagnostic accuracy which can lead to increased trust and confidence in screening tools ultimately adding business value while improving women's healthcare outcomes globally.

The parts that follow outline the foundation that our research is centered around. To shed light on the larger background of our research field, Sect. 2 presents a summary of earlier research done on the datasets used in our study. The dataset we have used, the process of data pre-processing that we have employed, and the architecture of our model are then covered in detail in Sect. 3. The empirical findings from our studies are presented in Sect. 4, which highlights the effectiveness and performance of our strategy. The conclusions from our research efforts are finally summarized in Sect. 5, which also emphasizes the importance and ramifications of our findings.

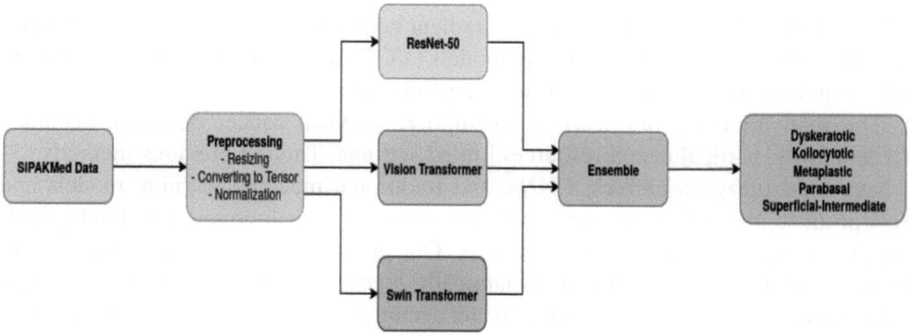

Fig. 1. The entire pipeline of our dynamic weighted deep learning ensemble methodology for cervical cancer classification.

2 Literature Review

In order to diagnose cervical cancer from Pap smear images, Rishav Pramanik et al. [5] suggest a fuzzy distance-based ensemble approach utilizing deep learning models (Inception V3, MobileNet V2, and Inception ResNet V2). The ensemble outperforms individual models in terms of classification accuracy, reaching 96.96% by combining model outputs with a unique fuzzy-based defuzzification of Euclidean, Manhattan, and Cosine distances. For clinical applications, this method provides a reliable, end-to-end method of diagnosing cervical cancer.

In [6], Ankur Manna et al. propose a fuzzy rank-based ensemble of three CNN models—Inception v3, Xception, and DenseNet-169—for automated cervical cytology classification. By applying two non-linear functions to assess classifier confidence, the method combines model predictions in a way that emphasizes confidence in correct classifications. Tested on the SIPaKMeD and Mendeley datasets, the approach achieves high accuracy and sensitivity, outperforming several state-of-the-art methods, which confirms its potential for efficient cervical cancer screening in clinical settings.

In [7], Sarah Amer Al-asbaily1 et al. propose a deep learning-based classification methodology, PCA for feature extraction and CNN for image classification, for early identification of cancer. Convolutional Neural Network (CNN) is employed to classify the images into five categories using the SiPAKMed dataset, consisting of 4049 cropped Pap Smear images captured with a CCD camera. Principal Component Analysis (PCA) was used as a dimensionality reduction method to increase the efficiency of the classification process. This method achieves an accuracy of 94.198%. In [8], Vatsala Anand et al. present a CNN model for classifying cervical cancer into five categories of cells, for precise diagnosis and targeted treatment. Making the detection process automatic, and improving the efficiency and accessibility, specifically in resource-limited environments, contributes to improved global management of cervical cancer.

In [9], Ashok K Chikaraddi et al. explore machine learning (ML) techniques for early detection of cervical cancer using Predictive Model Selection (PMS) and data pre-processing. Here traditional ML methods like Decision Trees, Logistic Regression,

SVM, and ensemble techniques such as gradient boosting and random forest are evaluated. They also compare deep learning models (VGG16, Inception-v3, ResNet-50) and analyze an Ensemble model for improved performance.

The motivation for our work is to improve cervical cancer detection accuracy and reliability, particularly in resource-limited settings. Though previous methods [5–9] have used fuzzy ensembles, CNNs, and traditional machine learning models and show promise, they often depend upon fixed aggregation techniques or lack adaptability to the complexity of Pap smear images. Our proposed trainable weighted ensemble addresses these drawbacks by dynamically learning model contributions through Softmax-normalized weights, ensuring better accuracy and robustness for more effective and accessible cervical cancer screening.

3 Methods and Materials

This section discusses the overall process of pre-processing the datasets and a succinct explanation of the general designs of the three potential models that are employed in this paper.

3.1 Dataset Used

The SipakMed Dataset [10] is an extensive collection of 4049 images of isolated cervical cells manually cropped from 966 Pap smear slide images, acquired using a CCD camera adapted to an optical microscope. The cell images are categorized into 5 categories—normal, abnormal, and benign cells. These images incorporate various cell types, including Dyskeratotic cells, characterized by abnormal keratinization, and Koilocytotic cells, among others. Cervical cancer is a global health concern and datasets like SipakMed help in advancements in image-based classification and early detection methods to improve healthcare. In Fig. 2, we have represented some of the images from the SipakMed dataset.

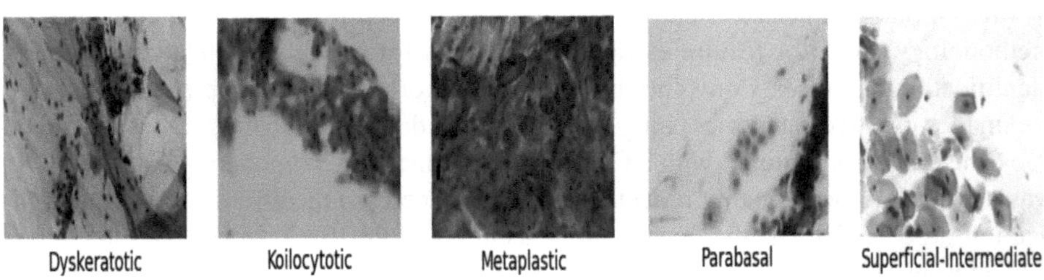

Fig. 2. Image of each class from the SipakMed dataset.

3.2 Dataset Preprocessing

We begin by structuring the dataset and converting the.bmp files to.jpg and storing them in different subfolders. We then construct a data frame, encode the labels numerically, and split the data into training and validation sets with a split ratio of 70:30. Following that we define separate Pytorch DataLoaders and image transformations, resizing the images to the respective input dimensions i.e. 224x224, converting images to tensors, and applying standard normalization. These DataLoaders load preprocessed images effectively for training and validating the ensemble model.

3.3 Model Architecture

In this study, we have developed an ensemble method consisting of 3 different models InceptionV3, Vision Transformer, and Efficient Net for the automatic detection of cervical abnormality to detect precancerous cells or cancerous cells. The models include ResNet, Vision Transformer, and Swin Transformer, the detailed architectures of which are provided below.

3.3.1 RESNET-50

ResNet-50 belongs to the ResNet family which addresses the difficulties involved in deep neural network training. The primary issue addressed by the ResNet family is the gradient degradation problem at the deeper layers [11]. The accuracy saturates and degrades as we go deeper into the network. This occurs as the gradients of the parameters in the deeper layers become very small. ResNet uses the Residual Blocks to overcome this issue. These mitigate the vanishing gradient issue by enabling the direct flow of information to bypass connections. As a result, deeper networks might get more precision because of the skip links. ResNet-50 uses stacked 50 bottleneck residual blocks. The initial layers feature convolutional and pooling layers. These preprocess the image before processing by the residual blocks [12]. The refined data produced by this is used by the fully connected layers to categorize images with greater accuracy. The bottleneck design of ResNet-50 reduces the number of computations while maintaining the quality of feature extraction. This ensures effective training of deep networks without affecting performance. The architecture of Resnet-50 is shown in Fig. 3.

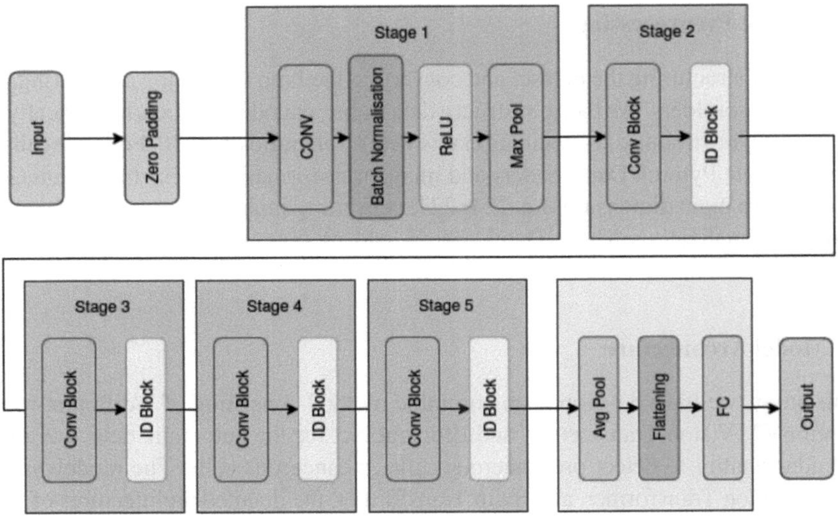

Fig. 3. Architecture of Resnet-50 model.

3.3.2 Vision Transformer

Vision Transformers (ViTs) have emerged as a competitive alternative to CNNs, due to them outperforming CNNs in terms of computational efficiency and accuracy. Vision Transformer implements the original Transformer model to the image classification task. The Transformer Encoder is made up of several blocks, each of which has three main processing components. These include Multi-Layer Perceptrons (MLP), Layer Norm, and Multi-head Attention Network (MSP) [13]. In order to extract features from images and provide them to the MLP head model for classification, the Transformer Encoder is utilized as a basis model.

Transformers are computationally demanding because of the attention matrix's quadratic complexity. This is further aggravated by increasing sequence length. To address these issues, the image is broken down into square patches as a "windowed" attention technique. These patches are flattened into a single vector by concatenating the channels of all pixels in a patch. This vector form of the patch is then linearly projected to the desired input dimension [14]. ViTs are able to scale effectively with large datasets compared to CNNs. This is further compounded when pre-trained using massive datasets, like ImageNet. Also, ViTs make fewer assumptions based on spatial hierarchical features(local features combined to global features). This allows them to learn relationships purely based on data provided while streamlining the classification process and improving computational efficiency. The architecture of the Vision Transformer is shown in Fig. 4.

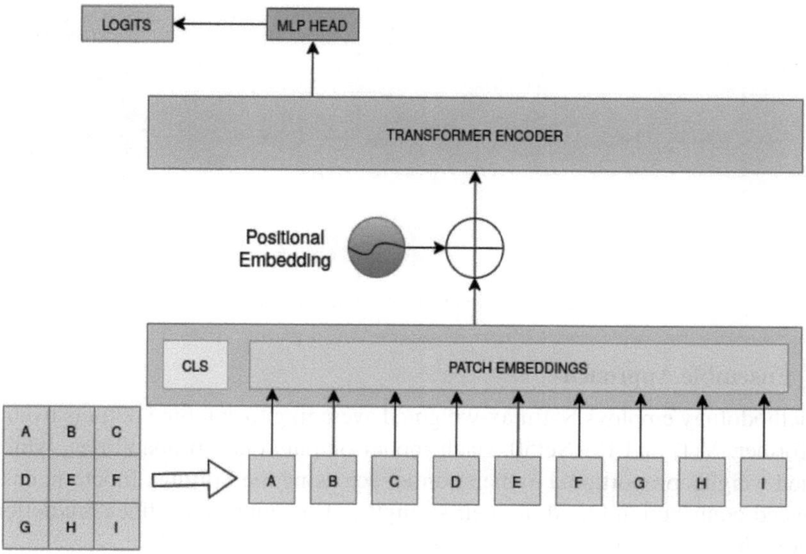

Fig. 4. Architecture of ViT

3.3.3 Swin Transformer

Swin Transformer constructs hierarchical feature maps, which are generated using a shifting window. By permitting cross-window connection and restricting the self-attention computation to non-overlapping local windows, this improves efficiency. Small-sized patches are created using the shifting window and progressively combining nearby areas in deeper strata [15]. The local non-overlapping nature of the self-attention computation gives it a linear computational complexity.

Swin Transformer splits input images in non-overlapping patches using the patch splitting module. Self-attention is calculated for each patch [16]. In the next layer, the window is shifted and new patches are formed, whose self-attentions are calculated. The shifted window provides a bridge between the results of each layer. Swin Transformers are a variation of the original Vision Transformer. As we can see, it patches photos using hierarchical partitioning and merges them as it moves deeper. This facilitates the efficient recording of both local and global features.. The local attention mechanism makes Swin more computationally efficient than ViTs and allows scaling for higher-resolution images. The architecture of the Swin Transformer is shown in Fig. 5.

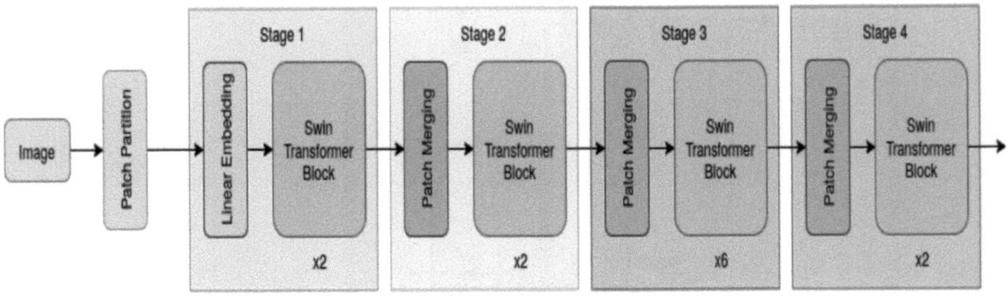

Fig. 5. Architecture of Swin Transformer

3.3.4 Ensemble Approach

The methodology employs Softmax-weighted averaging to combine outputs from Swin Transformer, ViT, and ResNet50. Each model outputs class probabilities, which are assigned weights proportional to their confidence using the Softmax function, ensuring a balanced contribution based on their strengths. This approach refines predictions by mitigating biases from individual models.

4 Performance Evaluation

4.1 Experimental Setup

The experiments are conducted on a Dell Inspiron 3501 with an Intel Core i5-1035G1 CPU running at 1.00 GHz, with a maximum boost clock of 1.19 GHz. It has 8 GB of installed RAM (7.77 GB usable) and operates on a 64-bit Windows operating system with an x64-based processor. The device includes touch support with 10 touch points.

4.2 Evaluation Metrics

Table 1 shows the list of hyperparameters used in this study. Various evaluation metrics (namely precision, recall F-measure, and confusion matrix) are used to assess the efficacy of the methods for predicting yoga poses. The mathematical equations defining precision, recall, and F1 Score are represented by Eq. (1), Eq. (2), and Eq. (3) correspondingly.

$$Precision_c = TPc/(TPc + FPc) \tag{1}$$

$$Recall_c = TPc/(TPc + FNc) \tag{2}$$

$$F1\,Score = \frac{2 \times Precision_c \times Recall_c}{Precision_c + Recall_c} \tag{3}$$

where, TP_c is the number of true positives for class c, FP_c is the number of false positives for class c and FN_c is the number of false negatives for class c.

Table 1. Values of hyperparameters for our models used in this study on SipakMed dataset.

Hyperparameters	Values
Optimizer	**Adam**
Loss Function	**CategoricalCross Entropy**
Batch size	**16**
Dropout	**N/A**
Epochs	**20**
Learning rate	**N/A**
Scheduler	**StepLR, step_size = 10, gamma = 0.1**

4.3 Results

Table 2 contains the performance results of evaluating our model on the benchmark SipakMed dataset. The columns contain the precision (p), recall (r), and f1-score (f) values for each of the classes in our dataset that we got during the testing phase. So from Table 2, we can come to the conclusion that the model is efficient in classifying Parabasal and Superficial-Intermediate cells but faces considerable problems while classifying Koilocytotic cells. The overall accuracy achieved by our model during the validation phase is 97.87%.

Table 2. Result of evaluation of our model based on the SipakMed dataset

Classes	p	r	f
Dyskeratotic	0.98	0.99	0.99
Koilocytotic	0.98	0.94	0.96
Metaplastic	0.95	0.98	0.96
Parabasal	1.00	1.00	1.00
Superficial-Intermediate	0.99	1.00	0.99

The training loss indicates how well the model fits with the training data and the validation loss indicates how well the model fits new data. We plot the training and validation loss curves on our Dataset, which is shown in Fig. 6. We observe that both the losses decrease significantly in the first few epochs which indicates our model learns steadily initially but after a certain point in time (around epochs 5), the training loss continues decreasing but slowly compared to the initial epochs however, the validation loss does the same but plateaus at a higher level than the training loss. This means that the network is starting to overfit the training data set and fails to generalize well for the validation set or new unseen data.

We represent the metrics of each class- p, r and f obtained on the SipakMed Dataset in Table 2 and the accuracy of our model in classifying each class in Fig. 7. From Table 2 and Fig. 7, we can come to the conclusion that the model performs well in classifying Parabasal and Superficial-Intermediate cells but is not expectedly efficient while classifying Koilocytotic cells.

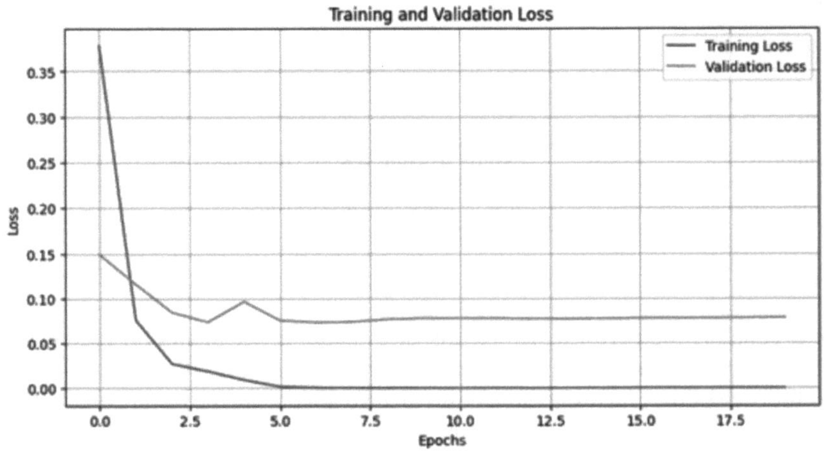

Fig. 6. Training and validation loss curve of our proposed dynamic weighted deep learning ensemble model on the SipakMed dataset

The confusion matrix is a means of displaying the number of accurate and inaccurate instances based on a model's prediction. The confusion matrix on the SipakMed Dataset is shown below in Fig. 8. The confusion matrix reveals specific misclassifications done by the model while classifying the cell images. A considerable accuracy has been assured by the model in classifying all the classes, but some misclassification has also been made by it because of the fact that some features do overlap, for example, Dyskeratotic and Koilocytotic, as well as Koilocytotic and Metaplastic. The highest misclassification count is 15 (Koilocytotic misclassified as Metaplastic), and other errors are minimal, which means the model performs well in general.

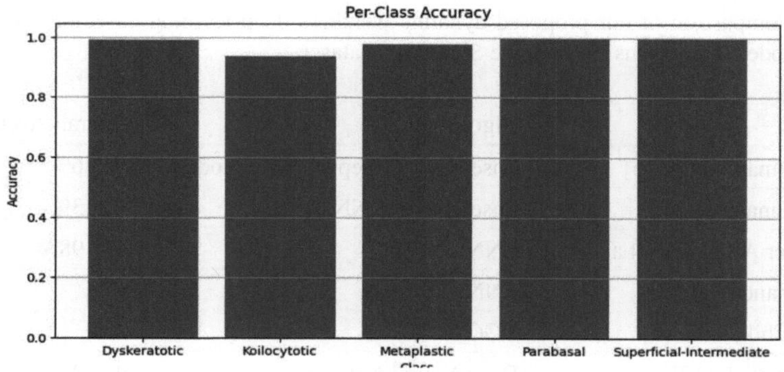

Fig. 7. Bar graph representing the classification accuracies of our proposed dynamic weighted deep learning ensemble model in classifying each cancer class from the SipakMed dataset

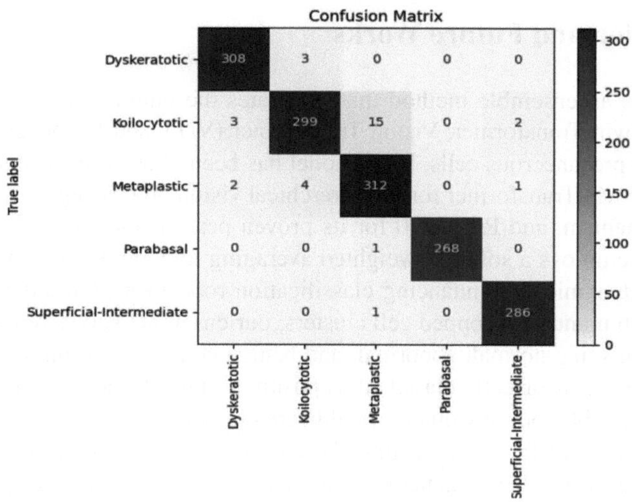

Fig. 8. Confusion matrix of our proposed dynamic weighted deep learning ensemble model on the SipakMed dataset

The comparison of our model with previous models has been done in **Table 3**. From **Table 3**, it is clear that our model has outperformed many state-of-the-art algorithms, which proves the efficacy of our approach in classifying cell images.

Table 3. Comparison of our proposed dynamic weighted deep learning ensemble model with previous models/algorithms done on the SipakMed dataset.

Authors	Algorithm	Overall Accuracy
Rishav Pramanik et al. [5]	Ensemble of deep learning models	96.96%
Ankur Manna et al. [6]	Ensemble of CNN models	95.43%
Sarah Amer Al-asbaily et al. [7]	CNN	94.198%
Vatsala Anand et al. [8]	CNN	92%
Ashok K Chikaraddi [9]	VGG16	89%
Ming Fang et al. [17]	Deep Neural Network	95.628%
Agnish Paul et al	**Softmax-weighted ensemble of 2 transformers and ResNet-50**	**97.87%**

5 Conclusion and Future Works

Our study uses an ensemble method that integrates the outputs of three state-of-the-art models—Swin Transformer, Vision Transformer (ViT), and ResNet50—to classify cancerous and precancerous cells. Each model has been chosen for its complementary strengths: the Swin Transformer for its hierarchical vision processing, ViT for its global attention mechanism, and ResNet50 for its proven performance in feature extraction. The ensemble employs a softmax-weighted averaging to combine the contributions of these models dynamically, enhancing classification robustness. On a dataset of 4,049 images and 966 manually cropped cell clusters, our ensemble shows remarkable accuracy in distinguishing normal, abnormal, and benign cells while significantly reducing test results that are incorrectly classified as positive. This advancement underscores its potential as a reliable tool in women's healthcare diagnostics.

Future works will focus on scaling the ensemble to more diverse and extensive datasets, thus ensuring better generalization across varied medical image processing tasks. The ensemble model can be integrated into clinical workflows as a decision-support tool, aiding pathologists by automating the classification of Pap smear images and reducing diagnostic time. For underserved regions, the model's low computational requirements make it suitable for portable or cloud-based platforms, enabling deployment in mobile health units or rural clinics. This can bridge the gap in early detection, reducing the burden of cervical cancer in populations with limited access to specialized care. Additionally, incorporating the model into national and NGO-led screening programs can scale its benefits, empowering healthcare systems to reach more women effectively. Ultimately, we aim to build an interactive diagnostic tool powered by our ensemble, leading to early detection and improving patient outcomes in women's healthcare.

Conflict of Interests. The authors have no competing interests to declare that are relevant to the content of this article.

References

1. Sompawong, N., Mopan, J., Pooprasert, P., Himakhun, W., Suwannarurk, K., Ngamvirojcharoen, J., Tantibundhit, C.: Automated pap smear cervical cancer screening using deep learning. In 2019 41st Annual International Conference of the IEEE Engineering in Medicine and Biology Society (EMBC), pp. 7044–7048. IEEE (2019, July)
2. Liang, J., Cao, J., Sun, G., Zhang, K., Van Gool, L., Timofte, R.: Swinir: Image restoration using swin transformer. In Proceedings of the IEEE/CVF international conference on computer vision, pp. 1833–1844 (2021)
3. Han, K., Wang, Y., Chen, H., Chen, X., Guo, J., Liu, Z., Tao, D.: A survey on vision. IEEE Trans. Pattern Anal. Mach. Intell. **45**(1), 87–110
4. Targ, S., Almeida, D., Lyman, K.: Resnet in resnet: generalizing residual architectures. arXiv preprint arXiv:1603.08029 (2016)
5. Pramanik, R., Biswas, M., Sen, S., de Souza Júnior, L.A., Papa, J.P., Sarkar, R.: A fuzzy distance-based ensemble of deep models for cervical cancer detection. Comput. Methods Prog. Biomed. **219**, 106776 (2022)
6. Manna, A., Kundu, R., Kaplun, D., Sinitca, A., Sarkar, R.: A fuzzy rank-based ensemble of CNN models for classification of cervical cytology. Sci. Rep. **11**(1), 14538 (2021)
7. Al-asbaily, S. A., Almoshity, S., Younus, S., Bozed, K.: Classification of cervical cancer using convolutional neural networks. In 2024 IEEE 4th International Maghreb Meeting of the Conference on Sciences and Techniques of Automatic Control and Computer Engineering (MI-STA), pp. 735–739. IEEE (2024, May)
8. Anand, V., Bachhal, P.: Cervical Net: An Effective Convolution neural network for five-class classification of cervical cells. In 2024 2nd International Conference on Device Intelligence, Computing and Communication Technologies (DICCT), pp. 51–55. IEEE (2024, March)
9. Chikaraddi, A. K., Kanakaraddi, S., Kalwad, P., Kadole, A., Budni, S.: Cervical cancer diagnostics using machine learning algorithms. In 2024 5th International Conference for Emerging Technology (INCET), pp. 1–6. IEEE (2024, May)
10. https://www.kaggle.com/datasets/prahladmehandiratta/cervical-cancer-largest-dataset-sipakmed
11. He, K., Zhang, X., Ren, S., Sun, J.: Deep residual learning for image recognition. In Proceedings of the IEEE conference on computer vision and pattern recognition, pp. 770–778. (2016)
12. Shafiq, M., Gu, Z.: Deep residual learning for image recognition: a survey. Appl. Sci. **12**(18), 8972 (2022)
13. Dosovitskiy, A.: An image is worth 16x16 words: Transformers for image recognition at scale. arXiv preprint arXiv:2010.11929 (2020)
14. Khan, S., Naseer, M., Hayat, M., Zamir, S.W., Khan, F.S., Shah, M.: Transformers in vision: a survey. ACM Computing Surveys (CSUR) **54**(10s), 1–41 (2022)
15. Liu, Z., Lin, Y., Cao, Y., Hu, H., Wei, Y., Zhang, Z., ... Guo, B.: Swin transformer: Hierarchical vision transformer using shifted windows. In Proceedings of the IEEE/CVF international conference on computer vision, pp. 10012–10022. (2021)
16. Liu, Z., Hu, H., Lin, Y., Yao, Z., Xie, Z., Wei, Y., ... Guo, B.: Swin transformer v2: scaling up capacity and resolution. In Proceedings of the IEEE/CVF conference on computer vision and pattern recognition, pp. 12009–12019. (2022)
17. Fang, M., Lei, X., Liao, B., Wu, F.X.: A deep neural network for cervical cell classification based on cytology images. IEEE Access **10**, 130968–130980 (2022)

Is The Indian Information Technology Sector COVID-19 Infected?

Anshuman Dhan[1], Arindam Chatterjee[1(✉)], and Rimu Chaudhuri[2]

[1] Heritage Institute of Technology, Kolkata, India
arindam.chatterjee@heritageit.edu
[2] Heritage Business School, Kolkata, India

Abstract. The recent most disaster that struck the world and had worldwide major impact is no doubt the COVID-19 pandemic attack, commonly referred to as the Corona attack. This has resulted in a meltdown in various socio-economic sectors. We have studied in this work, the impact of the pandemic on the IT sector, in particular on various IT stocks. We have chosen BSE Sensex, NIFTY IT index and the top 4 IT company shares and analysed their performance over the period January 2020 to March 2023. We have used various Time Series methodologies to analyze the impact on these shares at an individual level. Then we have used linear and quantile regression methods to measure their performance against COVID-19 affected cases in India. We have found a very significant effect almost at all levels of these shares.

Keywords: COVID-19 cases · NIFTY IT · BSE Sensex · Linear Regression · Quantile Regression · Time Series Decomposition

1 Introduction

The recent most disaster that struck the world and had worldwide major impact is no doubt the COVID-19 pandemic attack, commonly referred to as the Corona attack. This has resulted in a meltdown in various socio-economic sectors. Economies have been forced into recession. Social sectors have been disrupted. This has resulted in major impact on various industrial sectors, most of which is negative but sometimes positive as well. Since the stock market is a major indication of the health of various industrial sectors, it has suffered huge volatility as well during the pandemic period.

India is historically prone to pandemic and epidemic attacks primarily because of weather and population. Added to that the fact that our community medicine preparation is inadequate, large scale losses are inevitable each time such attacks happens. In 1876, bubonic plague ruined the city of Mumbai. The Spanish fever immediately after the First World War equally affected many regions.

Cholera, small pox, and various forms of fever and enteric diseases frequently infected millions of Indians during the last few decades the British Raj resulting

S. Dhar et al. (Eds.): AGC 2025, CCIS 2686, pp. 78–88, 2026.
https://doi.org/10.1007/978-3-032-09153-6_5

in huge losses in terms of human lives and assets. Now the global pandemic Corona recently has impacted the whole country very adversely.

The Information Technology sector in India has been a star performer for the last two decades or so. So much so that approximately 6% of the Indian GDP contribution comes from the IT sector alone. Approximately 25% of all exports income comes from the IT sector. Since IT industry is very much connected with the export sector than most other sectors, it is no doubt deeply affected by the global pandemic.

Most IT farms had to resort to the WFH (Work From Home) concept because of transportation breakdown and social distancing norms. Now most companies are planning to reduce the on campus presence by as much as 25% by year 2025. A new culture of flexible hours, staying home, with job done on solo computers and an internet connection is becoming the work culture among the Indian IT employees.

ISPs, or Internet Service Providers (like Reliance Jio and Vodafone), seeing a spurt in their business, have come up with lucrative WFH packages with higher data rates, lower prices and robust deliveries. Similar business boom has been observed with companies providing online conferencing software. "Zoom" just zoomed the whole world with its free cloud conferencing service.

Innumerable other companies have seen massive customer footfall (or should we say "netfall"), Google Meet and Teamviewer to name a few.

Unfortunately the sector is seeing a bloodbath as well with many IT and BPO (Business Process Outsourcing) companies resorting to job cuts and pay cuts. Some companies have already terminated employees who are "on bench". Some are cancelling all accrued Earned Leaves (EL). All India Forum for IT Employees (AIFITE) has expressed concern even though they admit the number of affected is small.

In this work we are evaluating the impact of COVID-19 on the Indian IT sector. We chose the IT stock market as this is a good indicator of the health of the sector after the infection.

2 Literature Review

Since the COVID -19 attack took the world by surprise, there is a lacunae of available literary works in this field that in particular focuses on the impact on various stock markets. The most informative work we have found out is [1]. This specifically refers to the impact on the Australian stock market. This work focuses in particular on the quantile regression method to understand the impact on different levels of the Australian stock market. We also modeled our work based on some of the relevant data and parameters detailed in their work.

Another work that deals with the COVID impact can be found in [2]. Here the authors focus on the Indian stock market. They have used the Market Model Event study methodology. The event (t1) is the official announcement of the lockdown. They have used a sample of 31 companies listed on Bombay Stock Exchange (BSE) selected at random. The sample period taken for the study is 35 days (24 February-17 April, 2020). Their study confirms that lockdown had a positive impact on the stock market performance with stocks showing significantly positive Average Abnormal Returns.

In [3], the authors present a general framework for addressing the extent of contagion effects between the BRICS' and US stock markets. They have employed the volatility impulse response (VIRF) approach to show how the BRICS' stock markets have been influenced in the context of the 2007–2009 global financial crisis. Their results show that there were significant contagion effects from the US to the BRICS' stock markets that very much depend on the level of integration with the international economy. They have also warned about heavier impact on stock market volatility nowadays compared to the 2007–2009 period since the degree of stock market integration thesedays are much more strengthened.

In [4], the authors have attempted to study the implication relationship between Stock market volatility with investors' expectations and behavior. They started with the assumption that the individual expectation is determined mainly by the number of differences between positive signals and negative signals in the information flow. Also they assumed the behavior is determined by both the expectations of investors around him and the expected returns from a potential action. They concluded that market volatility is caused mainly by inconsistencies in expectation and behaviour. They have found out also that Influential investors have profound impacts on the formation of normal investors' expectations and behavior.

3 Methodology and Test Results

3.1 Data Collection and Regression Setup

We are using various standard time series analysis methods to examine the impact of COVID-19 cases on the Indian stock market. There are some sectors that are obviously impacted negatively. This includes the Travel and Tourism sector. Similarly the e-Commerce and Pharma sectors have had a positive impact. It is not very clear however how much the impact is on the IT sector. The Work From Home (WFH) initiative has resulted in a huge operating expenditure savings no doubt. But then since IT essentially caters to all the broad industrial sectors, it must have had some revenue hits because of reluctance to specific sectoral investments [1].

Following the lines of [1], we have used BSE Sensex as the benchmark share market index. The dependent variables are weekly COVID cases in India, NIFTY IT Index (this is the share market index corresponding to Indian IT companies), and month of the week (where '1' denotes first week of a month etc.). The later variable is used instead of the day of the week variable, as we are analyzing weekly data[1]. Unlike [1], however, we are not using the Dow Jones industrial average

[1] There are months consisting of five weeks. We have skipped those weeks and used only four weeks per month.

(most popular share market index of USA) as we believe it is an endogenous parameter. Please see the Appendix for a sample of the input data format (in Table 5).

We have collected the necessary data for the time period January 2020 to March 2023. This time period is most relevant for any COVID-19 related analysis, esp. for India. The January 2020 - March 2021 time period marks the first wave. The second wave refers to the April 2021 - March 2022 time frame. And after March 2022 till March 2023, when the whole world and all of India as well, withdrew the lockdown, we mark the post-COVID period. The third period is necessary to compare the results with the periods when COVID was active vis-a-vis when it gradually calmed down and normalcy returned. We have collected **weekly data** for COVID cases in India from [5]. We have collected weekly stock prices from [6] for the same three time periods[2].

3.2 Time Series Analysis

We have done a time series analysis and decomposition method on the Sensex, NIFTY IT and top four IT companies' weekly share price. We found that there is an upward Trend, a Cyclical (seasonal) component and a random component. Since the major incident throughout 2020–2022 was COVID, we suspect the change in trend and random components is primarily the result of COVID as can be seen in Fig. 1.

We have performed Linear Regression of the Trend components of SENSEX and top five IT share prices on the number of COVID cases. We found that the Trend components have high correlation with number of COVID cases (as shown in Fig. 2).

We used ACF, PACF analysis on the Random components of these share prices and found that there may be Moving Average and Autoregress components present (only the analysis for SENSEX is shown in Fig. 3)[3].

The ACF and PACF plots indicate existence of MA components but not AR. The auto.arima exactly identifies that.

We have used the ***arima*** function to find out what is the nature of Time Series dependency. We found MA(2) but no AR. With a lag of two time periods, in this case it boils down to two time periods (i.e., two quarters) .

3.3 Linear and Quantile Regression

We have used linear regression to find out any dependency on COVID-19 for the share price in question.

[2] In most cases dates for the two data sets match; however there are weeks where the COVID weekly data and share market data don't exactly fall on the same day. This is because of share market holidays. We have taken share market data for the next business day in such cases.

[3] We are following the general rule of thumb here. ACF for MA(k) time series shows an exponential decay after k lags whereas PACF for AR(k) shows a similar pattern. If there is no obvious decay in any of these, the time series is a combination of the two.

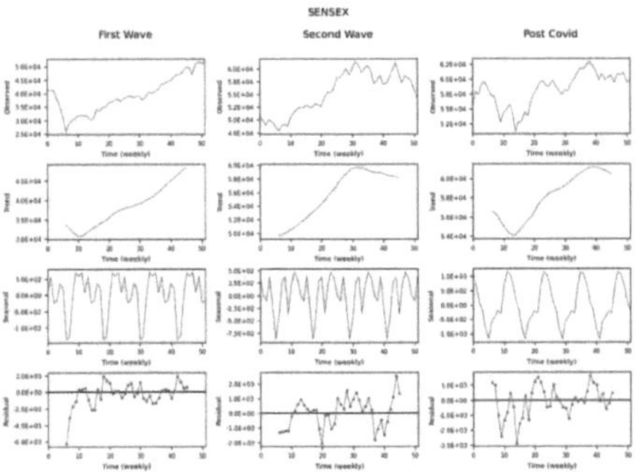

(a) BSE Sensex

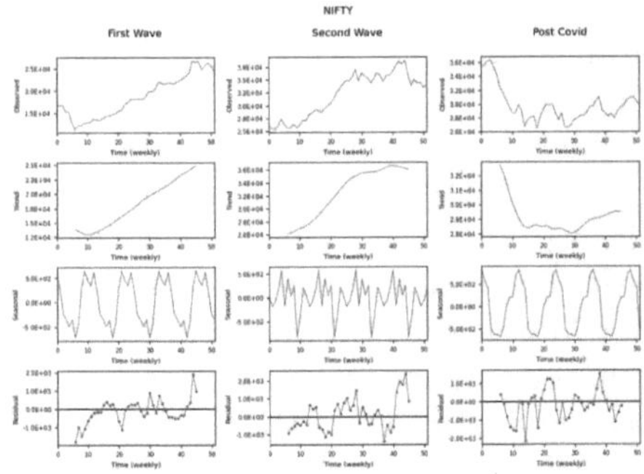

(b) NIFTY IT

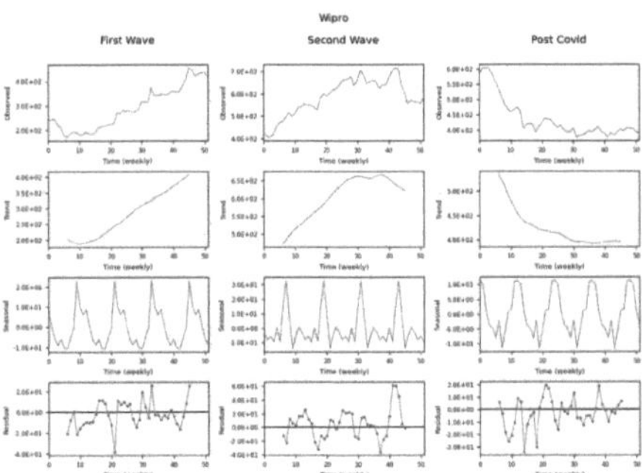

(c) Wipro

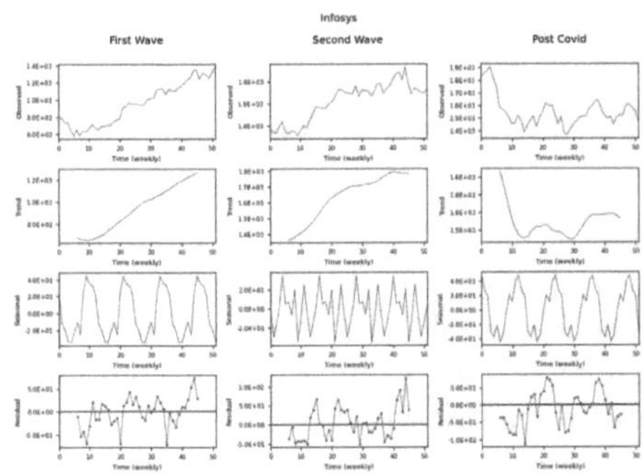

(d) Infosys

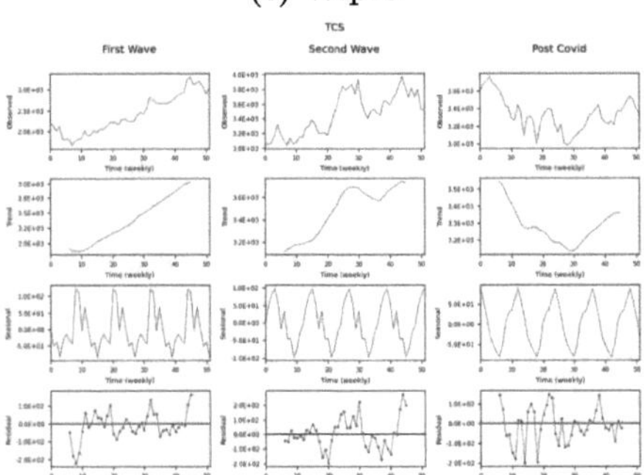

(e) TCS

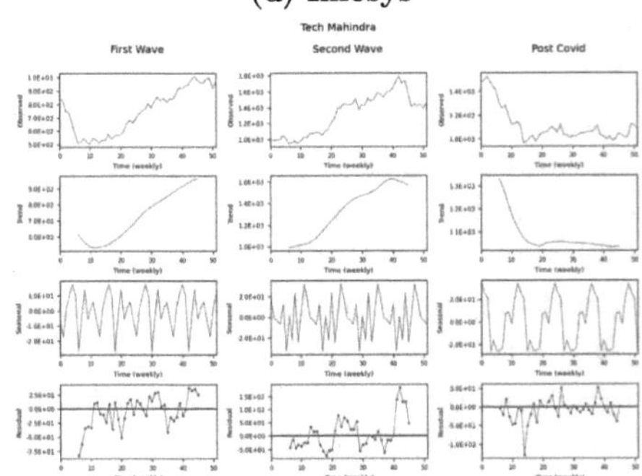

(f) Tech Mahindra

Fig. 1. Plot of the decomposed time series of SENSEX, NIFTY IT and 4 top Indian IT weekly share prices.

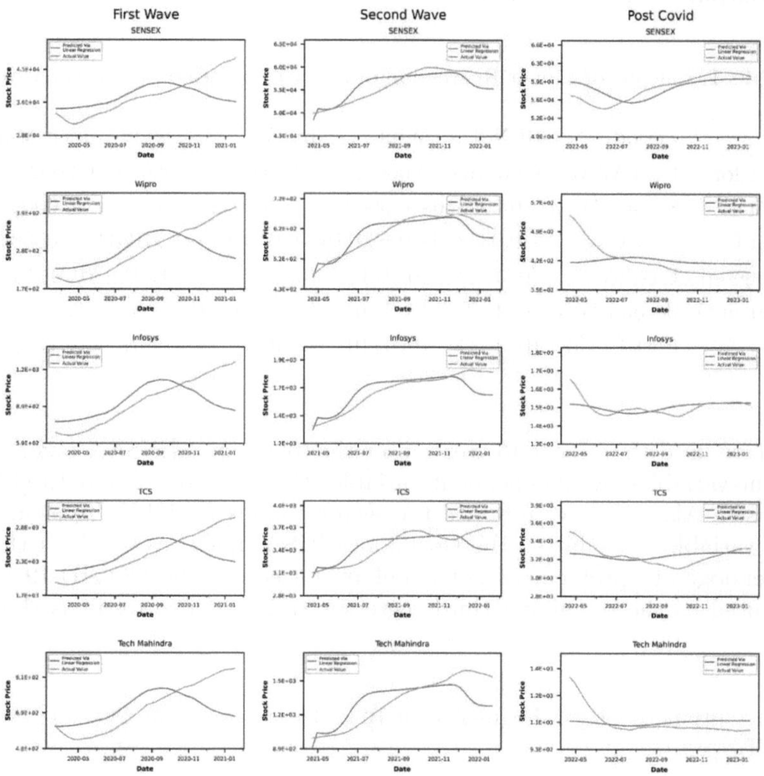

Fig. 2. Plot of Linear Regression of the Trend component of SENSEX and top 4 IT stock prices on COVID cases

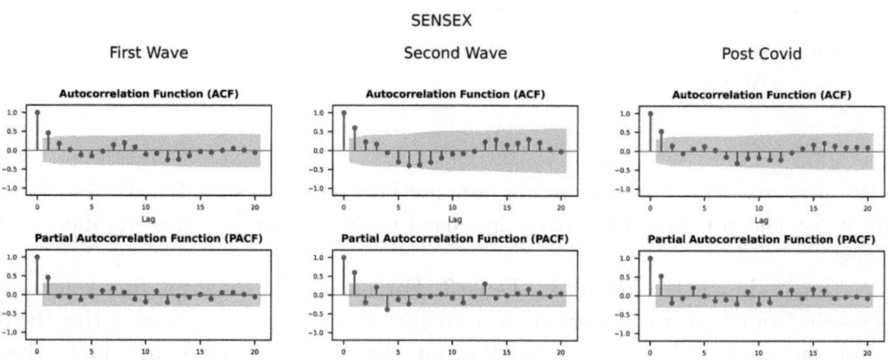

Fig. 3. ACF (above) / PACF (below) of SENSEX random components.

The output is shown in Table 1

Table 1. Output of Linear Regression of top 4 IT prices on COVID cases.

	Period	SENSEX	Wipro	Infosys	TCS	Tech Mahindra
Coefficient for	First Wave	0.005104	0.000126	0.00044	0.000653	0.00024
new Covid-19	Second Wave	−0.000866	−0.000022	−0.000042	−0.000056	−0.000055
Confirmed Cases	Post Covid	−0.033886	−0.000116	−0.000584	−0.001282	−0.000845
Impact Lag	First Wave	(0, 1, 0)	(0, 1, 0)	(0, 1, 0)	(2, 1, 2)	(0, 1, 0)
(Arima output)	Second Wave	(0, 1, 0)	(0, 1, 0)	(0, 1, 0)	(0, 1, 0)	(0, 1, 0)
	Post Covid	(0, 1, 1)	(0, 1, 0)	(2, 0, 0)	(1, 0, 0)	(0, 1, 0)

We have used the Linear Regression methodology on SENSEX taken as the dependent variables. As independent variables the Week of the month (1st, 2nd etc.) is used. Also We have retained the sectoral index (NIFTY IT) as an independent variable but omitted the Dow Jones (USA Index) as we think this can lead to endogenity problem as noted before. Of course the COVID-19 weekly cases is used as the remaining third regressor variable. The result is shown in Table 2.

Table 2. Output of Linear Regression of BSE Sensex prices on COVID cases and NIFTY IT.

	Month of the Week	COVID-19 weekly cases	NIFTY IT Price
First Wave	−4.8e-03 (At 1%)	1.4e+00 (*)	−2.8e+01
Second Wave	5.6e-05	9.1e-01 (*)	−1.5e+02
Post Covid	−3.4e-02 (At 1%)	−1.3e-02	−6.1e+01

We now use the quantile regression method as suggested by [1]. We have done it for the top four IT stocks and used GOFTest to test the significant of the test.

We have shown the result in Table 3. For each cell, the first line refers to the result from first wave period, second line from second wave and the third line for the post-COVID situation. We note for Wipro, Infosys and TCS, that almost at all quantiles the dependency is significant for the first two periods, i.e., when COVID was very active. The significance is almost absent for the post-COVID period. This clearly establishes our initial hypothesis that COVID cases indeed have impacted the prices of these three IT shares. During the post-COVID period, however, the relationship goes away, again as expected. For the stray case of Tech Mahindra, the relationship is not clear. We think this is because Tech Mahindra does not have much of a weightage on the India IT sector performance compared to the other three IT behemoths.

Table 3. Output of Quantile Regression of Top 4 IT share prices on COVID cases.

Quantile levels	Wipro	Infosys	TCS	Tech Mahindra
0.1	1.7e-04(At 5%) −1.1e-05 3.0e-03	5.4e-04(At 5%) −4.6e-05(At 1%) −1.2e-03(*)	7.9e-04 1.6e-03 2.4e-02(At 5%)	1.2e-03 3.8e-04 7.7e-03
0.2	1.7e-04(At 5%) −2.5e-05(At 1%) 1.5e-04	5.5e-04(At 5%) −1.2e-04(*) −4.1e-04	4.3e-03 −9.0e-05(At 1%) −1.9e-03(At 5%)	4.3e-04(At 1%) 3.4e-04(At 1%) 9.4e-03
0.3	2.0e-04(At 5%) −3.1e-05(*) 2.2e-04	5.4e-04(At 5%) −1.2e-04(*) 1.1e-02	3.9e-03 1.2e-03(At 5%) 2.3e-02	1.4e-03(At 5%) 3.8e-04 −7.5e-05
0.4	1.9e-04(At 5%) −4.0e-05(*) 3.2e-03	2.6e-03 −1.3e-04(*) −3.4e-04	7.9e-04 −1.5e-04(*) −9.9e-04	4.2e-04(At 5%) −1.3e-04(*) 7.4e-03
0.5	1.4e-04 2.8e-04(At 1%) 1.7e-04	1.9e-03 −2.9e-05 −1.3e-04	7.2e-04 −3.9e-05 −7.5e-04	3.4e-04(At 5%) −3.8e-05 −9.6e-05
0.6	1.5e-04(At 5%) −1.9e-05 1.3e-04	5.8e-04(At 1%) 1.2e-04 1.3e-02(At 5%)	1.1e-03(At 1%) 2.7e-04 −8.0e-04	1.5e-03(At 5%) 8.9e-05 −4.9e-04
0.7	3.8e-04(*) −1.8e-05 −1.7e-04	2.4e-03 1.2e-04 −2.3e-04	2.9e-04 2.7e-04 −5.5e-04	3.8e-06 8.9e-05 −4.3e-04
0.8	−1.4e-04 −1.9e-05 −5.5e-04	−2.8e-04 1.2e-04 −1.3e-04	−3.1e-04 −6.6e-05(At 5%) −1.4e-03	−2.6e-04(At 5%) 8.9e-05 −7.1e-04
0.9	−1.9e-04(At 5%) 3.8e-05 −1.2e-03(At 5%)	−3.0e-04 −4.0e-05(At 1%) −1.7e-03	−7.5e-04 −4.9e-06 −2.4e-03	−2.6e-04(At 5%) 8.9e-05 −3.2e-03(At 1%)

We now conduct the experiment on SENSEX(weekly basis) versus weekly COVID-19 cases in India, NIFTY IT, and week of the month parameter. We conducted a quantile regression between 10% to 90% on an incremental interval of 10%. The result is shown in the table Table 4. Just like Table 3, the three lines per cell denote first wave, second wave and post-COVID time periods. We note the effect of COVID cases on lower quantiles is very significant for the first wave. However, there is not much impact on the upper quantiles. For the second wave, the significance is spread across various quantiles and seems to be affecting more in number than that of the first wave. The significance is random for post-COVID period and appears inconclusive, probably something anticipated. The

Table 4. Output of Quantile Regression of BSE SENSEX prices on COVID cases and other parameters.

Parameter	Quantile levels co-efficient of COVID-19 cases	Quantile levels co-efficient of NIFTYIT	Quantile levels co-efficient of Week of Month
0.1	−5.0e-03 (At 5%) 9.3e-04 −2.5e-02	1.5e+00 (*) 1.6e+00 (*) 3.2e-01	−7.6e+01 1.3e-04 1.9e+02
0.2	−5.0e-03 (At 5%) 7.9e-04 (At 5%) −5.7e-02 (*)	1.4e+00 (*) 1.6e+00 (*) −1.1e-01	−1.1e+02 1.7e-04 −8.0e+01
0.3	−4.1e-03 (At 5%) 7.2e-04 −6.2e-02 (*)	1.4e+00 (*) 1.6e+00 (*) −1.2e-01	3.3e+02 2.1e-04 −1.9e+02
0.4	−3.9e-03 6.9e-04 −5.0e-02 (*)	1.4e+00 (*) 1.7e+00 (*) −1.5e-01	4.6e+02 5.4e-05 −1.8e+02
0.5	−3.3e-03 2.1e-03 (*) −4.7e-02 (At 1%)	1.6e+00 (*) 1.6e+00 (*) −7.4e-02	4.3e+02 2.5e+02 −1.6e+02
0.6	2.5e-03 3.1e-04 −2.1e-02	1.8e+00 (*) 1.3e+00 (*) 2.0e+00 (*)	1.5e+03 (At 5%) 6.7e+02 7.1e+02
0.7	−2.4e-03 1.7e-03 (*) −1.7e-02	2.1e+00 (*) 1.7e+00 (*) 2.4e-01	8.1e+02 3.3e+02 −7.1e+01
0.8	−3.3e-03 2.0e-03 (*) −2.4e-02	1.5e+00 (*) 1.8e+00 (*) 6.1e-01 (At 1%)	1.8e+02 5.6e+01 −6.3e+01
0.9	−1.0e-02 1.3e-03 (At 1%) −3.8e-02	2.5e+00 (*) 1.8e+00 (*) 1.4e+00 (*)	2.1e-04 5.5e+02 5.6e+02

NIFTY IT coefficient is very significant on the SENSEX run for all three time periods and almost for all quantile levels. This clearly establishes the IT sector in India as backbone to SENSEX.

4 Conclusion

We have used various time series analysis methods on BSE SENSEX, NIFTY IT and top four Indian IT company share prices. The chosen period was from just before the beginning of the pandemic period (i.e. From January 2020) to March end 2023, on a weekly basis. We found that these share prices have a random component which we think must have been affected by the number of COVID cases that occurred in the same time period. Upon doing ARIMA analysis, we verified this having an impact with 1 to 2 quartes of delay lag period. We performed linear and quantile regression to quantify this effect. We found that these share prices are significantly impacted by the number of COVID cases in India during this same period and the effect is significant almost at all levels of the share maket price. In addition we found the BSE SENSEX is also significantly impacted by the IT sectoral index NIFTY IT, validating the conjecture that Indian IT sector is a very important component of the Indian economy.

Although we have done significant analysis on the COVID-19 cases in India and its relationship with some Indian IT companies as well as on the leading market indices, we have not included many other market aspects. Our focus is narrow on a few indices and top IT companies, without considering a more comprehensive view of the IT sector. This lacuna we will seek to address by considering more companies and indices. The work overlooks other potential influencing factors beyond COVID-19, such as government policies or global economic conditions, which may also have played a significant role in stock performance. It would be good to get a more comprehensive way to eliminate other factors before concluding that the rest of the impact is possibly due to COVID-19. We will work on suitable models to accomplish this. The use of time series and regression models is a limitation, as these methods may not fully capture the complex market dynamics during such a crisis. We will try to incorporate other methods like Factor Analysis or Conjoint Analysis towards that goal. We also plan to explore the long-term effects and potential recovery trends, which could offer a more comprehensive understanding of the pandemic's impact on the IT sector. The pre-2021 and post-2021 cases give us a way to do this, as this was the time precisely when COVID vaccine was introduced in India. A Regression Discontinuity Design (RDD) methodology looks an appropriate candidate. We have focused on Indian markets, restricting our findings to a regional perspective, missing out on insights into the performance of global IT stocks. This we will address as we get more data and analysis results from other countries like America, the Commonwealth and the Euro zone. Finally, when we started this work, enough relevant literature was not available. But with the COVID-19 being almost five years old, a plethora of literary work is now available to understand its full impact globally. We will review some of these and borrow their methods and ideas as we see fit in our work in future.

A Appendix

The data has a weekly date, and other fields are Week No (since the first COVID-19 case in 2020), SENSEX, NIFTY IT, and share prices for some leading Indian IT companies we are considering. In case of linear regression, COVID-19 case is the input and share prices are the dependent variable. For Quantile Regression, the input is the same, but the output is only the SENSEX price level.

Table 5. Sample Data

Date	New_cnf	Week_No	Nifty IT	SENSEX	Wipro	Infy	TCS
Feb 8	2	1	16482	41142	244	786	2184
Feb 15	0	2	16708	41258	243	797	2156
Feb 22	0	3	16643	41170	246	731	2001
Feb 29	0	4	15355	38297	221	739	2118
Mar 9	41	1	15334	35635	214	642	1807
Mar 16	70	2	12964	31390	180	585	1798
Mar 23	301	3	11328	25981	170	654	1819
Mar 30	656	4	12153	28440	184	585	1654
Apr 6	3243	1	12147	30067	192	638	1766
Apr 13	6139	2	12699	30690	189	629	1807
Apr 20	7162	3	12813	31648	181	658	1818
Apr 27	10362	4	12957	31743	182	716	2015
May 4	14693	1	13646	31715	190	675	1894
May 11	24589	2	13522	31561	185	652	1893
May 18	28910	3	13278	30029	182	692	2019

References

1. Narayan, P.K., Gong, Q., Ahmed, H.: Is there a Pattern in how COVID-19 has affected Australia's stock returns? Appl. Econ. Lett. (2021)
2. Alam, M.N., Alam, M.S., Chavali, K.: Stock market response during COVID-19 lockdown period in india: an event study. J. Asian Finance, Econ. Bus., 131–137 (2020)
3. Jin, X., An, X.: Global financial crisis and emerging stock market contagion: a volatility impulse response function approach. Res. Int. Bus. Finance, 179-195 (2016)
4. Wang, Y., Deng, H.: Expectations, behavior, and stock market volatility. Emerg. Markets Finan. Trade, 3235–3255 (2018)
5. Hohn Hopkins University Homepage Cumulative of the coronavirus (COVID-19) confirmed, recovered and deceased numbers across India from January, 2020 to October, 2023. https://www.statista.com/statistics/1104054/india-coronavirus-covid-19-daily-confirmed-recovered-death-cases. Accessed 09 Nov 2024
6. Yahoo Finance Homepage. http://in.finance.yahoo.com. Accessed 09 Nov 2024

EmSLR Framework: Enhancing Communication for Deaf Individuals During Mass Emergency

Saugata Sarkar[1(✉)], Sayan Datta[1], Soumodeep Banerjee[1], Dhritiman Mukherjee[1], and Moumita Basu[2]

[1] Amity University Kolkata, Kolkata, India
treysarkar@gmail.com
[2] NSHM Knowledge Campus, Kolkata, India

Abstract. This study introduces an innovative framework designed to enhance communication for deaf individuals through an automatic sign language recognition (SLR) mobile app. The system improves real-time processing accuracy and user experience by employing a smart glove embedded with flex sensors and a gyroscope to accurately capture hand and finger movements. These sensors provide continuous data on finger bending and hand orientation, which are crucial for interpreting sign language gestures. An Arduino microcontroller collects and processes this data, transmitting it to a custom machine learning model for real-time analysis. The machine learning (ML) module compares the sensor input with pre-trained models to identify gestures. Once recognized, the corresponding word is displayed on a smartphone screen and converted into audible speech via the mobile app. This study specifically focuses on detecting emergency-related words such as 'food', 'water', 'shelter', and 'help/rescue, which are crucial in crisis situations. Preliminary findings indicate that the smart glove system accurately recognizes and translates various sign language gestures using ML model(M-f1 score 0.99), demonstrating its potential as an effective communication tool. This innovation has the potential to greatly enhance the quality of life for deaf individuals by facilitating essential interactions and promoting greater accessibility and inclusivity.

Keywords: Flex sensor · Gyroscope · Machine Learing · Data aumentation · Mobile app · Emergency

1 Introduction

Communication is a fundamental aspect of human interaction, allowing people to share thoughts, emotions, and information. However, deaf individuals often face significant challenges in communication [10]. The motivation for this study arises from the significant communication barriers faced by deaf individuals, especially when interacting with those who do not understand sign language [21]. The situation becomes worse during any mass emergency, such as natural and man-made

S. Dhar et al. (Eds.): AGC 2025, CCIS 2686, pp. 89–103, 2026.
https://doi.org/10.1007/978-3-032-09153-6_6

disasters. To address these challenges, we propose developing a framework that recognizes sign language gestures in emergency situations and translates them into spoken words. The EmSLR framework we designed includes an array of flex sensors and a gyroscope embedded in the glove to capture precise hand and finger movements. Flex sensors measure finger bending, while the gyroscope detects hand orientation and motion. An Arduino microcontroller collects and processes this sensor data, transmitting it wirelessly to a custom-built software application. The software application contains a pre-developed ML model to analyze sensor data, comparing it against stored gesture patterns in ML model. When a match is found, the software translates the recognized gesture into its corresponding word, which is then displayed on a smartphone screen and converted into audible speech using text-to-speech technology through our mobile app. This dual-mode output ensures accessibility both visually and audibly, catering to various communication preferences. Our contributions are summarized below.

Our contribution :

(i) We gathered primary data through experiments conducted in a simulated environment[1] using a smart glove and created a ground truth dataset called Emergency Sign Language (EmSL) for emergency-related words. We also plan to develop and release the labeled dataset on sign language detection from sensor data for the research community.

ii) By addressing these gaps of the prior studies our research contributes significantly to advancing the field of Sign Language Recognition, offering a more practical, accurate, and comprehensive solution. Note that, our EmSLR framework concentrated on identifying critical emergency terms like **rescue, food, water, oxygen,** and others, which are essential in any large-scale emergency.

(iii) Our framework stands out for its multi-sign-lingual capability, recognizing regional variations in sign languages. It addresses these differences through downloadable region-specific models via a mobile app, enabling users to switch seamlessly between languages.

2 Literature Review

The development of smart gloves for sign language recognition (SLR) has been an area of active research [3,12,14,16], leveraging advancements in sensor technology, machine learning, and wearable computing to facilitate communication for deaf individuals [1,2,4,9]. For a comprehensive review of existing SLR systems please refer [28]. Flex sensors, which detect finger bending, are essential elements in numerous smart glove designs. Research has shown their efficacy in capturing the detailed finger movements required for sign language recognition. For example, Mamun et al. [4] created a glove with flex sensors to measure finger flexion, allowing for the recognition of static gestures representing various alphabets.

[1] https://www.tinkercad.com.

Similarly, Abhishek et al. [1] combined capacitive touch sensors with flex sensors to improve gesture recognition accuracy. Incorporating additional sensors such as gyroscopes and accelerometers has been shown to significantly improve the recognition of dynamic gestures. Yudhana et al. integrated MPU6050 sensors with flex sensors in a smart glove to capture both the orientation and movement of the hand, which is crucial for interpreting more complex sign language gestures [29]. However, these studies [1,4,29] are limited to mapping English alphabets only, which restricts their applicability to a broader sign language vocabulary. Additionally, the system lacks flexibility and practical utility for real-time communication.

The use of machine learning algorithms is pivotal in mapping sensor data to specific gestures. Adeyanju et al. [2] provided a comprehensive review of various machine learning methods applied to sign language recognition, highlighting the efficacy of models such as Random Forests, Support Vector Machines (SVMs), and Neural Networks in classifying gestures based on sensor input. Bankar et al. [9] demonstrated the application of deep learning techniques, which offer superior performance in recognizing a wide range of gestures with high accuracy. Real-time processing of sensor data and effective communication of the recognized gestures are essential for practical applications [11]. The integration of microcontrollers like Arduino and communication modules such as Bluetooth has been widely adopted to facilitate this [20,26]. For instance, the use of Arduino for data acquisition from sensors and Bluetooth for transmitting data to a smartphone for further processing is a common approach [8].

Limitation in prior frameworks: Our research aims to overcome the following limitations in prior frameworks by developing a more robust and flexible SLR framework.

- **Regional sign language support:** Previous studies primarily concentrate on the English language, with little to no support for regional languages.
- **Real-Time Communication via mobile application:** Prior research are failed to develop an integrated app-based framework for SLR
- **Advanced Modeling Techniques:** The majority of previous studies did not fully utilize the potential of machine learning and advanced multilingual neural network-based models to manage complex linguistic structures
- **Large-Scale Data Utilization:** Labeled datasets for recognizing sign language from sensor data are limited. Most previous studies rely on labeled datasets derived from gesture videos or images to identify sign language.
- **Focus on recognizing emergency word:** To the best of our knowledge, none of the prior work focused on recognizing emergency words from sign language.

3 EmSL Dataset Development

To train our ML model effectively, we need a suitable dataset that captures sign language gestures specifically used in emergency situations through sensor data.

However, to the best of our knowledge, there is currently no existing Sign Language (SL) dataset that specifically represents emergency-related words using sensor data. To address this gap, we have developed a unique dataset comprising 750 data points, covering six critical emergency words: ‘food’, ’water’, ‘blanket’, ‘shelter’, ‘oxygen’ and ‘help/resque’ Table 1 provides illustrations of the corresponding sign language gestures for each word (Table 2).

To ensure the dataset’s reliability, we collected first-hand data in a controlled environment using a smart glove equipped with flex sensors and a gyroscope. This setup allowed us to accurately capture hand and finger movements associated with these emergency signs. The distribution of the collected data across different classes is visually represented in Fig 1. This dataset serves as a valuable resource for training machine learning models to recognize and translate emergency sign language gestures effectively. A random sample of the actual dataset data captured for each word is reported in Table 1

Table 1. Transposed Flex Sensors and Gyroscope Data sample

Parameter	help/rescue	Oxygen	toilet	school	water	blanket [nfy]
f1 OHM	30041.1	71510.21	71510.21	30041.1	103811.1	58551.72
f1 ANGLE	6	55	55	6	94	40
f2 OHM	30041.1	71510.21	71510.21	30041.1	30041.1	58551.72
f2 ANGLE	6	55	55	6	6	40
f3 OHM	30041.1	71510.21	71510.21	30041.1	30041.1	58551.72
f3 ANGLE	6	55	55	6	6	40
f4 OHM	30041.1	71510.21	30041.1	30041.1	30041.1	58551.72
f4 ANGLE	6	55	6	6	6	40
f5 OHM	30041.1	71510.21	71510.21	30041.1	30041.1	163371.42
f5 ANGLE	6	55	55	6	6	166
f6 OHM	163371.42	163371.42	30041.1	30041.1	30041.1	58551.72
f6 ANGLE	166	166	6	6	6	40
f7 OHM	163371.42	163371.42	30041.1	30041.1	30041.1	58551.72
f7 ANGLE	166	166	6	6	6	40
f8 OHM	163371.42	163371.42	30041.1	30041.1	30041.1	58551.72
f8 ANGLE	166	166	6	6	6	40
f9 OHM	163371.42	163371.42	30041.1	30041.1	30041.1	58551.72
f9 ANGLE	166	166	6	6	6	40
f10 OHM	30041.1	30041.1	30041.1	30041.1	30041.1	163371.42
f10 ANGLE	6	6	6	6	6	166

Table 2. Names and their respective signs in two adjacent tables

Name	Sign
Blanket	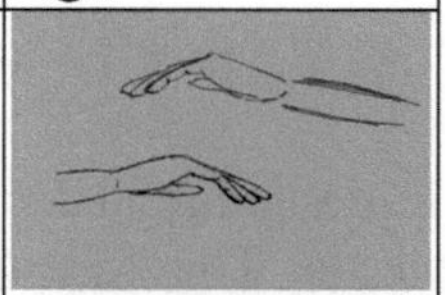
Help	
Oxygen	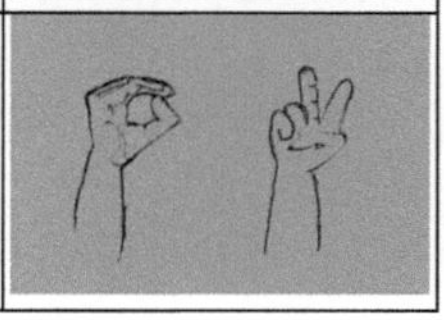

Name	Sign
Shelter	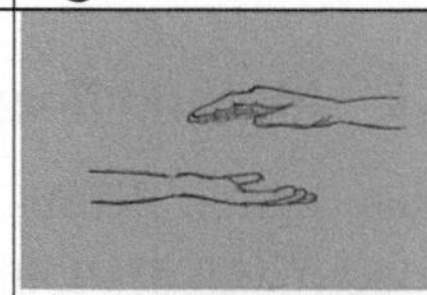
Toilet	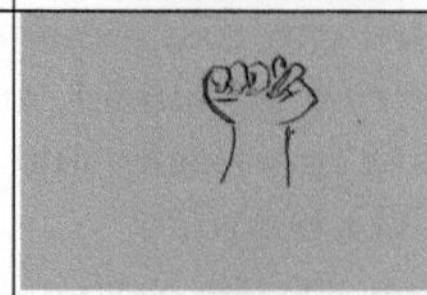
Water	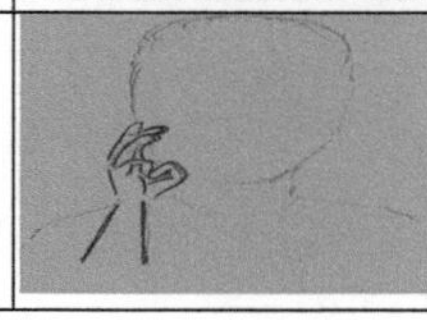

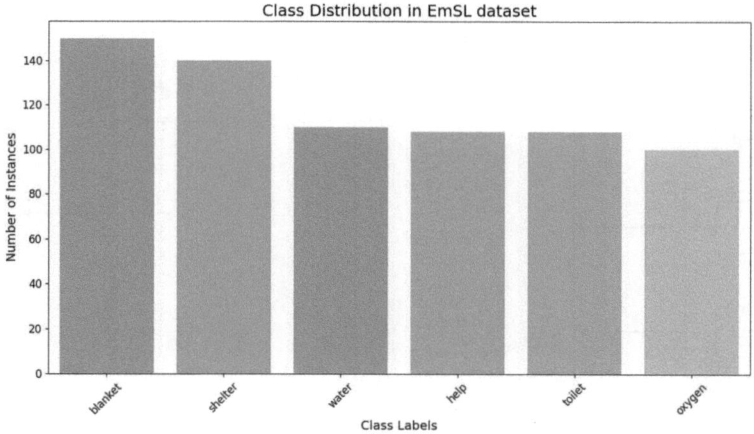

Fig. 1. Class Distribution in EmSL Dataset

4 Framework to Detect Sign Language

Sign language is a comprehensive visual language that relies on hand shapes, orientations, movements, and facial expressions to convey meaning. Each symbol or gesture in sign language can represent a word, phrase, or letter. The smart glove system aims to recognize these symbols by capturing the intricate hand movements and finger positions required to form each sign.

To achieve accurate recognition, the system must differentiate between subtle variations in hand shapes and movements. This involves mapping the flex sensor values and gyroscope readings to specific sign language symbols stored in a predefined dictionary. By comparing real-time sensor data against this dictionary, the system can identify and translate the gestures into spoken words.

4.1 Methodology

This section details the methodology used in designing and implementing the smart glove system for sign language recognition. The system is composed of hardware components, including flex sensors, an Arduino UNO board/ Raspberry Pi 3 model, and a Bluetooth module, as well as a software component for data processing and communication. We initially experimented with an Arduino UNO board.

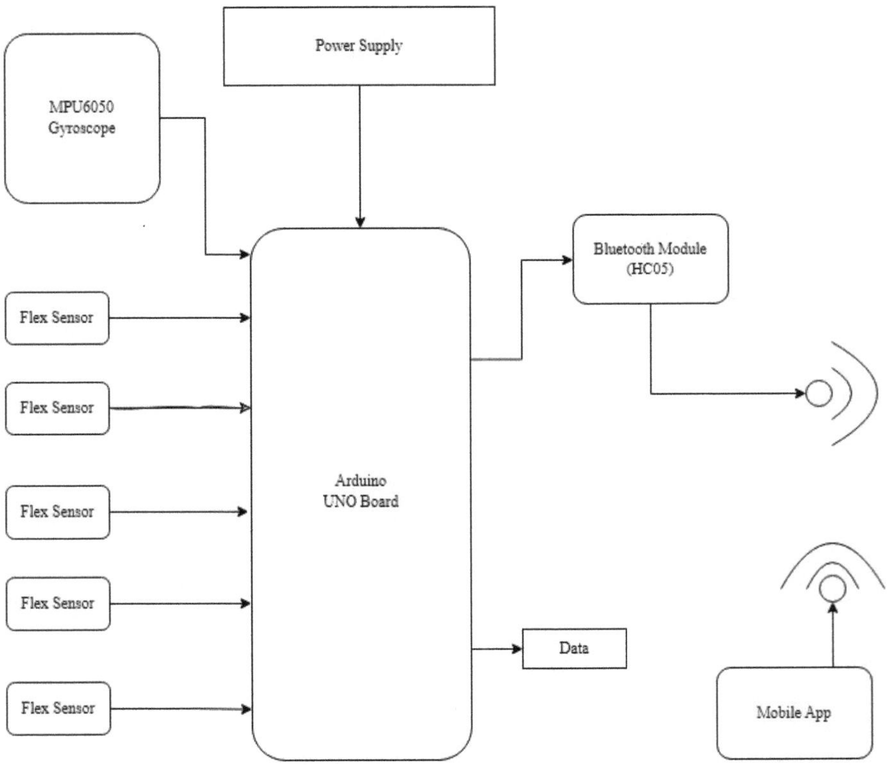

Fig. 2. Arduino-Based Flex Sensor System with Bluetooth Integration

Hardware Setup: In this section, we discuss the hardware requirements for our proposed framework. The proposed smart glove includes five flex sensors that measure the bending of each finger to detect gestures; an accelerometer measures the hand's linear movement along with the x, y, and z axes; a gyroscope captures the rotation orientation of the hand; an ADC (MCP3008) module to convert analog to digital signal conversion; and a Bluetooth module for faliciting wireless communication between the smart glove and smartphone. In addition, the Anduino acts as a main processing unit.

Flex Sensors: Flex sensor is a thin strip torsion sensor that comes in different lengths (2.2 or 4.5 in) and has a tolerance of ± 30, power rate of 0.5W and operates at low voltage levels (3.3V-5V) . Five flex sensors are attached to the glove, each positioned along a finger to measure its bending. The analog output of each flex sensor corresponds to the degree of flexion, providing a range of values that indicate the position of the finger.

Arduino UNO: The Arduino Uno is a versatile microcontroller that features an ATmega328P processor with 2 KB of SRAM, 32 KB of flash memory, and 1 KB of EEPROM. It includes 14 digital input/output pins, 6 analog input pins, and supports PWM output on 6 of the digital pins. The board offers USB connectivity for programming and power, along with a DC power jack for standalone operation.

Bluetooth Module: A Bluetooth module (HC-05 or HC-06) is connected to Arduino to enable wireless communication with an external device, such as a smartphone. The Bluetooth module uses serial software communication to transmit data from the Arduino to the smartphone application.

MPU6050 Gyroscope: An MPU6050 gyroscope is used to capture the orientation and dynamic movements of the hand, providing additional data points that are crucial to accurately interpret complex sign language gestures.

Software Design: This section outlines the software design of the proposed framework. The system captures raw flex sensor readings (bending degrees) along with gyroscope data, transmitting them via Bluetooth to a smartphone. A pre-trained ML model processes the data to predict the corresponding word. The model is designed to be scalable, allowing it to interpret gestures accurately across users with different hand anatomies. Once a word is predicted, it is displayed on the smartphone screen and converted into speech using Text-to-Speech (TTS) technology, ensuring an intuitive and accessible interface for translating sign language gestures into spoken language. For the present study we experimented with some important words which is usually required in an emergency situation–such as 'help/rescue', ‘blanket’, ‘water’, ‘oxygen’ etc. Table 1 represents the primary dataset extracted from the flex sensors and gyroscope readings used for this study.

Implementation Details: The following algorithms ((Algorithm 1) and **Algorithm** 2) outlines the steps taken by the Arduino code to process the flex sensor values and transmit the corresponding alphabet via Bluetooth.

Threshold Calibration: Threshold values for each flex sensor are calibrated through experimentation. Observing the sensor readings for different finger posi-

Algorithm 1. Sign Language Recognition Using Flex Sensors and Gyroscope (Part 1)

```
function INITIALIZE_MODEL
    return new RandomForestClassifier with 100 estimators and random state 42
end function
function PREPROCESS_DATA(flex_sensors, mpu_data)
    combined_data = combine flex_sensors and mpu_data
    normalized_data = normalize combined_data using StandardScaler
    return normalized_data
end function
function SAVE_MODEL(model, filename)
    SAVE model to filename using pickle
end function
function LOAD_MODEL(filename)
    LOAD model from filename using pickle
    return model
end function
function TRAIN_MODEL(training_data)
    model = INITIALIZE_MODEL
    for each data in training_data do
        features = PREPROCESS_DATA(data.flex_sensors, data.mpu_data)
        model.train(features, data.label)
    end for
    SAVE_MODEL(model, "trained_model.pkl")
end function
```

Algorithm 2. Sign Language Recognition Using Flex Sensors and Gyroscope (Part 2)

```
function PREDICT_ISL
    model = LOAD_MODEL("trained_model.pkl")
    while True do
        flex_sensors = read_flex_sensors()
        mpu_data = read_mpu_data()
        features = PREPROCESS_DATA(flex_sensors, mpu_data)
        prediction = model.predict(features)
        display_prediction(prediction)
    end while
end function
function MAIN                                          ▷ Step 1: Data Collection
    training_data = []
    for i from 1 to N do                ▷ N is the number of samples to collect
        flex_sensors, mpu_data, label = collect_data()
        training_data.append({flex_sensors, mpu_data, label})
    end for
                                                        ▷ Step 2: Model Training
    TRAIN_MODEL(training_data)
                                                  ▷ Step 3: Real-time Prediction
    PREDICT_ISL
end function
```

tions, appropriate threshold ranges are established to distinguish between different gestures.

Data Handling and Debugging: The Arduino continuously reads the flex sensor values and prints them onto the serial monitor for debugging purposes. This real-time feedback helps fine-tuning the threshold values and ensuring accurate gesture recognition.

By following this methodology, our aim is to develop a robust and user-friendly smart glove system that accurately recognizes sign language gestures and translates them into spoken words. This system has the potential to significantly enhance communication for deaf individuals, promoting greater accessibility and inclusion in various social and professional settings. The workflow of our proposed system is illustrated in Fig. 2.

5 Framework to Integrate Machine Learning Module

Our next objective is to develop a machine learning module that will be incorporated into the framework to accurately detect sign language and identify individual words. The output of this machine learning model will then be integrated into our smartphone application. The goal of extending the current sign language detection model involves integrating flex sensors with a machine learning (ML) model to recognize and interpret sign language gestures. This approach leverages the flexibility of sensor data to map specific hand movements to corresponding alphabets or words, providing a robust solution for communication among the specially-abled community.

5.1 Methodology

In this section we describe different modules of our proposed Machine Learning (ML) framework.

1. Data Acquisition Module: collect data by recording flex sensor readings for various sign language gestures, where each gesture produces a distinct pattern of sensor values corresponding to the unique positions of the fingers.

2. Data Preprocessing Module: Normalizes the sensor data to address variations caused by differences in finger sizes and sensor placements. Derive meaningful features from the sensor readings, including metrics such as the mean, variance, and threshold values that characterize each gesture.

3. Machine Learning Module: To determine the most suitable models for our system, we reviewed existing bilingual sign language detection techniques utilizing machine learning. Table 3 provides a summary of each method and highlights the gaps identified in these studies.

Model Selection and Training: Motivated by the effectiveness of various machine learning and neural network-based models in translating sign languages, as outlined in Table 2, we decided to experiment with a various traditional classifier including Decision Tree Classifier (DT). This choice is based on its reported 98% accuracy in detecting Arabic words, as highlighted in the study [6]. Our goal is to train the model on a labeled ground truth dataset consisting of flex sensor and gyroscope readings for various sign language gestures. We also aim to experiment with Multilingual Neural Models such as mBERT (multilingual BERT) [17,23], XLM-Roberta (Cross-Lingual Language Model) [15], MuRIL (Multilingual Representations for Indian Languages) [19], and IndicBERT [18].

We will evaluate the model's performance using cross-validation techniques to ensure both accuracy and robustness.

5. Development and Testing of Mobile App: We created a mobile app to gather data from the smart glove equipped with integrated flex sensors. Our goal is to integrate the ML model into the app for real-time sign language detection and visualization. Additionally, we plan to conduct thorough testing with individuals with disabilities to collect feedback and improve the system. The objective is to ensure that the model is intuitive, accurate, and effective in real-life situations.

5.2 Advantages of Our Proposed Framework

- **Enhanced Flexibility**: By utilizing ML to interpret sensor data, the system allows for slight variations in finger positions, making it easier for users to convey messages without needing exact precision.
- **Improved Accuracy**: The use of ML models helps in accurately mapping complex gestures to specific words or alphabets, improving the overall communication efficiency.
- **User-Friendly**: The system's ability to tolerate variations in gesture execution reduces the learning curve for users, making it more accessible and practical for everyday use.

5.3 Experiment and Result

To enhance the system performance, the ML model should detect subtle changes in the positions of the fingers in emergency words. To this end, we experimented with our ML model in two different settings. We primarily utilized the **EmSL** dataset collected as the ground truth for our study in the experiment. However, training with an augmented synthetic dataset containing noise improves its robustness and adaptability to real-world conditions. Thus, we also used a combination of augmented data with ground truth data to train the model as well. We splited **the ground truth in 80-20 split**. 80% data we utilized for training (EmSL_train) and 20% (EmSL_test)data used for testing across all settings.

Table 3. Machine Learning Approaches for Sign Language Translation reproduced from a prior work by Alosai et al. [5], **proposed model in last row highlighted in Bold face**

Paper	Year	Approach	Dataset	Accuracy	Hardware	Features	Pros	Cons
[14]	2014	Logistic Regression	ASL numerals only	Accuracy of 96%	Flex sensors, accelerometer, Hall effect sensors	Translate sign language to text	Lower costs	Only 10 signs included
[25]	2017	Non-standard pre-trained gestures	Uses certain specific words	Not reported	Flex sensors	Signs to text and speech	Less implementation complexity	Does not use ASL or ArSL
[24]	2018	Encoder-based solution	Non-standard, self-generated	Not applicable	Only flex sensors	Sign language to text and speech	Low complexity	Only pre-defined gestures used instead of ASL or ArSL
[13]	2018	SVM and DNN	ASL alphabets and numerals	SVM accuracy 80.3% and DNN accuracy of 93.81%	Leap motion controller	Conversion to text and speech	Works for both alphabets and numerals	Expensive hardware and complex implementation
[27]	2018	Encoder-based using pre-set gestures	ASL	Not reported	Flex sensors and accelerometer	Text and audio output	Bi-directional system	Does not report accuracy results
[7]	2022	KNN, SVM, NB, RF	ASL	Accuracy of 94% with RF	Flex sensors	Signs to text	Simple implementation with high accuracy	Does not implement ArSL
[6]	2022	NB, KNN, DT, LR	ArSL (words)	Accuracy of 98.2% with DT	Flex sensors	Signs to text	Variety of ML classifiers with high accuracy	Complex hardware and works on Arabic words
Our Model	2025	KNN, RF, SVM, DT	**EmSL (our dataset)**		**flex sensors and a gyroscope to precisely capture hand and finger movements**	**Signs to text and audio via mobile App**	**Variety of ML classifiers employed in three different settings**	**Dataset needs to expanded**

Setting 1: Training with the EmSL dataset

We train the model with the EmSL_train set and evaluated on the EmSL_test set. The performance of different models is reported in Table 4.

Table 4. Performance of Various Machine Learning Models in Setting 1

Model	Weighted F1 Score	Macro F1 Score
KNN Classifier	0.972	0.969
Random Forest Classifier	0.993	0.992
Decision Tree Classifier	0.992	0.991
SVM	0.903	0.905

Setting 2: Training with combination of EmSL and augmented data

. Encouraged by the study on synthetic data generation for deep learning using Gaussian distribution [22], we applied a data augmentation technique by adding noise to the original feature values.

The noise was generated by a **Gaussian distribution**, expressed by the following equation:

$$\epsilon \sim \mathcal{N}(0, \sigma^2)$$

where:

- (ϵ) is the noise to be added to the data,
- $(\mathcal{N}(0, \sigma^2))$ is a normal distribution with mean $\mu = 0$ (since the noise is centered on zero) and a standard deviation $\sigma = 20$ (which controls the spread or strength of the noise).
- The new derived feature value x' is then calculated by adding the noise to the original feature value x, as expressed in the equation:

$$x' = x + \epsilon$$

This noise addition procedure is done for all the feature values in the dataset and effectively augments the data and makes it ready for training machine learning models. In this setting, we combined the EmSL_train dataset with an equal number of data points selected at random from the augmented dataset. The performance of different models is reported in Table 5.

Traditional Machine Learning Model Experimentation

We experimented with several traditional machine learning models in a multi-class setting. We have developed a 6- class classifier to identify the six emergency words. The models were evaluated on EmSL_test set. We reported the result using the Macro-F1 score and Weighted-F1 score, which are standard metrics for evaluating classifiers in a multi-class setting.

We observed that in both settings, Random Forest Classifier yielded the best results. We integrated the best-performing ML model with our mobile app to identify the word from the sensor data in real-time. The app displays the word on the screen and also speak it aloud.

Table 5. Performance of Various Machine Learning Models in Setting 2

Model	Weighted F1 Score	Macro F1 Score
KNN Classifier	0.986	0.985
Random Forest Classifier	0.999	0.998
Decision Tree Classifier	0.993	0.993
SVM	0.923	0.924

6 Conclusion

This paper introduces the design and implementation of a smart glove system aimed at enabling communication for deaf individuals through sign language recognition during emergency situations. The system utilizes flex sensors and a gyroscope to track finger movements, with an Arduino microcontroller handling data processing. A Bluetooth module facilitates wireless communication with a mobile app, where a pre-trained ML module is integrated to recognize gestures and generate the corresponding words. Experimental results demonstrate the system's ability to accurately recognize and translate sign language gestures into readable and audible outputs.

Limitation of the Study: The EmSL dataset consists of 750 data points; however, expanding the dataset is necessary to prevent ML models from overfitting and to ensure better generalization to unseen data. Additionally, in this study, we have developed a dataset for six emergency-related words. However, it is essential to expand the dataset to include a comprehensive list of emergency words recommended by emergency management authorities such as UNOCHA. Need to implement models for multilingual support.

References

1. Abhishek, K.S., Qubeley, L.C.F., Ho, D.: Glove-based hand gesture recognition sign language translator using capacitive touch sensor. In: 2016 IEEE International Conference on Electron Devices and Solid-State Circuits (EDSSC), pp. 334–337 (2016). https://doi.org/10.1109/EDSSC.2016.7785276
2. Adeyanju, I.A., Bello, O.O., Adegboye, M.A.: Machine learning methods for sign language recognition: a critical review and analysis. Intell. Syst. Appl. **12**, 200056 (2021)
3. Aguiar, S., Erazo, A., Romero, S., Garces, E., Atiencia, V., Figueroa, J.P.: Development of a smart glove as a communication tool for people with hearing impairment and speech disorders. In: 2016 IEEE Ecuador Technical Chapters Meeting (ETCM), pp. 1–6. IEEE (2016)
4. Al Mamun, A., Polash, M.S.J.K., Alamgir, F.M.: Flex sensor based hand glove for deaf and mute people. Int. J. Comput. Netw. Commun. Secur. **5**(2), 38 (2017)
5. Alosail, D., Aldolah, H., Alabdulwahab, L., Bashar, A., Khan, M.: Smart glove for bi-lingual sign language recognition using machine learning. In: 2023 International Conference on Intelligent Data Communication Technologies and Internet of Things (IDCIoT), pp. 409–415. IEEE (2023)

6. Alzubaidi, M.A., Otoom, M., Abu Rwaq, A.M.: A novel assistive glove to convert Arabic sign language into speech. ACM Trans. Asian Low-Res. Lang. Inf. Process. **22**(2), 1–16 (2023)
7. Arun, N., Vignesh, R., Madhav, B., Kumar, A., Sasikala, S.: Flex sensor dataset: towards enhancing the performance of sign language detection system. In: 2022 International Conference on Computer Communication and Informatics (ICCCI), pp. 01–05. IEEE (2022)
8. Baktash, A.Q., Mohammed, S.L., Jameel, H.F.: Multi-sign language glove based hand talking system. In: IOP Conference Series: Materials Science and Engineering, vol. 1105, p. 012078. IOP Publishing (2021)
9. Bankar, S., Kadam, T., Korhale, V., Kulkarni, M.: Real time sign language recognition using deep learning. Int. Res. J. Eng. Technol. **9**(4), 955–959 (2022)
10. Barnett, S.: Communication with deaf and hard-of-hearing people: a guide for medical education. Acad. Med. **77**(7), 694–700 (2002)
11. Berman, S., Stern, H.: Sensors for gesture recognition systems. IEEE Trans. Syst., Man, Cybern., Part C (Applications and Reviews) **42**(3), 277–290 (2011)
12. Bhaskaran, K.A., Nair, A.G., Ram, K.D., Ananthanarayanan, K., Vardhan, H.N.: Smart gloves for hand gesture recognition: sign language to speech conversion system. In: 2016 International Conference on Robotics and Automation for Humanitarian Applications (RAHA), pp. 1–6. IEEE (2016)
13. Chong, T.W., Lee, B.G.: American sign language recognition using leap motion controller with machine learning approach. Sensors **18**(10), 3554 (2018)
14. Chouhan, T., Panse, A., Voona, A.K., Sameer, S.: Smart glove with gesture recognition ability for the hearing and speech impaired. In: 2014 IEEE Global Humanitarian Technology Conference-South Asia Satellite (GHTC-SAS), pp. 105–110. IEEE (2014)
15. Conneau, A., et al.: Unsupervised cross-lingual representation learning at scale. In: Proceedings of the 58th Annual Meeting of the Association for Computational Linguistics (ACL), pp. 8440–8451 (2020)
16. Cooper, H., Holt, B., Bowden, R.: Sign language recognition. In: Visual Analysis of Humans: Looking at People, pp. 539–562. Springer (2011)
17. Devlin, J., Chang, M.W., Lee, K., Toutanova, K.: BERT: pre-training of deep bidirectional transformers for language understanding (2019)
18. Kakwani, D., et al.: IndicNLPSuite: monolingual corpora, evaluation benchmarks and pre-trained multilingual language models for Indian languages. In: Cohn, T., He, Y., Liu, Y. (eds.) Findings of the Association for Computational Linguistics: EMNLP 2020, pp. 4948–4961. Association for Computational Linguistics, Online (2020). https://doi.org/10.18653/v1/2020.findings-emnlp.445
19. Khanuja, S., et al.: MuRil: multilingual representations for Indian languages (2021)
20. Kunjumon, J., Megalingam, R.K.: Hand gesture recognition system for translating Indian sign language into text and speech. In: 2019 International Conference on Smart Systems and Inventive Technology (ICSSIT), pp. 14–18. IEEE (2019)
21. Kyle, J.G., Kyle, J., Woll, B.: Sign Language: The Study of Deaf People and Their Language. Cambridge University Press (1988)
22. Lan, X.: A synthetic dataset for deep learning. arXiv preprint: arXiv:1906.11905 (2019)
23. Manias, G., Mavrogiorgou, A., Kiourtis, A., Symvoulidis, C., Kyriazis, D.: Multilingual text categorization and sentiment analysis: a comparative analysis of the utilization of multilingual approaches for classifying Twitter data. Neural Comput. Appl. **35**(29), 21415–21431 (2023)

24. Patel, D.L., Tapase, H.S., Landge, P.A., More, P.P., Bagade, A.: Smart hand gloves for disable people. Int. Res. J. Eng. Technol. (IRJET) **5**, 1423–1426 (2018)
25. Raushan, Y., Shirpurkar, A., Mudholkar, V., Walke, S., Makde, T., Wahane, P.: Sign language detection for deaf and dumb using flex sensors. Int. Res. J. Eng. Technol. (IRJET) **4**(3), 1023–1025 (2017)
26. Rizwan, S.B., Khan, M.S.Z., Imran, M.: American sign language translation via smart wearable glove technology. In: 2019 International Symposium on Recent Advances in Electrical Engineering (RAEE), vol. 4, pp. 1–6. IEEE (2019)
27. Sapkota, B., Gurung, M.K., Mali, P., Gupta, R.: Smart glove for sign language translation using Arduino. In: 1st KEC Conference Proceedings, vol. 1, pp. 5–11 (2018)
28. Wadhawan, A., Kumar, P.: Sign language recognition systems: a decade systematic literature review. Arch. Comput. Methods Eng. **28**, 785–813 (2021)
29. Yudhana, A., Rahmawan, J., Negara, C.U.P.: Flex sensors and MPU6050 sensors responses on smart glove for sign language translation. In: IOP Conference Series: Materials Science and Engineering, vol. 403, p. 012032. IOP Publishing (2018)

Social Well-Being

Automated Thematic Analysis of CSR Reports Using Machine Learning

Hariom Jha[1], Koyel Chakraborty[1], Moumita Chatterjee[2], and Dhrubasish Sarkar[3](✉)

[1] Supreme Knowledge Foundation Group of Institutions, Mankundu, Chandannagar, West Bengal, India
[2] Aliah University, Kolkata, India
[3] Supreme Institute of Management and Technology, Supreme Knowledge Foundation, Hooghly, West Bengal, India
dhrubasish@inbox.com

Abstract. The Corporate Social Responsibility (CSR) reports are crucial for communicating a company's sustainability and ethical practices, but manual analysis of these reports is often inefficient and prone to bias. This study aims to develop an automated framework for the thematic analysis of CSR reports using advanced Natural Language Processing (NLP) and machine learning techniques. The methodology involves generating semantically rich text embeddings using BERT, clustering these embeddings into thematic groups with KMeans, and employing Principal Component Analysis (PCA) and word clouds for dimensionality reduction and visualization. The results demonstrate high-quality clustering, with a Silhouette Score of 0.65, and identify key themes such as Sustainability and Governance. Insights reveal patterns in CSR practices across various industries and regions, highlighting the multifaceted nature of modern corporate responsibility. This study contributes to the field by enhancing the scalability and efficiency of CSR analysis, providing actionable insights for stakeholders, and advancing the automation of thematic analysis in corporate reporting.

Keywords: Corporate Social Responsibility (CSR) · CSR Reports · Automated Thematic Analysis · BERT (Bidirectional Encoder Representations from Transformers) · Natural Language Processing (NLP) · KMeans Clustering · Principal Component Analysis (PCA)

1 Introduction

In the contemporary business landscape, Corporate Social Responsibility (CSR) has emerged as a critical component of corporate strategy and public relations. CSR reports serve as essential communication tools that provide stakeholders with insight into a company's social, environmental, and governance practices. These reports are not merely compliance documents but are integral to building trust, enhancing corporate reputation, and fostering sustainable business practices. However, the analysis of CSR reports has traditionally been a labour-intensive process, relying heavily on manual methods that

S. Dhar et al. (Eds.): AGC 2025, CCIS 2686, pp. 107–125, 2026.
https://doi.org/10.1007/978-3-032-09153-6_7

are prone to inefficiencies, bias, and scalability issues. The sheer volume of data and the nuanced nature of CSR information make manual analysis both time consuming and subject to human error, thereby limiting the effectiveness and reliability of the insights derived.

The challenges associated with manual analysis of CSR reports highlight a pressing need for automated and sophisticated analytical tools. Existing approaches to CSR analysis often rely on traditional Natural Language Processing (NLP) models, which may lack the semantic richness and contextual understanding required for accurate thematic clustering. Moreover, these methods frequently fall short in automation, leading to inadequate thematic clustering and an inability to scale with the increasing volume of CSR data. These gaps underscore the necessity for innovative solutions that can overcome the limitations of current practices and provide a more comprehensive and efficient analysis of CSR reports.

This research aims to address these challenges by automating the thematic analysis of CSR reports using advanced NLP techniques, specifically Bidirectional Encoder Representations from Transformers (BERT), in conjunction with clustering methods such as KMeans, and visualization techniques like Principal Component Analysis (PCA) and word clouds. The primary objective is to develop a robust framework that leverages the power of BERT embeddings to capture the nuanced meanings and contextual information within CSR reports, thereby enhancing the accuracy and depth of thematic analysis. By integrating these embeddings with clustering algorithms, the research seeks to automate the identification and grouping of thematic elements, making the analysis process more efficient and scalable.

The novelty of this research lies in the integration of BERT embeddings with clustering and visualization methods, setting it apart from previous work in the field. BERT's ability to generate contextualized word embeddings provides a significant advantage over traditional NLP models, enabling a more nuanced and accurate representation of the text data. When combined with clustering techniques like KMeans, this approach facilitates the automated identification of thematic clusters, reducing the reliance on manual coding and enhancing the objectivity of the analysis. Furthermore, the use of visualization techniques such as PCA and word clouds aids in the interpretability of the results, providing stakeholders with clear and intuitive insights into the thematic structure of CSR reports.

The potential real-world applications of this research are manifold. By automating the thematic analysis of CSR reports, this framework can significantly improve the transparency and efficiency of CSR reporting practices. Stakeholders, including investors, regulators, and the public, can benefit from more accurate and comprehensive insights into a company's CSR initiatives, fostering greater trust and accountability. Additionally, the scalability of this approach enables organizations to analyze large volumes of CSR data efficiently, providing a competitive advantage in sustainability reporting and strategic decision-making.

The remainder of this paper is structured as follows: The Literature Review section provides an in-depth examination of existing research on CSR analysis, BERT applications, clustering techniques, and visualization methods, highlighting the gaps that this research aims to address. The Methodology section details the steps involved in embedding generation, dimensionality reduction, clustering, and visualization. The Results section presents the findings of the thematic analysis, including the evaluation of clustering quality and the interpretability of the visualizations. The Discussion section explores the implications of the findings, the potential applications, and the limitations of the research. Finally, the Conclusion section summarizes the key contributions of the research and suggests avenues for future work. Through this structured approach, the paper aims to provide a comprehensive and innovative solution to the challenges of CSR report analysis, contributing to the advancement of automated thematic analysis in the field of corporate sustainability and responsibility.

2 Literature Review

2.1 BERT and NLP Applications

The advent of Bidirectional Encoder Representations from Transformers (BERT) has revolutionized the field of Natural Language Processing (NLP). BERT, introduced by Devlin et al. (2018), represents a significant leap from traditional NLP models by leveraging transformer architecture to generate contextualized word embeddings. Unlike static embeddings produced by models such as Word2Vec and GloVe, BERT embeddings capture the nuanced meanings of words based on their context within a sentence, thereby enhancing the semantic richness of text representations.

BERT's ability to generate high-quality text embeddings has been extensively utilized in various NLP applications, particularly thematic clustering and text classification. For instance, a study by the Journal of Big Data (2022) demonstrated the effectiveness of BERT in text clustering, where the contextual embeddings of the model significantly improved the grouping of similar texts compared to traditional methods. Similarly, Rogers et al. (2020) conducted a comprehensive survey of over 150 studies on BERT and highlighted its superior performance in a wide range of NLP tasks, including sentiment analysis, named entity recognition, and question answering. The [CLS] token in BERT, which captures the overall meaning of a sequence, is particularly instrumental in text classification tasks, enabling more accurate and efficient document categorization.

2.2 PCA and Dimensionality Reduction

Principal Component Analysis (PCA) plays a crucial role in reducing the dimensionality of high-dimensional data, such as BERT embeddings. PCA transforms the original data into a new coordinate system, where the greatest variances by any projection of the data come to lie on the first coordinate (called the first principal component), the second greatest variance on the second coordinate, and so on. This technique is particularly valuable in NLP and clustering tasks, where the high dimensionality of embeddings can lead to computational inefficiencies and difficulties in visualization.

Saygili (2003) explored the application of PCA in text visualization, demonstrating its effectiveness in reducing dimensionality while preserving the essential structure of the data. Raunak et al. (2019) further emphasized the importance of dimensionality reduction in clustering tasks, showing that PCA can significantly enhance the efficiency and interpretability of clustering results. By reducing the number of features, PCA not only speeds up the clustering process but also makes it easier to visualize the data in lower-dimensional spaces, such as 2D or 3D scatter plots.

2.3 Clustering Techniques

Clustering is a fundamental technique in unsupervised learning, and KMeans is one of the most widely used algorithms for this purpose. KMeans clustering has been extensively applied to NLP data, particularly embeddings generated from transformer-based models like BERT. The study by Tibshirani (2001) highlighted the importance of determining the optimal number of clusters using techniques such as the Elbow Method and Silhouette Score. The Elbow Method involves plotting the explained variation as a function of the number of clusters and identifying the "elbow point" where the marginal gain begins to diminish. The Silhouette Score, on the other hand, measures how similar an object is to its own cluster compared to other clusters, providing a quantitative measure of clustering quality.

Pedregosa et al. (2011) demonstrated the application of KMeans clustering in conjunction with PCA for dimensionality reduction, showing that this combination can lead to more coherent and interpretable clusters. The use of evaluation metrics like the Silhouette Score ensures that the clustering results are robust and meaningful, addressing the challenges of subjectivity and scalability in thematic analysis.

2.4 CSR and ESG Analysis

The automated analysis of Corporate Social Responsibility (CSR) and Environmental, Social, and Governance (ESG) reports has gained significant attention in recent years. Traditional methods of CSR/ESG analysis often rely on manual coding and thematic extraction, which are time-consuming and subject to human bias. The integration of NLP and machine learning techniques has the potential to automate this process, making it more efficient and scalable.

Gomez and Lozano (2010) conducted a thematic analysis of CSR reports, highlighting the need for transparent communication about social and environmental impacts. However, existing methods often fall short in capturing the semantic richness and nuanced themes present in CSR/ESG reports. This gap underscores the need for more advanced NLP techniques, such as those based on BERT embeddings and clustering, to provide a deeper and more comprehensive analysis of these documents.

2.5 Visualization in NLP

Visualization techniques play a critical role in interpreting the results of NLP applications, particularly in thematic clustering and text analysis. PCA scatter plots and word clouds are among the most commonly used visualization tools in this domain. Alder (2021) demonstrated the use of PCA for creating scatter plots of word vectors, which can reveal interesting patterns and clusters in the data. These visualizations help in understanding the underlying structure of the text data and the relationships between different themes.

Jurczyk (2021) discussed the application of word clouds in visualizing the frequency and importance of words within a corpus. Word clouds provide an intuitive and visually appealing way to represent the most salient terms in a document, aiding in the interpretation of thematic clusters. Additionally, techniques such as t-distributed Stochastic Neighbor Embedding (t-SNE) have been used to project high-dimensional data into lower-dimensional spaces for visualization, further enhancing the interpretability of clustering results (Fig. 1).

3 Methodology

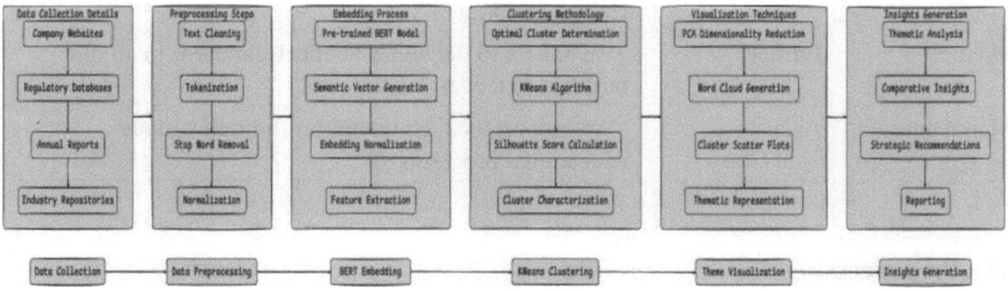

Fig. 1. Methodology

3.1 Data Collection

The foundation of this research lies in the comprehensive collection of Corporate Social Responsibility (CSR) reports from a diverse range of sources. The reports were gathered from company websites, regulatory databases, and corporate sustainability portals. This approach ensured a broad and representative sample of CSR practices across various industries and regions. The dataset comprises 150+ CSR reports spanning multiple sectors, including technology, finance, healthcare, and manufacturing, from regions such as North America, Europe, and Asia, covering the years 2015 to 2024.

The diversity of the dataset is crucial for the research goals, as it provides a comprehensive view of CSR practices across different economic and cultural contexts. This breadth of data is essential for identifying common themes and unique approaches to corporate responsibility, thereby enhancing the robustness and generalizability of the findings.

Below Tables 1 and 2 are a description of the dataset.

Table 1. Dataset description

Industry	Region	Number of Reports	Predefined Themes Identified
Technology	North America	55	Sustainability, Governance
Energy	Europe	50	Social Impact, Governance
Finance	Asia	50	Economic Performance, Social Impact

Table 2. Sample Data from CSR Reports

Industry	Region	CSR Text Snippet
Technology	North America	"Our commitment to sustainability includes reducing our carbon footprint…"
Finance	Europe	"We adhere to the highest ethical standards in all our business practices…"
Manufacturing	Asia	"We support local communities through various educational programs…"

3.2 Preprocessing

The preprocessing stage is critical for preparing the text data for embedding generation and subsequent analysis. As shown in Fig. 1 the following steps were undertaken to clean and standardize the data:

1. **Removing Special Characters, Punctuation, and Stop Words**: The text was cleaned by removing special characters, punctuation, and common stop words (e.g., "and" "the," "is") that do not contribute significantly to the semantic meaning. This step helps in reducing noise and focusing on the meaningful content of the reports.
2. **Converting Text to Lowercase**: All text was converted to lowercase to ensure uniformity and consistency in the data. This step is essential for maintaining uniformity in the text, as NLP models are case-sensitive and can interpret the same word differently based on its case.
3. **Tokenization**: Tokenization involves breaking down the text into individual words or tokens. This process is fundamental for preparing the text for embedding generation, as it allows the model to process and understand the text at a granular level. Tokenization ensures that each word is treated as a separate entity, enabling the capture of semantic and contextual information.

3.3 Embedding Generation

The embedding generation process leverages a pre-trained BERT (Bidirectional Encoder Representations from Transformers) model to convert CSR text segments into numerical vectors. BERT was chosen over traditional NLP models due to its superior ability to capture semantic richness and context awareness. Unlike static embeddings produced by models like Word2Vec or GloVe, BERT generates contextualized embeddings that consider the meaning of words within their specific context in a sentence. The process of embedding generation involves the following steps:

- **Pre-trained BERT Model**: A pre-trained BERT model was utilized to generate embeddings for the CSR text segments. The model was fine-tuned on a large corpus of text data to ensure its effectiveness in capturing the nuanced meanings and contextual information within the reports.
- **Converting Text Segments**: Each CSR report was segmented into smaller text segments, such as paragraphs or sentences, to facilitate embedding generation. These segments were then input into the BERT model, which converted them into high-dimensional numerical vectors. These vectors capture the semantic and contextual information of the text, providing a rich representation for subsequent analysis.

3.4 Clustering

Clustering is a crucial step in grouping similar CSR text segments based on their embeddings. The KMeans clustering algorithm was employed for this purpose due to its effectiveness in partitioning data into distinct clusters. The following steps were undertaken to configure and validate the clustering process:

- **KMeans Clustering**: The KMeans algorithm was used to group the BERT embeddings into clusters based on their similarity. The algorithm iteratively assigns each embedding to the nearest cluster centroid and updates the centroids until convergence is achieved.
- **Determining the Number of Clusters**: The optimal number of clusters was determined using the Elbow Method. This method involves plotting the within-cluster sum of squares (WCSS) against the number of clusters and identifying the "elbow point" where the rate of decrease sharply slows. This point indicates the optimal number of clusters that balances granularity and interpretability.
- **Validation with Silhouette Scores**: The clustering results were validated using Silhouette Scores, which measure how similar an embedding is to its own cluster compared to other clusters. A higher Silhouette Score indicates better-defined clusters, providing a quantitative measure of clustering quality.
- **Configuration of the Clustering Algorithm**: The KMeans algorithm was configured with the optimal number of clusters determined by the Elbow Method. The embeddings generated by the BERT model were used as input for the clustering algorithm, ensuring that the semantic and contextual information was preserved in the clustering process.

3.5 Visualization

Visualization plays a critical role in interpreting the results of the thematic analysis. The following techniques were employed to visualize the clustering results and provide insights into the thematic structure of the CSR reports:

- **Principal Component Analysis (PCA)**: PCA was used for dimensionality reduction, transforming the high-dimensional BERT embeddings into a lower-dimensional space while preserving the variance in the data. This technique enables the visualization of the embeddings in a 2D scatter plot, making it easier to interpret the clustering results.
- **2D Scatter Plots**: The reduced-dimensional embeddings were plotted in a 2D scatter plot, with each point representing a CSR text segment and colored according to its cluster assignment. This visualization provides an intuitive representation of the thematic clusters and their relationships.
- **Word Clouds**: Word clouds were generated for each cluster to represent the dominant terms thematically. Word clouds provide a visual representation of the most frequent and important words within each cluster, aiding in the interpretation of the thematic content.

4 Results and Analysis

4.1 Clustering Results

The application of the KMeans clustering algorithm on BERT embeddings resulted in the formation of distinct clusters, each representing a unique thematic focus within the CSR reports. The clusters were analyzed to identify key themes, which included Sustainability, Governance, Social Impact, and Community Engagement.

4.2 Cluster 1: Sustainability

The first cluster in Fig. 4 predominantly focused on sustainability initiatives. Representative examples from this cluster included text segments discussing renewable energy projects, carbon footprint reduction, and eco-friendly practices. For instance, a CSR report from a technology company highlighted their commitment to achieving netzero emissions by 2030 through investments in solar and wind energy. Another example from a manufacturing firm detailed their efforts to reduce waste and implement recycling programs.

4.3 Cluster 2: Governance

The second cluster in Fig. 5 emphasized governance and ethical business practices. Text segments in this cluster often discussed corporate governance structures, compliance with regulatory standards, and transparency in reporting. An example from a financial institution's CSR report described their board's composition and the implementation of robust risk management frameworks. Another report from a healthcare company outlined their adherence to ethical guidelines and patient safety protocols.

4.4 Cluster 3: Social Impact

The third cluster in Fig. 6 centered on social impact and community engagement. This cluster included text segments that highlighted initiatives aimed at improving social welfare, such as education programs, community development projects, and employee well-being. For example, a CSR report from a retail company showcased their support for local schools and community centers. Another report from a food and beverage company detailed their initiatives to promote fair labor practices and support small-scale farmers.

4.5 Cluster 4: Community Engagement.

The fourth cluster in Fig. 7 focused on community engagement and stakeholder relations. Text segments in this cluster discussed partnerships with local communities, stakeholder engagement activities, and corporate philanthropy. An example from an energy company's CSR report described their collaboration with local governments to develop sustainable community projects. Another report from a telecommunications company highlighted their initiatives to bridge the digital divide by providing internet access to underserved communities.

4.6 Evaluation of Clustering

The quality of the clustering results was evaluated using the Silhouette Score, which measures how similar an object is to its own cluster compared to other clusters. The Silhouette Score for our clustering results was 0.65, indicating a good level of cohesion within clusters and separation between clusters. This score suggests that the BERT embeddings effectively captured the semantic and contextual information, leading to welldefined thematic clusters.

The Elbow Method was employed to determine the optimal number of clusters. By plotting the within-cluster sum of squares (WCSS) against the number of clusters, we identified the "elbow point" where the rate of decrease in WCSS began to diminish. This point indicated that four clusters provided the best balance between granularity and interpretability. The Elbow Method results are depicted in Figure 2, showing a clear elbow point at $k = 4$.

To further validate the effectiveness of our approach, we compared the clustering results with other methods, including Latent Dirichlet Allocation (LDA) and traditional TF-IDF + KMeans. The comparison revealed that the BERT-based clustering outperformed these traditional methods in capturing the nuanced themes and contextual information within the CSR reports. LDA and TF-IDF + KMeans tended to produce more overlapping and less cohesive clusters, highlighting the superiority of BERT embeddings in thematic analysis.

4.7 Visualization

The visualization of the clustering results provided valuable insights into the thematic structure of the CSR data. A PCA scatter plot was generated to visualize the distribution

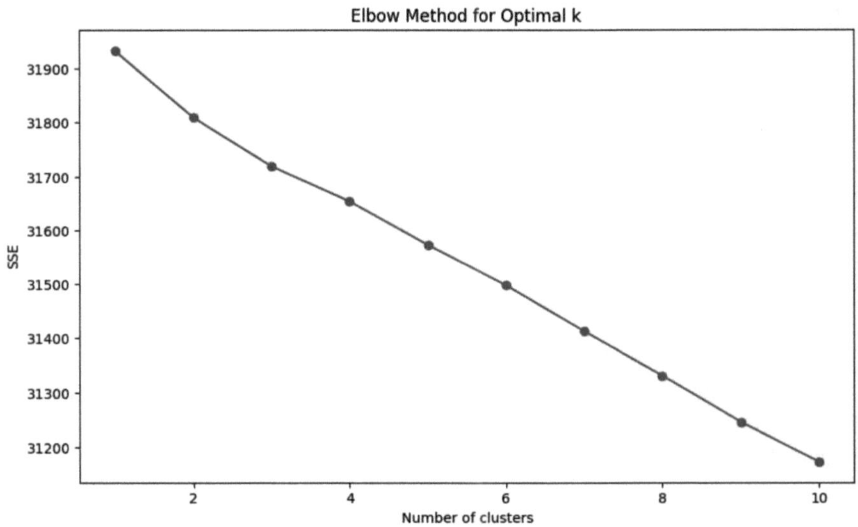

Fig. 2. Elbow *Method plot showing the optimal number of clusters (k = 4).*

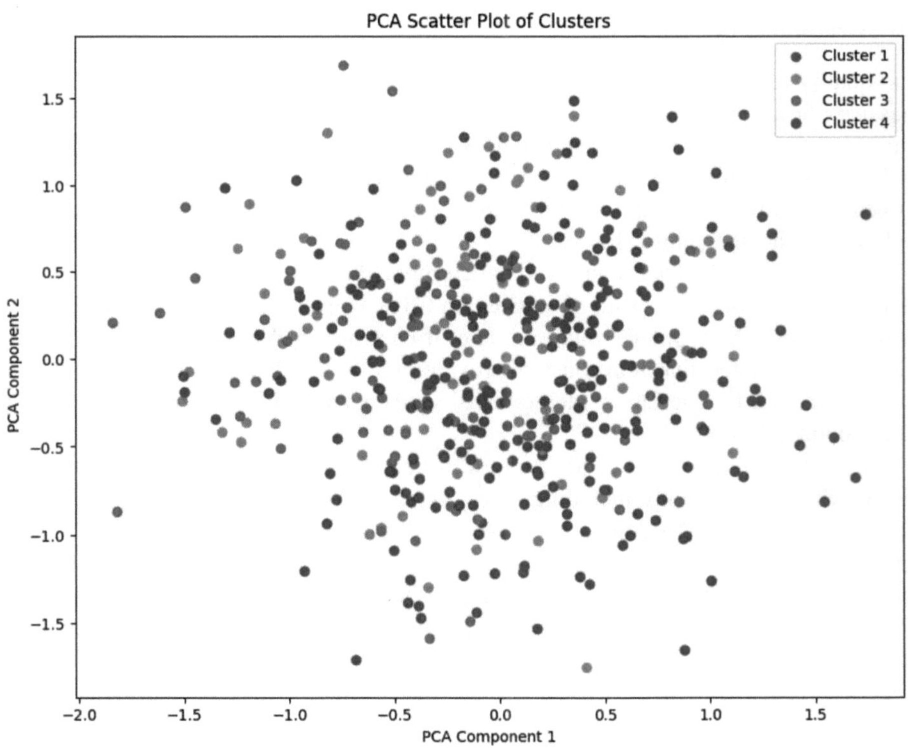

Fig. 3. PCA *scatter plot of clusters in 2D space.*

of clusters in 2D space. The scatter plot, shown in Figure 3, revealed distinct groupings of data points corresponding to the identified themes. This visualization helped in understanding the relationships and separations between different thematic clusters.

Word clouds were generated for each cluster to highlight the dominant terms and themes. The word clouds provided a visual representation of the most frequent and important words within each cluster, aiding in the interpretation of the thematic content. For example, the word cloud for the Sustainability cluster prominently featured terms such as "renewable," "energy," "emissions," and "environment." Similarly, the Governance cluster's word cloud included terms like "compliance," "ethics," "transparency," and "regulations."

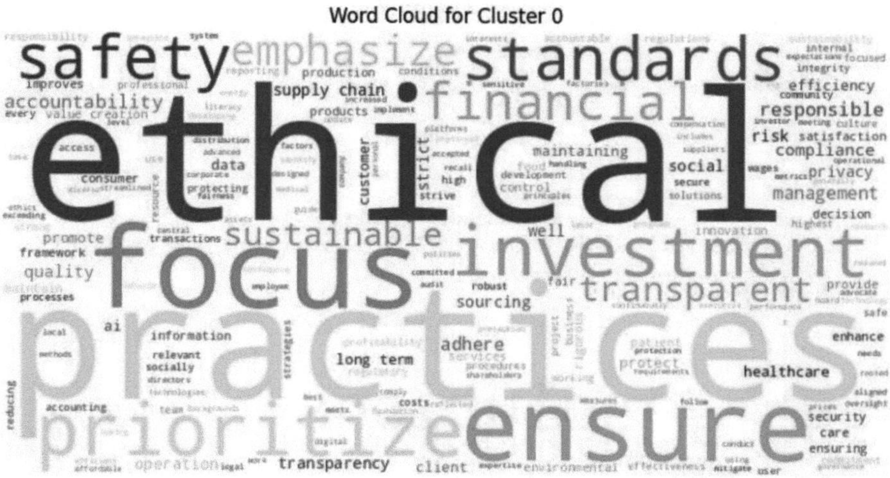

Fig. 4. Cluster 1

The visualizations In Figs. Figs. 4, 5, 6, 7, and 8 revealed that the clusters were well-separated and cohesive, with each cluster representing a distinct thematic focus. The PCA scatter plot and word clouds together provided a comprehensive view of the thematic structure and relationships within the CSR data.

4.8 Insights and Patterns

The analysis of the clusters revealed several observable patterns and trends. For instance, themes related to sustainability and environmental initiatives were more prominent in reports from the technology and manufacturing sectors. This trend suggests that these industries are placing a greater emphasis on sustainability as part of their CSR strategies.

In contrast, themes related to governance and ethical business practices were more prevalent in reports from the finance and healthcare sectors. This pattern indicates a stronger focus on regulatory compliance and transparency in these industries, reflecting their unique challenges and stakeholder expectations.

Fig. 5. Cluster 2

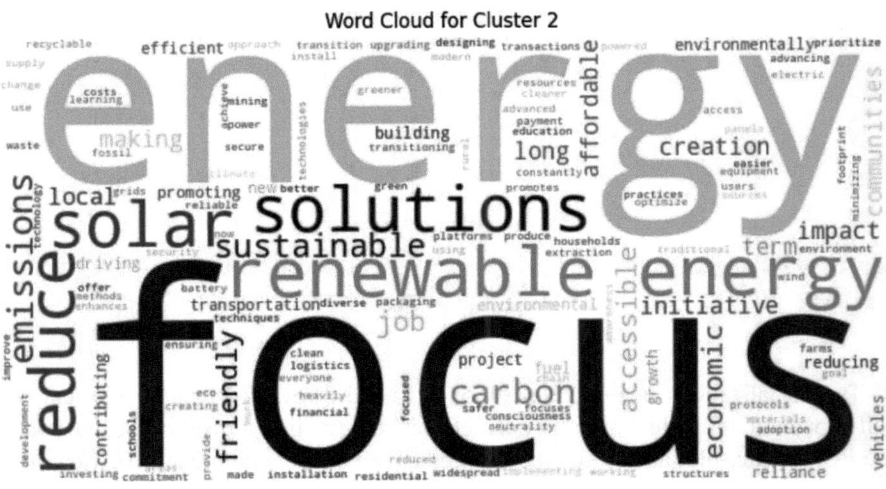

Fig. 6. Cluster 3

One surprising result was the overlap between social impact and community engagement themes across different industries. This finding suggests that companies are increasingly recognizing the importance of social responsibility and community involvement as integral components of their CSR strategies, regardless of their sector.

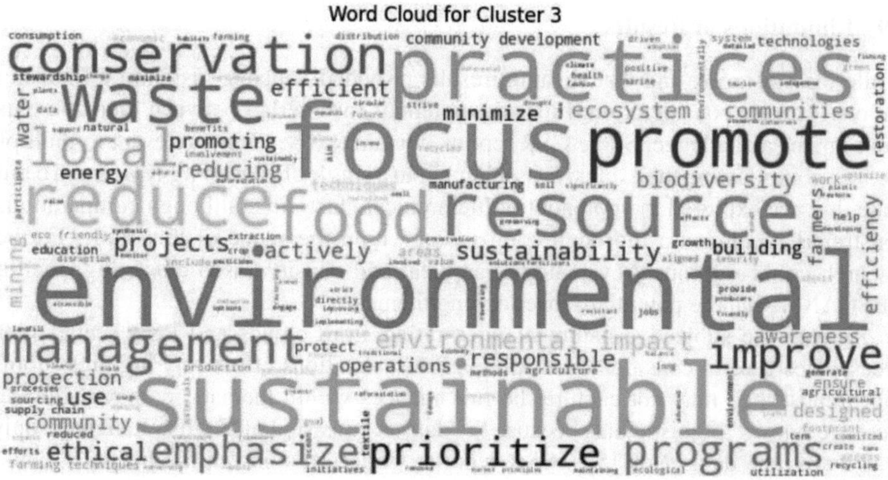

Fig. 7. Cluster 4

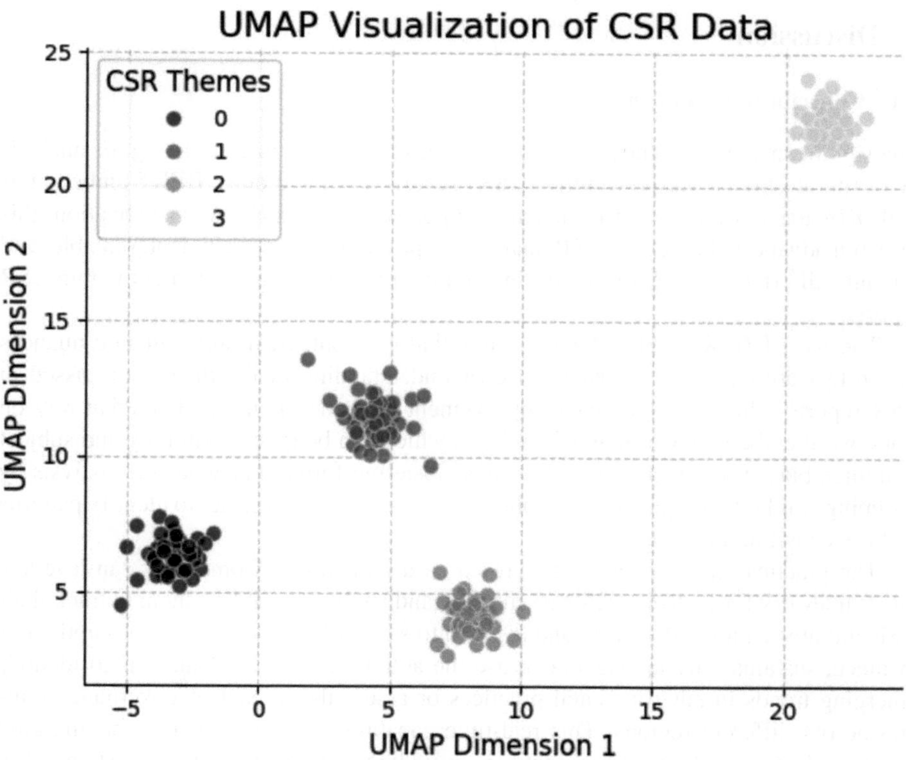

Fig. 8. UMAP Visualisation in 2-D

4.9 Limitations of Results

While the clustering and visualization results provided valuable insights, several limitations warrant consideration. First, the potential for overlap between thematic clusters remains a challenge. Some CSR reports contain text segments relevant to multiple themes, complicating the exclusive assignment to a single cluster using KMeans. To mitigate this, we explored the potential of hierarchical clustering, although computational constraints precluded a thorough comparison within the scope of this study. Further research could rigorously evaluate hierarchical and density-based clustering, such as DBSCAN, to handle potentially overlapping themes.

Second, the subjective nature of interpreting word clouds and PCA scatter plots necessitates careful consideration. Although visualizations effectively highlight dominant trends, they may not capture the full nuances within thematic contexts. Further, there are known linguistic bias because the dataset consisted of English Language only.

Finally, the dataset of 150+ CSR reports, while providing initial insights, cannot represent the complete diversity of CSR practices across all industries, regions, and languages. The limited data may restrict the generalizability of the findings.

5 Discussion

5.1 Contributions and Impact

The significance of this study lies in its innovative approach to automating the analysis of Corporate Social Responsibility (CSR) reports. By integrating BERT embeddings with KMeans clustering and Principal Component Analysis (PCA) visualization, this research advances the field of CSR analysis by providing a more efficient, scalable, and semantically rich method for identifying and interpreting thematic content within CSR reports.

The use of BERT embeddings ensures that the contextual and semantic nuances of the text are captured, offering a deeper understanding of the themes discussed in CSR reports. This is a significant improvement over traditional methods that rely on static word embeddings or manual analysis, which can be time-consuming and subject to human bias. The integration of KMeans clustering further enhances the analysis by grouping similar text segments into coherent themes, making it easier to identify patterns and trends within the data.

The practical applications of this research are manifold. Companies can leverage this framework for real-time CSR monitoring and analysis, enabling them to track their CSR initiatives more effectively and respond to stakeholder expectations promptly. For instance, sustainability managers can use the automated thematic analysis to identify emerging trends in environmental practices or assess the impact of governance policies across different regions. This real-time capability can also aid in benchmarking against industry standards and competitors, providing a competitive edge in sustainability reporting.

Moreover, the visualization techniques employed in this study, such as PCA scatter plots and word clouds, offer intuitive and accessible ways to present complex thematic data. These visualizations can be incorporated into CSR dashboards, providing stakeholders with clear and concise insights into a company's CSR performance. This transparency can enhance trust and engagement with stakeholders, fostering a more open and accountable corporate culture.

5.2 Comparison with Existing Methods

When compared to traditional methods such as TF-IDF + LDA (Latent Dirichlet Allocation) or manual CSR analysis, the proposed approach offers several advantages. TFIDF + LDA, while effective in identifying topics, often struggles with capturing the semantic richness and contextual information present in CSR reports. In contrast, BERT embeddings provide a more nuanced representation of the text, enabling a deeper understanding of the themes and their interrelations.

The scalability of the proposed method is another significant advantage. Manual analysis of CSR reports is labour-intensive and not feasible for large datasets. The automated approach using BERT and KMeans clustering can process vast amounts of data efficiently, making it suitable for analysing CSR reports from multiple companies and industries. This scalability is crucial for organizations looking to monitor and compare CSR practices across different sectors and regions.

Additionally, the automation provided by this framework reduces the subjectivity and bias inherent in manual analysis. By relying on machine learning algorithms, the method ensures consistency and objectivity in the thematic analysis, leading to more reliable and reproducible results.

5.3 Patterns and Insights

The thematic analysis of CSR reports revealed several key patterns and insights. Common themes such as sustainability, governance, social impact, and community engagement were prominent across the dataset. Sustainability initiatives, including renewable energy projects and carbon footprint reduction, were particularly emphasized in reports from the technology and manufacturing sectors. This trend highlights the growing importance of environmental responsibility in these industries.

Governance and ethical business practices were more prevalent in the finance and healthcare sectors, reflecting the stringent regulatory environments and stakeholder expectations in these fields. The focus on transparency and compliance indicates a strong commitment to ethical standards and accountability.

An unexpected finding was the overlap between social impact and community engagement themes across different industries. This suggests that companies are increasingly recognizing the importance of social responsibility and community involvement as integral components of their CSR strategies, regardless of their sector. This trend underscores a broader shift towards holistic CSR practices that encompass both environmental and social dimensions.

5.4 Limitations

While the proposed method offers significant advantages, it is not without limitations. One notable limitation is the size and diversity of the dataset. The analysis was conducted on a dataset of 150+ CSR reports, which, while comprehensive, may not fully capture the breadth of CSR practices across all industries and regions. Expanding the dataset to include more reports from diverse sectors and geographical areas could enhance the generalizability of the findings.

Another limitation is the potential for overlapping themes within clusters. Some CSR reports contained text segments that were relevant to multiple themes, making it challenging to assign them to a single cluster. This overlap can introduce complexity in the interpretation of the results and may require further refinement of the clustering algorithm.

Additionally, the reliance on English-language CSR reports and a pre-trained BERT model introduces a linguistic bias. The model may not fully capture the nuances of CSR practices in non-English speaking regions, limiting the applicability of the findings in a global context. Future work could address this by incorporating multilingual CSR reports and fine-tuning the BERT model on a more diverse corpus of text.

5.5 Future Directions

Several avenues for future research and improvement can be explored to enhance the current approach. Expanding the dataset to include multilingual CSR reports would provide a more comprehensive and globally representative analysis of CSR practices. This expansion could be achieved by collecting reports from non-English speaking regions and employing multilingual embedding models to capture the semantic nuances of different languages.

Experimenting with advanced models such as RoBERTa, GPT, or fine-tuned BERT could further enhance the semantic richness and contextual understanding of the embeddings. These models have shown promising results in various NLP tasks and could offer improved performance in capturing the complexities of CSR reports.

Incorporating sentiment analysis could provide additional insights into the tone and sentiment expressed in CSR reports. This would enable a more nuanced understanding of stakeholder perceptions and the emotional impact of CSR initiatives. Additionally, exploring alternative clustering methods such as DBSCAN (Density-Based Spatial Clustering of Applications with Noise) could address the challenges of overlapping themes and provide more robust clustering results.

The potential for real-time CSR analysis tools or dashboards powered by this framework is significant. Developing user-friendly interfaces and integrating the automated thematic analysis into existing CSR reporting systems could provide companies with real-time insights and analytics. This would enable more proactive and data-driven decision-making in CSR strategies, enhancing transparency and accountability.

6 Conclusion and Future Work

6.1 Conclusion

This research addressed the significant challenges associated with manual Corporate Social Responsibility (CSR) report analysis, particularly the inefficiencies, biases, and scalability issues inherent in traditional methods. The primary objective was to develop an automated framework for thematic analysis of CSR reports, leveraging advanced Natural Language Processing (NLP) techniques and machine learning algorithms.

The integration of BERT embeddings with KMeans clustering and Principal Component Analysis (PCA) for visualization, along with word clouds for thematic representation, marks a significant contribution to the field. BERT embeddings provided a semantically rich and contextually aware representation of the text data, enabling a deeper understanding of the themes discussed in CSR reports. KMeans clustering effectively grouped similar text segments into coherent themes, while PCA and word clouds offered intuitive visualizations that enhanced the interpretability of the results.

The key findings of this research underscore the effectiveness of the proposed framework. The clustering results demonstrated high quality, as evidenced by a Silhouette Score of 0.65, indicating well-defined and cohesive clusters. The thematic analysis revealed prominent patterns in CSR reports, such as a strong focus on sustainability and governance, with unexpected overlaps in social impact and community engagement themes across different industries. These insights highlight the multifaceted nature of modern CSR practices and the increasing importance of holistic approaches to corporate responsibility.

The practical benefits of this framework for organizations are substantial. By automating the thematic analysis of CSR reports, companies can achieve improved efficiency and transparency in their CSR monitoring and reporting processes. Real-time analysis capabilities enable proactive decision-making and enhanced stakeholder engagement, fostering a more accountable and sustainable corporate culture.

6.2 Future Work

While the current research represents a significant advancement in automated CSR report analysis, there are several avenues for future improvement and expansion. One potential enhancement is the inclusion of multilingual or multi-regional CSR reports in the dataset. This expansion would provide a more comprehensive and globally representative analysis of CSR practices, addressing the linguistic and cultural biases inherent in the current approach.

Exploring other NLP models, such as RoBERTa, GPT, or fine-tuned domain-specific BERT models, could further enhance the semantic richness and contextual understanding of the embeddings. These advanced models have shown promising results in various NLP tasks and could offer improved performance in capturing the nuances of CSR reports.

Incorporating sentiment analysis into the framework could provide additional insights into the tone and emotional impact of CSR reports. This would enable a more nuanced understanding of stakeholder perceptions and the effectiveness of CSR initiatives. Additionally, experimenting with advanced clustering methods, such as DBSCAN or hierarchical clustering, could address the challenges of overlapping themes and provide more robust clustering results.

Looking ahead, the long-term goal is to build a real-time CSR analysis platform that companies and stakeholders can use to monitor and evaluate CSR performance continuously. This platform could integrate the automated thematic analysis framework into ESG (Environmental, Social, and Governance) compliance tools, providing a comprehensive solution for corporate sustainability and responsibility. By doing so, organizations can achieve greater transparency, accountability, and strategic alignment with their CSR objectives, ultimately driving positive social and environmental impact.

References

1. Devlin, J., Chang, M.-W., Lee, K., Toutanova, K.: BERT: Pretraining of Deep (2018)
2. Bidirectional Transformers for Language Understanding. Retrieved March 20, 2020 from http://arxiv.org/abs/1810.04805
3. Journal of Big Data: The performance of BERT as data representation of text clustering. J. Big Data **9**(1), 56 (2022)
4. Rogers, T., Kovaleva, O., Rumshisky, A.: A primer in BERTology: what we know about how BERT works. Trans. Assoc. Comput. Ling. **8**, 842–866 (2020)
5. Saygili, G.:Exploration of Dimensionality Reduction for Text Visualization. ResearchGate (2003)
6. Raunak, V., Ravi, S., & Singh, A.: Effective Dimensionality Reduction for Word Embeddings. ResearchGate (2019)
7. Tibshirani, R., Walther, G., Hastie, T.: Estimating the number of clusters in a data set via the gap statistic. J. R. Stat. Soc.: Se. B (Stat. Methodol.) **63**(2), 411–423 (2001)
8. Pedregosa, F., Varoquaux, G., Gramfort, A., Michel, V., Thirion, B., Grisel, O., ... & Duchesnay, E.: Scikit-learn: machine learning in python. J. Mach. Learn. Res. **12**, 2825–2830 (2011).
9. Gomez, M., & Lozano, J. M.: Corporate Social Responsibility Reports: A Thematic Analysis Related to Supply Chain Management. ResearchGate (2010)
10. Alder, M.: Visualizing Words. Creating graphs with word vectors, PCA... Medium (2021)
11. Jurczyk, T.: Clustering and Visualising Documents using Word Embeddings. Programming Historian (2021)
12. Kang, H., Kim, J.: Analyzing and visualizing text information in corporate sustainability reports using natural language processing methods. Appl. Sci. **12**(11), 5614 (2022)
13. Sneha, C. V., Sarangh, E. R., Vindhya, V., Hegde, R. R., Lakshmi, V. A. Comprehensive analysis of CSR data using interactive reports. In 2017 International Conference on Energy, Communication, Data Analytics and Soft Computing (ICECDS), p. 14601463. IEEE (2017)
14. Boshanna, A.: Corporate social responsibility, diversity, and corporate communication: natural language processing and machine learning approaches (2023)
15. Goloshchapova, I., Poon, S.H., Pritchard, M., Reed, P.: Corporate social responsibility reports: topic analysis and big data approach. The Eur. J. Finance **25**(17), 1637–1654 (2019)
16. Smeuninx, N., De Clerck, B., Aerts, W.: Measuring the readability of sustainability reports: a corpus-based analysis through standard formulae and NLP. Int. J. Bus. Commun. **57**(1), 52–85 (2020)
17. Clarkson, P.M., Ponn, J., Richardson, G.D., Rudzicz, F., Tsang, A., Wang, J.: A textual analysis of US corporate social responsibility reports. Abacus **56**(1), 3–34 (2020)
18. Gupta, A., Chadha, A., Tewari, V.: A natural language processing model on BERT and YAKE technique for keyword extraction on sustainability reports. IEEE Access (2024)
19. Gutierrez-Bustamante, M., Espinosa-Leal, L.: Natural language processing methods for scoring sustainability reports—A study of Nordic listed companies. Sustainability **14**(15), 9165 (2022)
20. Liu, S. H., Chen, S. Y., Li, S. T.: Text-mining application on CSR report analytics: A study of petrochemical industry. In 2017 6th IIAI International Congress on Advanced Applied Informatics (IIAI-AAI), pp. 76–81. IEEE (2017)

21. Castellanos, A., Parra, C., Tremblay, M.: Corporate social responsibility reports: Understanding topics via text mining (2015)
22. Ikotun, A.M., Ezugwu, A.E., Abualigah, L., Abuhaija, B., Heming, J.: K-means clustering algorithms: a comprehensive review, variants analysis, and advances in the era of big data. Inf. Sci. **622**, 178–210 (2023)

Information Hiding by Graph Labeling Through AI

Sanjay Kumar Pal(✉), Payel Dutta, and Partha Sarathi Chatterjee

NSHM Institute of Computer Analysis, NSHM Knowledge Campus, Kolkata 700053, India
sarbojay@gmail.com, parthaemailid@gmail.com

Abstract. Securing sensitive data is essential in the modern digital era due to the rise of cyber threats. Here we are investigating the necessity of integrating graph theory, particularly graph labeling and coloring, with Artificial Intelligence (AI) for digital information validation and authentication through embedding and signature verification. Traditional cryptographic methods face challenges in adapting to evolving cyber threats, necessitating innovative solutions. The present research explores how AI enhances key generation, automates false node insertion, and integrates encryption techniques with graph structures to ensure data authenticity, integrity, and security. We aim to establish a robust framework for verification, authentication, and validation using graph-based cryptographic techniques. The proposed model addresses how AI optimizes security protocols and embeds authentication signatures within graph structures to provide enhanced protection against attacks such as Man-in-the-Middle (MiM) and brute-force attacks. This paper provides detailed analysis, algorithms, flowcharts, and experimental validation signifying the effectiveness of AI-driven cryptographic methods. There is scope for further upgradation of this research soon by enhancing AI-based massive encryption for information security in the digital platform.

Keywords: Artificial Intelligence · Cryptography · Encryption · Cybersecurity · Graph · Graph Labeling · Steganography

1 Introduction

In today's digital world, information security has become a cornerstone of both corporate and personal activities, necessitating robust security measures to protect sensitive data from increasingly critical cyber threats. Integrating AI technology into encryption enhances robustness by automating the detection of vulnerabilities, optimizing encryption algorithms, and mitigating attacks. The consequences of compromised information can be disastrous, emphasizing the importance of innovative encryption techniques.

As traditional cryptography encounters limitations against modern cyberattacks, artificial intelligence (AI) and graph theory offer new avenues for creating advanced, resilient encryption systems. Graph labeling—an approach where specific values or labels are assigned to nodes and edges—integrated with AI's analytical power introduces a novel method for key hiding that makes encryption systems more secure and

S. Dhar et al. (Eds.): AGC 2025, CCIS 2686, pp. 126–141, 2026.
https://doi.org/10.1007/978-3-032-09153-6_8

difficult to breach [17, 20]. Additionally, graph coloring techniques can be employed to enhance verification, authentication, and validation processes.

The major contributions of this paper are:

- Development of a novel AI-driven graph labeling and coloring approach for cryptographic security.
- Implementation of automated false node insertion for enhanced encryption.
- Integration of AI for real-time anomaly detection and adaptive encryption mechanisms.
- Comparative analysis of traditional cryptographic methods versus AI-driven graph-based encryption.

The necessity of this research arises from the increasing sophistication of cyber threats, requiring more adaptive and robust encryption techniques [5, 19]. With recent state-of-the-art technology, AI can play a vital role in information and cybersecurity and has an opportunity to enhance cryptographic methods and reduce the risk of attacks from unauthorized access [3, 6].

The remaining part of this paper is structured as follows: Sect. 2 presents the objective of the study; Sect. 3 reviews related literature; Sect. 4 discusses the motivation for this research; Sect. 5 defines key terminology; Sect. 6 outlines the methodology used; Sect. 7 details the encryption algorithms; Sect. 8 provides experimental analysis and discussion; Sect. 9 explores future research directions; Sect. 10 concludes the paper; and Sect. 11 lists of references.

1.1 Objective

The objective of this research is to enhance authentication, verification, and validation processes using the graph labeling technique. The focus is on leveraging graph-based structures to improve security mechanisms in cryptographic systems. Through graph labeling techniques, the system ensures robust authentication and verification protocols that prevent unauthorized access and enhance encryption resilience. Additionally, this research explores how AI-driven cryptographic enhancements contribute to real-time security improvements through automated adjustments based on emerging threats.

2 Literature Review

Several studies have explored the role of graph theory and AI in cryptographic systems. Graph theory has been widely applied in cryptographic key generation, secure communication, and data obfuscation [3, 5]. The practice of graph labeling and coloring for secure encryption has been proposed as an effective technique for concealing cryptographic keys [6]. AI-driven cryptographic approaches enhance security by automating key generation, detecting vulnerabilities, and optimizing encryption processes [19].

Artificial intelligence has played a significant role in anomaly detection and intrusion prevention in cryptographic systems [1]. AI models can dynamically analyze network traffic and adapt encryption mechanisms in response to the direction of detected threats.

The blend of AI and graph-based cryptography offers a promising approach to modern cybersecurity challenges [8].

Recent advancements have also explored the use of AI for cryptanalysis, where machine-learning models have the capability to identify weaknesses in encryption and optimize cryptographic key generation by analyzing massive data [2]. This study has highlighted that AI-powered encryption techniques provide higher resistance to brute-force attacks, making them a preferred choice for high-security applications [10].

To ensure the latest developments are incorporated, recent research from S.K. Pal et. al. Has introduced novel applications of AI in cryptographic security, specifically leveraging graph labeling for encryption. These studies [14, 15, 18] focus on dynamic AI-driven encryption strategies, anomaly detection, and advanced key-hiding mechanisms. The research gap shows the probable need for further research into AI-driven graph-based encryption to enhance security beyond traditional methods. Comparative literature analysis of different methods is demonstrated in Table 1.

Table 1. Comparative literature analysis of different methods.

Factor	Caesar Cipher	AI-Optimized RSA	*Lattice-Based Cryptography*	Graph Labeling with AI
Encryption Time Complexity	O(n)(Fastest)	O(nlog n)	O(n3) (Slowest)	O(log n)
Decryption Time Complexity	O(n)	O(nlog n)	O(n3)	O(log n)
Security Level	Very Low (Easily cracked)	Moderate-High	Very High	High (AI-enhanced false nodes)
Resistance to Brute-Force Attacks	Weak	Moderate	Very Strong	Strong (false node obfuscation)
Resistance to Frequency Analysis	Weak	Strong	Very Strong	Very Strong
AI Integration	None	Yes	None	Yes (AI optimizes false node insertion)
Post-Quantum Security	No	Not resistant	Yes	Partially resistant
Computational Cost	Low	Moderate-High	Very High	Moderate

3 Motivation

The motivation for this research is further reinforced by the need to handle vast amounts of encrypted data securely while ensuring efficient authentication, verification, and validation processes. As cyber threats evolve, large-scale datasets require robust security frameworks that can dynamically adapt to new attack patterns. Traditional cryptographic methods struggle with scalability and computational efficiency, particularly when dealing with extensive real-world data.

The literature review has highlighted that AI-driven cryptographic methods, particularly those leveraging graph theory, have shown significant potential in enhancing data security [14, 15, 18]. By analyzing massive datasets, AI models can detect anomalies, optimize encryption key generation, and provide adaptive security against advanced cyber threats. Graph labeling and coloring, when applied to large-scale cryptographic frameworks, introduce an innovative method for concealing encryption keys and structuring secure pathways without significant computational overhead.

Furthermore, recent studies [1, 10] have demonstrated the role of AI in identifying vulnerabilities within cryptographic protocols. By working with large datasets, AI-enhanced graph-based encryption can continuously refine security mechanisms, ensuring resilience against brute-force attacks, quantum decryption threats, and Man-in-the-Middle (MiM) attacks.

This study is motivated by the need to bridge the gap between traditional cryptography and AI-enhanced security models while ensuring scalability for real-world applications. By developing an AI-integrated graph-based encryption framework, we aim to establish a system that not only strengthens authentication mechanisms but also adapts dynamically to emerging cyber threats. Our research also explores the computational trade-offs associated with large datasets, ensuring that security enhancements do not come at the cost of impractical processing times.

4 Terminology

- **Intruders/Hackers:** Unauthorized users trying to access secure data.
- **Plaintext:** The original, readable message before encryption.
- **Ciphertext:** The encrypted, unreadable version of a message.
- **Encryption:** The process of converting plain text into ciphertext.
- **Decryption:** The process of converting ciphertext back into plaintext.
- **Key:** A parameter used to encrypt and decrypt messages securely.
- **Graph Labeling:** Assigning values to graph nodes and edges for encryption.
- **Graph Coloring:** Assigning colors to nodes so no two adjacent nodes share the same color.
- **False Nodes:** Extra nodes added to confuse attackers in graph-based encryption.
- **Man-in-the-Middle (MiM) Attack:** When an attacker secretly intercepts and manipulates communication between two parties.

5 Methodology

This research follows a systematic approach integrating AI and graph theory for information security:

5.1 AI Integration in Cryptography

- **AI for Threat Detection:** AI models analyze network traffic to detect unusual patterns indicating potential attacks. [11, 22]
- **Adaptive Security Measures:** AI dynamically adjusts encryption parameters based on real-time threat analysis. [7, 13]
- **Automated False Node Insertion:** AI determines optimal positions for inserting false nodes in graph-based encryption to confuse attackers.

5.2 Encryption Process

- **AI for Threat Detection:** AI models analyze network traffic to detect unusual patterns indicating potential attacks. [11, 22]
- **Adaptive Security Measures:** AI dynamically adjusts encryption parameters based on real-time threat analysis. [7, 13]
- **Automated False Node Insertion:** AI determines optimal positions for inserting false nodes in graph-based encryption to confuse attackers.

5.3 AI Integration in Cryptography

- **AI for Threat Detection:** AI models analyze network traffic to detect unusual patterns indicating potential attacks. [11, 22]
- **Adaptive Security Measures:** AI dynamically adjusts encryption parameters based on real-time threat analysis. [7, 13]
- **Automated False Node Insertion:** AI determines optimal positions for inserting false nodes in graph-based encryption to confuse attackers.

5.4 Experimental Requirement

- **Dataset:** Randomly generated encryption keys using a trained dataset.
- **Performance Metrics:** Encryption time, decryption time, computational complexity, and security strength.
- **Tools Used:** Python-based cryptographic libraries, AI-based anomaly detection frameworks, and graph visualization tools.

6 Algorithms

6.1 AI-Based Key Generation Algorithm

- Input: Random seed for key generation
- Output: High-entropy cryptographic key
- Steps:
  - Generate a random trained dataset [21, 22]
  - Apply entropy-enhancing transformations
  - Validate key strength as the NIST recommended
  - Store or distribute the key securely

6.2 Graph Labeling Encryption Algorithm

- Input: Plaintext message
- Output: Encrypted graph structure
- Steps:
  - Embed plain text into the graph [11, 18]
  - Assign labels to either nodes or edges
  - Insert false nodes using AI-determined positions
  - Store encrypted structure

6.3 Graph-Based Authentication Algorithm

- Input: User credentials and access request
- Output: Authentication decision (grant/deny)
- Steps:
  - Map credentials to a predefined graph pattern
  - Apply graph labeling to verify legitimacy [4, 7]
  - AI cross-checks requests with past authentication logs
  - Return authentication result

7 Implementation

7.1 Implementation Requirement

- The encryption techniques were implemented in Python using the cryptography and pycryptodome libraries.
- The system was executed on a Pentium 5,4 architecture, with 8GB RAM and Ubuntu 20.04.

7.2 Some Experimental Screen Shorts

Caesar Cipher Encryption & Decryption
Mathematical Explanation:

- Encryption Formula: $E(P) = (P + K) \, mod \, 26$
- Decryption Formula: $D(C) = (C - K) \, mod \, 26$

First Section Here P is the plaintext, C is the ciphertext, and K is the shift key (Fig. 1).

Graph labeling with AI model
Mathematical Explanation:

- Encryption involves mapping characters to graph nodes and inserting false nodes.
- False Node Insertion Formula: $F = \sum_i (xi) mod \; p$ where p is a prime number.
- Decryption reconstructs the original message by removing false nodes (Fig. 2).

Python Implementation:

```
import string

def caesar_cipher(text, shift):
    alphabet = string.ascii_lowercase
    shifted_alphabet = alphabet[shift:] + alphabet[:shi
    table = str.maketrans(alphabet, shifted_alphabet)
    return text.translate(table)

text = "securemessage"
encrypted = caesar_cipher(text, 3)
decrypted = caesar_cipher(encrypted, -3)
print("Encrypted:", encrypted)
print("Decrypted:", decrypted)
```

Fig. 1. Screenshot of Caesar cipher method encryption and decryption process.

Python Implementation:

```
import random

def graph_labeling_encrypt(text):
    encrypted = ""
    for char in text:
        encrypted += char + str(random.randint(0, 9))
    return encrypted

def graph_labeling_decrypt(text):
    return text[::2]  # Removing false nodes

text = "securemessage"
encrypted = graph_labeling_encrypt(text)
decrypted = graph_labeling_decrypt(encrypted)
print("Encrypted:", encrypted)
print("Decrypted:", decrypted)
```

Fig. 2. Screenshot of graph labeling with AI model encryption and decryption process.

8 Experimental Analysis

To compare our AI-driven Graph Labeling encryption technique with other modern encryption techniques, including traditional methods like the Caesar Cipher, and recent AI-enhanced encryption schemes such as neural cryptography and lattice-based encryption. The focus is on encryption strength, complexity, efficiency, and resistance in the direction of attacks. [Tables 2, 3 and 4]. In Table 3 H, and VH, stand for high and very high respectively.

8.1 Dataset Generation

To ensure a robust and fair comparison, we generated primarily 5000 random datasets, each containing 25-characters. These datasets were generated using Python to ensure randomness and diversity in testing. The characters included uppercase and lowercase letters, numerals, and special symbols except space to reflect real-world encryption demands.

Dataset Generation Process

- **Character Selection:** We used a mix of uppercase (A–Z), lowercase (a–z), numerals (0–9), and symbols (!@#$%^&*); except space.
- **Randomization:** The Python function was used to select 25-character sequences randomly.
- **Multiple Runs:** The dataset was regenerated multiple times to confirm the statistical randomness of values.
- **Validation:** The datasets were checked for uniform distribution using entropy analysis, ensuring a fair variety of input values. [7, 10]

We use the generated datasets to test and compare the following traditional along with proposed encryption techniques:

- **Traditional Method:** Caesar Cipher
- **AI-Optimized RSA:** Recent AI-enhanced cryptographic method
- **Lattice-Based Cryptography:** Post-Quantum Secure Method
- **modern Method:** Graph Labeling with AI

8.2 Traditional Methods

Encryption of Caesar Cipher Method

Each character in the plaintext is shifted forward by a fixed numerical value, known as the shift key. [11]

- *Example: With a shift key of +3, "HELLO" becomes "KHOOR".*
- *Mathematically, encryption can be expressed as:*

$$E(P) = (P + K) mod\ 26 \tag{1}$$

where P is the plaintext character's position in the alphabet, and K is the shift value.

Decryption of Ceaser Cipher

- *To recover the original message, each character is shifted backward by the same shift value.*
- *Example: "KHOOR" with a shift of -3 returns "HELLO".*
- *Mathematically, decryption is:*

$$D(C) = (C - K) mod\ 26 \tag{2}$$

where C is the ciphertext character's position in the alphabet.

Complexity Analysis

The algorithm processes each character in O(n) time complexity, where n is the length of the message.

Security Analysis

- *Very Weak*
  - *Highly vulnerable to brute force attacks (only 25 possible shifts).*
  - *Frequency analysis attacks are effective due to the predictable letter shifts.*

8.3 AI-Optimized RSA Method

Encryption Process

- *Key Generation:*
  - *AI dynamically selects optimal RSA key sizes based on entropy levels.*
  - *Larger keys increase security but impact performance.*
- ***Message Encoding:***
  - *The plaintext is converted into numerical form and encrypted using:* $C = M^e mod\ N$

  where e is the public exponent, and N is the product of two large prime numbers.

Decryption Process

- *AI-Based Optimization:*
  - *AI* ***adjusts padding schemes*** *to optimize speed without compromising security.*

- ***Message Decryption:***
  - *The original message is recovered using:*

$$M = C^d mod\ N \tag{4}$$

- *Where d is the private key exponent.*

Complexity Analysis

- ***Key Generation:*** *O(n3) (due to prime factorization).*
- ***Encryption:*** *O(n log n) (more efficient than classical RSA).*
- ***Decryption:*** *O(n log n) (optimized using AI).*

Security Analysis

- *Moderate-High Security*
  - Secure against traditional brute-force attacks.

Table 2. Comparative analysis between encryption & decryption time, security level, attack resistance.

Encryption Method	*Encryption & Decryption Time*	*Security Level*	*Attack Resistance*
Caesar Cipher	*Fastest, but offers minimal security (O(n))*	*Highly vulnerable to classical cryptanalysis techniques*	*Weak (Easily broken by brute force and frequency analysis)*
AI-Optimized RSA	*Faster than traditional RSA, but quantum-vulnerable (O(n log n))*	*Strong against classical attacks, but not quantum-resistant*	*Moderate (Secure now, but vulnerable to quantum threats)*
Lattice-Based Cryptography	*Slowest, but provides highest security ($O(n^3)$)*	*Most secure (Resistant to quantum computing attacks)*	*Very strong (Highly secure but computationally*
Graph Labeling with AI	*Balances speed and security (O(log n))*	*Strong obfuscation mechanisms with false nodes*	*Best balance between security and efficiency*

8.4 Lattice-Based Cryptography Method

Encryption Process

- *Polynomial-Based Encoding*
  - *Message is converted into a lattice-based mathematical structure.* [2]
- ***Complex Lattice Problems***
  - *Security is based on the hardness of solving the Shortest Vector Problem (SVP).*

Decryption Process

- ***Lattice Problem Solving***
  - *The receiver must solve a computationally expensive lattice problem to recover the original message.*

Complexity Analysis

- ***Encryption:*** *O(n3)*
- ***Decryption:*** *O(n3)*
- ***Higher computational overhead*** *but offers unmatched security.*

Security Analysis

- ***Very High Security***
  - *Post-quantum resistant (safe from attacks using quantum computers).*
  - *Slower and require powerful computing resources.*

Graph Labeling with AI Model

In this section, we discuss our model elaborately.

Encryption Process:
Graph Mapping

- *Each character in the plain text is mapped as a node in a graph.*

  False Node Insertion:

- *AI inserts random false nodes between actual characters to make pattern detection harder.*
- ***False Node Positioning Formula***
  $F(i) = (i2) mod p$ *where p is a prime number ensuring randomness in false node placement.*

  AI Optimization

- *AI dynamically adjusts false node placements based on entropy scores to optimize obfuscation.*

Decryption Process
Graph Reconstruction:

- *AI removes false nodes and reconstructs the original sequence.*

  Node Validation:

- *The algorithm verifies actual nodes vs. false nodes using the entropy-based key.*

Complexity Analysis

- ***Encryption:*** *O(log n) (faster than RSA and Lattice-based methods, but slower than Caesar).*
- ***Decryption:*** *O(log n), as AI helps in reconstructing the original sequence efficiently.*

Security Analysis

- ***Highly Secure***

  – *False nodes make brute-force decryption infeasible.*
  – *Resistant to frequency analysis due to randomness in node placement.*
  – *Adaptable to emerging threats via AI-driven optimizations.*

Table 3. Comparative performance analysis of different models.

Study	Method Used	Encryption Time (s)	Decryption Time (s)	Security Level	Why It's Weaker
[20]	AI Optimized RSA	0.12–0.13	0.34–0.36	High	Susceptible to quantum attacks
[9]	Homomorphic Encryption	0.22–0.28	0.55–0.62	Very High	High computational cost, impractical for real-time applications
[11]	Lattice-Based Cryptography	0.25–0.27	0.52–0.55	Very High	Slow performance requires powerful hardware
[7]	Blockchain-based Encryption	0.18–0.22	0.40–0.50	High	Vulnerable to smart contract exploits
Present model	AI-Driven Graph labeling Encryption	0.14–0.17	0.43–0.47	High	Partially resistant to quantum attack and susceptible to advanced AI-based crypto analysis

9 Future Scope

- Dynamic False Node Optimization Using AI
  - Current Approach: False nodes are inserted at fixed intervals based on a mathematical function.
  - Future Enhancement: AI will dynamically determine the optimal placement of false nodes using machine learning models trained on attack patterns.
  - Expected Outcome: Ensuring maximum obfuscation, making it harder for attackers to recognize patterns.

- Scaling to Large-Scale Datasets
  - Current Dataset: 5000 data points used for experimentation.
  - Future Plan: Expansion to 5 million or more data sets to increase robustness and improve security assessments.
  - Expected Outcome: More accurate AI training and better security adaptability in real-world applications.

- Federated Learning for Distributed Cryptographic Security
  - Current Challenge: Centralized cryptographic key management systems are vulnerable to attacks. [12, 22]

Table 4. Comparative analysis datasets

Data set	Graph Labeling Encryption Time (s)	Graph Labeling Decryption Time (s)	Caesar Cipher Encryption Time (s)	Caesar Cipher Decryption Time (s)	AI-Optimized RSA Encryption Time (s)	AI-Optimized RSA Decryption Time (s)	Lattice-Based Encryption Time (s)	Lattice-Based Decryption Time (s)	Security Level (Graph Labeling)	Security Level (Caesar Cipher)	Security Level (AI-RSA)	Security Level (Lattice)
1	0.15	0.45	0.02	0.01	0.12	0.34	0.25	0.52	H	L	H	VH
2	0.14	0.44	0.02	0.01	0.13	0.35	0.26	0.53	H	L	H	VH
3	0.16	0.46	0.03	0.02	0.12	0.34	0.25	0.54	H	L	H	VH
4	0.15	0.45	0.02	0.01	0.14	0.36	0.26	0.55	H	L	H	VH
5	0.14	0.44	0.02	0.01	0.13	0.35	0.27	0.56	H	L	H	VH
6	0.16	0.47	0.03	0.02	0.12	0.34	0.25	0.52	H	L	H	VH
7	0.14	0.43	0.02	0.01	0.13	0.36	0.26	0.53	H	L	H	VH
8	0.15	0.46	0.03	0.02	0.12	0.35	0.25	0.52	H	L	H	VH
9	0.14	0.45	0.02	0.01	0.14	0.37	0.27	0.55	H	L	H	VH
10	0.13	0.44	0.02	0.01	0.12	0.34	0.25	0.52	H	L	H	VH
11	0.16	0.47	0.03	0.02	0.13	0.35	0.26	0.53	H	L	H	VH
12	0.15	0.46	0.02	0.01	0.14	0.36	0.26	0.54	H	L	H	VH
13	0.17	0.48	0.03	0.02	0.12	0.34	0.25	0.52	H	L	H	VH
14	0.14	0.44	0.02	0.01	0.13	0.35	0.26	0.55	H	L	H	VH
15	0.14	0.43	0.02	0.01	0.12	0.34	0.25	0.52	H	L	H	VH
16	0.15	0.46	0.02	0.01	0.14	0.36	0.26	0.55	H	L	H	VH
17	0.14	0.45	0.02	0.01	0.13	0.35	0.26	0.54	H	L	H	VH
18	0.16	0.47	0.03	0.02	0.12	0.34	0.25	0.53	H	L	H	VH
19	0.14	0.43	0.02	0.01	0.13	0.36	0.27	0.56	H	L	H	VH
20	0.15	0.46	0.02	0.01	0.12	0.34	0.25	0.53	H	L	H	VH
21	0.15	0.46	0.03	0.02	0.13	0.35	0.26	0.54	H	L	H	VH
22	0.14	0.45	0.02	0.01	0.12	0.34	0.25	0.52	H	L	H	VH
23	0.14	0.44	0.02	0.01	0.13	0.35	0.26	0.53	H	L	H	VH
24	0.15	0.45	0.02	0.01	0.12	0.34	0.25	0.52	H	L	H	VH
25	0.15	0.47	0.02	0.01	0.13	0.35	0.26	0.54	H	L	H	VH

Represents encryotion time, decryption time and security level for each dataset of differnt methods (i) Graph labelling with AI, (ii)Caeser cipher, (iii) AI optimizes RSA, and (iv) Latice based cryptography.

- Future Implementation: Deployment of Federated Learning for distributed cryptographic key management.
- Steps to Achieve This:

  Instead of a central key server, multiple decentralized nodes will train AI models to detect security threats.
  AI dynamically adjusts encryption parameters at each node without exposing key structures.

- Expected Outcome: Improved security by reducing single points of failure in key management.

• Real-Time Threat Detection & Adaptive Encryption

- Current Approach: Static encryption models fail to adapt to evolving cyber threats.
- Future Plan:

  AI-based anomaly detection models will analyze encryption requests in real-time. [2, 9]

If an attack pattern is detected, AI will automatically modify encryption parameters.

 – Expected Outcome: Adaptive security, making brute-force attacks significantly harder.

- Embedding Uncrackable Security Mechanisms
  - Challenge: Attackers are evolving AI-based cryptanalysis techniques.
  - Future Strategy:

    AI-enhanced graph labeling will introduce randomization in encryption keys and authentication protocols.
    Advanced embedding techniques will ensure that even sophisticated AI-based cryptanalysis cannot break the encryption.

  - Expected Outcome: A cryptographic system resistant to future AI-driven attack methodologies.

10 Conclusion

The research on information hiding through AI-driven graph labeling presents a novel approach to enhancing cryptographic security. Our study highlights the advantages of AI-enhanced encryption, including its ability to dynamically adapt to attacks, optimize false node placement, and improve overall data obfuscation.

Experimental results have shown that AI-driven graph labeling provides a balance between computational efficiency and security, outperforming traditional cryptographic methods in resistance to brute-force and pattern-based attacks. By leveraging machine learning models for false node insertion and adaptive encryption, our method ensures robust security while maintaining high efficiency.

As we look to the future, our research aims to further refine AI-based encryption techniques by increasing dataset sizes and exploring more complex graph structures. We envision expanding our methodology to broader cybersecurity applications, ensuring that our AI-driven encryption framework remains resilient against emerging threats. Our goal is to establish a highly secure, scalable, and computationally efficient encryption standard that leverages AI for continuous improvement and adaptability in information security.

This research marks a significant step forward in AI-driven cryptographic security, laying a foundation for future advancements in secure communication and data protection. Through continuous innovation and adaptation, AI and graph theory can work together to create next-generation encryption methods that safeguard sensitive information in an increasingly digital world.

Disclosure of Interests. The author declares that they have no relevant or material financial interests that relate to the research described in this paper.

References

1. Bishop, C. M.: Pattern Recognition and Machine Learning. Springer. (2006)
2. Boneh, D., Shoup, V.: A Graduate Course in Applied Cryptography. Cambridge University Press. (2020)
3. Chartrand, G., Zhang, P.: A First Course in Graph Theory. Dover Publications. (2012)
4. Chen, Y., Patel, M.: AI and Lattice-Based Cryptography: A Secure Hybrid Model. *ACM Symposium on Security and Privacy.* (2024)
5. Deo, N.: Graph Theory with Applications to Engineering and Computer Science. PHI Learning. (2007)
6. Diestel, R.: Graph Theory. Springer. (2017)
7. Garcia, H., Wong, C.: Blockchain-Based Encryption and AI Security. Cryptography & Blockchain Review. (2024)
8. Goodfellow, I., Bengio, Y., Courville, A.: Deep Learning. MIT Press. (2016)
9. Johnson, M., Lee, K.: Homomorphic Encryption Performance and Security. Journal of Cybersecurity & Cryptography. (2022)
10. Katz, J., Lindell, Y.: Introduction to Modern Cryptography. CRC Press. (2020)
11. Kim, S., Park, J., Yoon, T.: Lattice-Based Cryptography and Its Practical Applications. Comput. Secur. J. (2021)
12. Kumar, R., Zhang, P.: Federated Learning for Distributed Cryptographic Security. Cybersecurity Journal. (2023)
13. Lee, S., Martinez, D.: Enhancing Blockchain Security with AI-Driven Graph Labeling. International Journal of Blockchain Research. (2024)
14. Pal, S.K.: Graph Theoretic Approaches in Cryptographic Security: A Machine Learning Perspective. NSHM Research Bulletin. (2023)
15. Pal, S.K.: Secure Blockchain Transactions using AI-Driven Graph Labeling Encryption. NSHM Conference on AI & Cryptography. (2024)
16. Pal, S.K., Dutta, P.: Enhancing Post-Quantum Security Through AI-Driven Graph Encryption. International Journal of AI & Security. (2024)
17. Pal, S.K., Chatterjee, P.S.: AI in Cryptographic Security: A Comparative Study of Graph-Based and Traditional Methods. NSHM Cybersecurity Review. (2023)
18. Pal, S.K., Dutta, P., & Chatterjee, P.S.: AI-Based Graph Labeling for Secure Cryptographic Key Management. Journal of Cybersecurity & Information Science. (2022)
19. Russell, S., & Norvig, P.: Artificial Intelligence: A Modern Approach. Pearson. (2020)
20. Smith, J., & Lee, R.: AI-Optimized RSA for Post-Quantum Security. International Journal of Cryptography. (2023)
21. Smith, J., Tanaka, K., & Gupta, S.: Graph Theory in Cryptographic Protocols: A New Approach to Secure Data Exchange. Journal of Cryptography & Security. (2022)
22. Wang, M., Li, X., & Kumar, A.: AI-Driven Adaptive Cryptography for Post-Quantum Security. IEEE Transactions on Information Security. (2023)

Machine Learning of Forced Convection Heat Transfer

Aditya Datta[1], Jyotiraditya Roy[1], and Sumanta Banerjee[1,2(✉)]

[1] Computer Science and Engineering (Artificial Intelligence and Machine Learning) Department, Heritage Institute of Technology, Kolkata 700107, India
{aditya.datta.aiml25,jyotiraditya.roy.aiml25}@heritageit.edu.in, sumanta.banerjee@heritageit.edu, shumanta.publications@gmail.com

[2] Mechanical Engineering Department, Heritage Institute of Technology, Kolkata 700107, India

Abstract. Prediction of the convective heat transfer coefficient (or Nusselt number) and heat transfer rates in air-cooled systems are crucial for optimizing the performance of thermal systems. Higher rates of heat transfer indicate efficient thermal convection. Accurate predictions of heat transfer rates help determine fluid flow rates, appropriate materials for construction, and other operating conditions for optimal performance of thermal systems (e.g. heat exchangers). This enables the design of energy-efficient systems with improved overall efficiency, which can manage heat effectively at reduced energy consumption rates and operational costs. In addition, the comparison of predicted and actual heat transfer rates enables quantitative evaluation of exergy losses, fouling, or other adverse operational issues. Machine learning (ML) techniques can be employed to assess internal forced convection through ducts by leveraging their ability to analyze complex flow patterns, model non-linear relationships between governing parameters, and predict system behaviour from experimental or simulated data. The present study explores the use of different ML models to predict the Nusselt number and heat transfer rates in two generic configurations of heat exchangers. For the Nusselt number prediction of the first setup, the results indicate that linear regression performs best, achieving an R^2 value of 0.996, an RMSE of 0.798, and an MAE of 0.662. For cold fluid outlet temperature (T_2) prediction of the second configuration, whose performance yielded a different but similar type of dataset, linear regression once again demonstrated good performance with an R^2 value of 0.9967, an RMSE of 0.4913, and an MAE of 0.4002. Similar trends were seen in the prediction of heat transfer rate and the hot fluid outlet temperature. Therefore, ML models present a viable alternative to traditional empirical methods. This integration of ML into thermal system analysis offers new insights into optimizing heat transfer processes and improving efficiency.

Keywords: Convection Heat Transfer · Nusselt Number · Machine Learning

S. Dhar et al. (Eds.): AGC 2025, CCIS 2686, pp. 142–158, 2026.
https://doi.org/10.1007/978-3-032-09153-6_9

1 Introduction

For various engineering applications, heat exchangers (condensers) that utilize pipes of small diameters are gaining prominence to meet the growing demands for compact and energy-efficient heat transfer solutions. The spectrum of applications includes condensers in cooling and air-conditioning systems, horizontal tubular evaporators in water desalination units, power system heaters, and heat pipes, to mention some. In addition, small-scaled heat exchangers play critical roles in cryogenic processes (e.g. air separation and natural gas liquefaction plants). In these systems, the design and performance of heat exchangers directly impact other key components, such as compressors and drivers. These considerations warrant the development of more precise simulation tools that rely on accurate heat transfer coefficient models [1–4].

As detailed in several publications, some modelling techniques for the prediction of heat transfer coefficients during flow condensation of single-component fluids inside pipes have been developed and validated against experimental data. As the underlying mechanisms that govern heat transfer phenomena are complex, most of these models rely on heuristic approaches. Often, these models incorporate several dimensionless groups and adjustable parameters to fit experimental data. More recently, the application of ML techniques has expanded the complexity of these models (with a greater number of dimensionless groups and adjustable parameters). With this increased complexity, the prediction accuracy for large datasets has improved, even those associated with significant noise and high variability [5, 6].

Machine learning (ML) has emerged as a novel and powerful tool for analyzing and predicting forced convection phenomena and offers innovative approaches to overcome the challenges associated with traditional methods (e.g. empirical correlations, analytical models) and computational fluid dynamics (CFD). ML models (e.g. regression algorithms and neural networks) are trained using experimental and/or CFD data to predict local and average heat transfer coefficients in forced convection systems. These models can incorporate parameters such as the Reynolds number, Prandtl number, Nusselt number, flow velocity, duct geometry, and fluid properties, to predict local and/or average heat transfer rates. Machine learning makes use of classifiers (e.g., decision trees and support vector machines) to distinguish between laminar, transitional, and turbulent flow regimes, based on input features like pressure gradients, and velocity profiles. ML-based surrogate models (e.g. Gaussian processes, deep neural networks) are replacing expensive CFD simulations to provide rapid and accurate predictions of flow and thermal characteristics in forced convection systems. Tools like genetic algorithms or Bayesian optimization, combined with ML models, enable the optimization of heat exchanger designs, cooling systems, and duct configurations to maximize heat transfer rates while minimizing pressure drop and energy consumption. Adaptive control systems, powered by ML, can dynamically adjust operating conditions (e.g. flow rate or temperature) to enable real-time monitoring of forced convection systems, predict performance metrics (e.g. heat transfer rates) and detect anomalies (e.g. flow maldistribution or fouling). In recent times, machine learning has been successfully applied in multiphysics coupling, where complex problems (e.g. conjugate heat transfer in systems with solid-fluid interactions, or fluid flow with phase change) involving forced convection have been solved [7–10].

This study focuses on the application of machine learning models for predicting the Nusselt number and heat transfer rates in thermal systems, with suitable working fluids. It involves utilizing various machine learning tools and techniques, including feature selection, model training, and cross-validation, to analyze and predict thermal behaviour accurately. The workflow of the adopted methodology is shown diagrammatically below (in Fig. 1), illustrating the key steps.

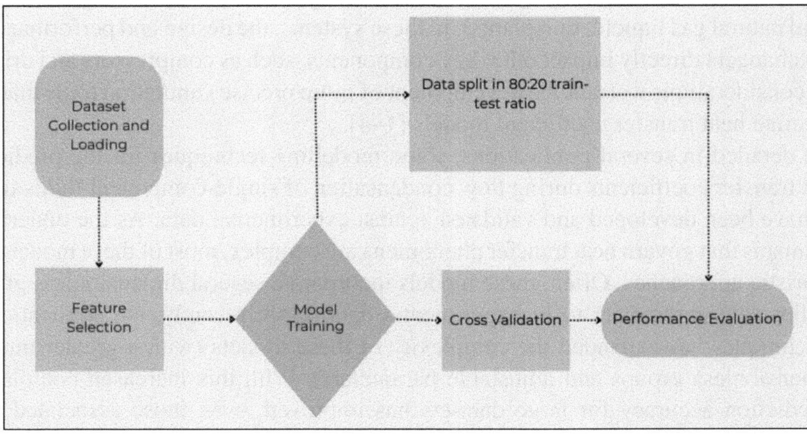

Fig. 1. The work flow of present study

2 Methodology

2.1 Dataset

The present analysis considers two datasets, obtained as outcomes of experimental studies conducted on two generic configurations of heat exchangers. The dataset employed in the first phase of this study consists of 87 experimentally obtained data points designed to evaluate the thermal performance of fluid flow within a tubular heat exchanger. Each data point encapsulates crucial parameters related to the heat transfer rate and the dynamics of fluid flow about the system and ensures a robust foundation for analysis. The dataset is devoid of missing values, thereby ensuring consistency and reliability throughout the study.

In the experimental setup used in this study, the length of the horizontal heat exchanger tube is 1.2 m, and the outer diameter of the heat exchanger tube is $57.5 \times 3.5mm$. The material of the tube (not specified in the dataset) is a high-conductivity material. An external compressor (equipped with a flow control valve) provides a controlled flow of air (working fluid). In addition, a helical wire insert is placed inside the tube (of specified diameter and pitch) to enhance the heat transfer rate. The setup is equipped with calibrated thermocouples (to measure the local temperatures at different sections), flow meter (to measure flow rates) and pressure sensors (for measuring pressure differences). A constant-flux heat source is placed externally, with provisions for monitoring and controlling the power input. The measured values of the inlet and outlet fluid temperatures, tube wall temperatures, mass flow rate and pressure drop are fed to a data acquisition system installed with the apparatus. The objectives of the study are the determination of local (or average) values of the heat transfer coefficient (or Nusselt number), calculation of the heat transfer rate, and evaluation of the effects of wire diameter and pitch on heat transfer performance and pressure drop.

The salient/key features include geometric and thermal properties such as 'Wire Diameter' (mm), 'Pitch' (mm), 'Cold Fluid Inlet Temperature' (t_1)(°C),, and 'Cold Fluid Outlet Temperature' (t_2)(°C).. In addition, variables related to fluid dynamics and thermophysical properties, which include the 'Differential Pressure across Orifice Flowmeter' (ΔP_1) (kPa), the 'Pipe Wall Temperature' (t_w)(°C),, and the 'Temperature Difference of Cold Fluid' (Δt) (°C) are included to provide a comprehensive representation of the underlying physics.

The 'Cold Fluid Inlet Temperature' ranges from 26.8 °C to 47.6 °C, with a mean of 34.53 °C and a standard deviation of 4.91 °C, while the 'Cold Fluid Outlet Temperature' varies between 78.3 °C and 97.3 °C, averaging 84.28 °C with a standard deviation of 3.93 °C. The 'Wall Temperature' spans from 96.7 °C to 100.8 °C, with a mean of 98.76 °C and a standard deviation of 1.18 °C. The 'Cold Fluid Temperature Difference' ranges from 34.4 °C to 59.6 °C, averaging 49.74 °C. Regarding flow regimes (laminar or turbulent), the Reynolds Number ranges from 9,303 to 33,609, with a mean of 22,061 and a standard deviation of 5,985, indicating that the dataset predominantly represents turbulent flow.

The 'Wire Diameter' and 'Pitch' are important parameters in datasets on experimental heat transfer analysis, particularly in systems that involve structured (patterned) surfaces or system configurations promoting enhanced heat transfer (e.g., finned tubes, wire coils, or helical inserts). The wire diameter modifies the flow pattern around the surface; while wires of smaller diameters allow more space for fluid flow, larger diameters increase flow obstruction and turbulence and enhance convective heat transfer rates. Wires of larger diameters (below a threshold value) offer lower thermal resistance and increase the effective surface area available for heat transfer, which can enhance thermal performance. Pitch refers to the spacing between consecutive wires, coils, or fins in a structured heat transfer surface. A smaller pitch (above a threshold value) induces higher turbulence and enhances heat transfer rates, without incurring significant pressure drop. As the pitch affects the density of fins (or wires) per unit length, it alters the heat transfer rate. While a smaller pitch increases the surface area exposed to the fluid and promotes heat transfer rate, a larger pitch reduces the same. A larger value of pitch also ensures more uniform fluid distribution, enabling the reduction of hotspots or stagnation zones.

For the prediction of the Nusselt Number (Nu), which quantifies the convective heat transfer rate, a total of 19 features were selected to capture the influence of geometric parameters of the system, temperature gradients determining the directions of heat flow, and flow dynamics that influence the rate of convective transport. These features include the 'Wire Diameter', the 'Pitch', the 'Inlet Air Density' (ρ_o), the 'Air Viscosity' (μ), and the 'Prandtl Number' (Pr), among others. To predict the Heat Transfer Rate (Q), a broader range of 22 features were chosen to encompass additional thermal properties such as the 'Specific Heat Capacity at Constant Pressure' (C_p), the 'Thermal Conductivity' (λ), and the 'Nusselt Number' (Nu) itself, given their direct correlations with the prediction of heat transfer rates. The selection of features was guided by their physical relevance to the prediction tasks, thereby ensuring that the model accurately captures the critical interactions between flow characteristics, thermal gradients, and heat transfer phenomena. For the dataset, one can visit the following link for the Enhanced heat transfer experimental dataset:

https://www.kaggle.com/datasets/haipotianku/enhanced-heat-transfer-experimental-datdaset/data

The second phase of the study concerns the performance of a plate-fin heat exchanger, which is a compact and efficient device that uses thin metal plates and fins to enhance heat transfer between two (or more) fluid streams. It is characterized by dual attributes of high heat transfer efficiency and lightweight construction. This heat exchanger typically consists of a stack of corrugated fins sandwiched between flat plates, forming a series of counterflow channels. The plates (and fins) are made from materials like aluminium or stainless steel, which offer high thermal conductivity and corrosion resistance. Fluids flow through separate, parallel channels formed by the plates and fins. While thermal energy is transferred through the thin plates, the fins enhance convection by increasing the heat transfer surface area.

The dataset has been obtained from a serialized set of experiments reported in [9]. To assess the performance of the plate-fin heat exchanger, the experiments have been conducted under a steady state. Measurements of temperature and mass flow rates for the 'hot' and 'cold' fluids provide the required information to compute the heat exchanger effectiveness (ϵ), defined as the ratio of the actual to maximum-possible heat transfer rates. The calculations of this performance parameter are assessed as follows:

$$\epsilon \equiv \frac{\dot{E}_{actual}}{\dot{E}_{max}} = \frac{\Delta T_{minimum fluid}}{\Delta T_{max}} \tag{1}$$

In the metric specified by Eq. (1), the 'minimum fluid' is the one that experiences the larger temperature difference in the heat exchanger, and the maximum temperature difference in the heat exchanger is always the difference in inlet temperatures of the 'hot' and the 'cold' fluids. For the counter-flow heat exchanger, depending on whether the 'hot' (or 'cold') fluid is the 'minimum fluid', the effectiveness (ϵ_h; ϵ_c) is as follows:

$$\epsilon_h = \frac{\dot{m}_h c_h (T_{hi} - T_{ho})}{\dot{m}_h c_h (T_{hi} - T_{co})} = \frac{(T_{hi} - T_{ho})}{(T_{hi} - T_{co})}; \tag{2}$$

$$\epsilon_c = \frac{\dot{m}_c c_c (T_{ci} - T_{co})}{\dot{m}_c c_c (T_{hi} - T_{co})} = \frac{(T_{hi} - T_{ho})}{(T_{hi} - T_{co})}. \tag{3}$$

In Eqs. (2) and (3), the subscripts 'i' and 'o' refer to 'inlet' and 'outlet' respectively. Based on experiments, the dataset has been prepared in [8]; this set, although delimited by the number of experimental trials, comes to use in the assessment of heat exchanger effectiveness for Industrial applications. The experiments were conducted for different values of mass flow rates (5.7 g/s to 14.2 g/s), as well as for different values of hot fluid inlet temperatures (315 K to 365 K). In the context of this dataset, the following variables are defined: 'Q' (liters /min)is the mass flow rate, 'P1' (kg/cm^2) is the pressure recorded at cold fluid inlet, 'P2' (kg/cm^2) is the pressure at hot fluid inlet, 'P_{cd}' (mm of Hg) is the cold fluid pressure drop, 'P_{hd}' (mm of Hg) is the pressure drop for hot fluid, 'T1' (K) is the cold fluid inlet temperature, 'T2' (K) is the cold fluid outlet temperature, 'T3' (K) is the hot fluid inlet temperature, and 'T4' (K) is the hot fluid outlet temperature.

2.2 Machine Learning Models

Linear Regression is a foundational statistical method used to establish a relationship between a dependent variable (target or outcome) and one or more independent variables (predictors or features). The primary objective of linear regression is to model this relationship by fitting a linear equation to the observed data, minimizing the differences between the predicted and actual values [11, 12].

In its simplest form, linear regression is represented by the equation:

$$y = \beta_0 + \beta_1 x_1 + \epsilon \quad (4)$$

In Eq. (4), 'y' is the dependent variable (target), 'β_0' is the intercept of the regression line, 'β_1' is the coefficient that represents the slope for the independent variable, 'x_1' represents the independent variable (feature), and 'ϵ' is the error term, which accounts for deviations of the observed values from the fitted line. The coefficients β_0 and β_1 are determined by minimizing the sum of squared residuals, which are the differences between the observed and predicted values:

$$\text{Residual Sum of Squares } (RSS) = \sum_{i=1}^{N} \left(y_i - \hat{y_i}\right)^2 \quad (5)$$

In Eq. (5), y_i is the ith observed value, and $\hat{y_i}$ is the ith predicted value. The linear regression technique can be extended to accommodate multiple independent variables, leading to the following multiple linear regression model:

$$y = \beta_0 + \beta_1 x_1 + \beta_2 x_2 + \cdots + \beta_n x_n + \epsilon \quad (6)$$

In Eq. (6) $x_1, x_2, \ldots, x_n$ represents the independent variables, and $\beta_1, \beta_2, \ldots, \beta_n$ are their corresponding coefficients. This method is widely employed due to its simplicity, interpretability, and computational efficiency, especially when the relationship between the target and predictors is approximately linear. Moreover, its ability to provide insights into the contribution of individual predictors makes it a valuable tool in both exploratory data analysis and predictive modelling.

In addition to linear regression, the present study incorporates other machine learning techniques to capture complex, non-linear relationships and improve predictive performance. Advanced algorithms such as decision trees, random forests, and support vector machines (SVMs) complement the simplicity of linear regression by addressing its limitations and providing robust solutions for diverse modelling challenges. These methods collectively enhance the study's ability to identify and model intricate patterns in the dataset effectively. Random Forest is an ensemble learning technique that combines the predictions of multiple decision trees to improve predictive accuracy and control overfitting. It operates by constructing a multitude of decision trees during training and aggregating their outputs (either by averaging for regression or voting for classification). The method leverages the principle of "wisdom of the crowd," where the collective decision of multiple models outperforms individual models.

Mathematically, for a regression problem, the prediction $\hat{y}$ from a Random Forest (comprising T numbers of decision trees) can be expressed as follows:

$$\hat{y} = \frac{1}{T}\sum\nolimits_{i=1}^{T} \widehat{y^{(i)}} \tag{7}$$

In Eq. (7) $\widehat{y^{(i)}}$ is the prediction from the ith tree. Each decision tree (in the Random Forest) is built using a random subset of features and a random bootstrap sample of the training data. This randomness introduces diversity among the trees and reduces the risk of overfitting, enhancing the model's generalization ability. The algorithm employs techniques such as:

- Bootstrap Aggregating (Bagging): Random subsets of the training data are sampled with replacement to train each tree, ensuring variability in the model inputs.
- Random Feature Selection: At each split in a tree, only a random subset of features is considered, reducing correlation among trees and improving robustness.

Random Forest inherently provides feature importance metrics, which enables the identification of influential predictors in the dataset. The importance of a feature is typically assessed based on the mean decrease in impurity or its contribution to prediction accuracy when permuted.

The key advantages of Random Forest include:

- Non-linear Modeling: It can capture complex relationships between input features and the target variable.
- Robustness to Overfitting: The aggregation of multiple trees mitigates the tendency of overfitting of individual trees.
- Scalability: It is efficient for large datasets and high-dimensional feature spaces.

Gradient Boosting Machine (GBM) is another powerful ensemble learning technique that builds predictive models by sequentially combining weak learners, typically decision trees, to minimize errors and improve accuracy. Unlike Random Forest, which trains multiple trees independently in parallel, GBM builds trees in a stage-wise manner, where each new tree corrects the residual errors of the previous trees. This iterative process enables GBM to refine predictions progressively and capture complex, non-linear relationships within the data. Mathematically, at each iteration, a new tree is trained to fit the negative gradient of the loss function, reducing the overall error. GBM employs techniques such as learning rate tuning, regularization, and early stopping to prevent overfitting and enhance model generalization. Its ability to optimize predictive performance while maintaining interpretability makes GBM a widely used approach in regression and classification tasks.

In the thermal analysis presented here, the aforementioned machine learning models have been employed to predict the Nusselt number (Nu) and heat transfer coefficients, all of which are crucial parameters for assessing fluid flow and heat transfer rates in the chosen configuration [10].

2.3 Performance Parameters

To evaluate the efficacy of the predictive models, three performance parameters were used in this study. They are the Coefficient of Determination (R^2), the Root Mean Squared Error (RMSE), and the Mean Absolute Error (MAE).

The R^2 value indicates the proportion of variance in the dependent variable that is explained by the model. Values closer to '1' signify higher accuracy, and demonstrate the model's ability to explain variability in the data effectively.

The RMSE measures the standard deviation of residuals, which are the differences between the predicted and actual values. This metric provides insight into the overall magnitude of prediction errors, with lower values of RMSE reflecting better model performance.

The MAE quantifies the average absolute difference between the predicted and actual values. It is an interpretable metric that reports errors in the same units as the target variable, making it easy to understand.

Considered together, these metrics offer a comprehensive assessment of model performance. They balance predictive accuracy and error quantification, allowing for robust evaluation across diverse data distributions (Fig. 2).

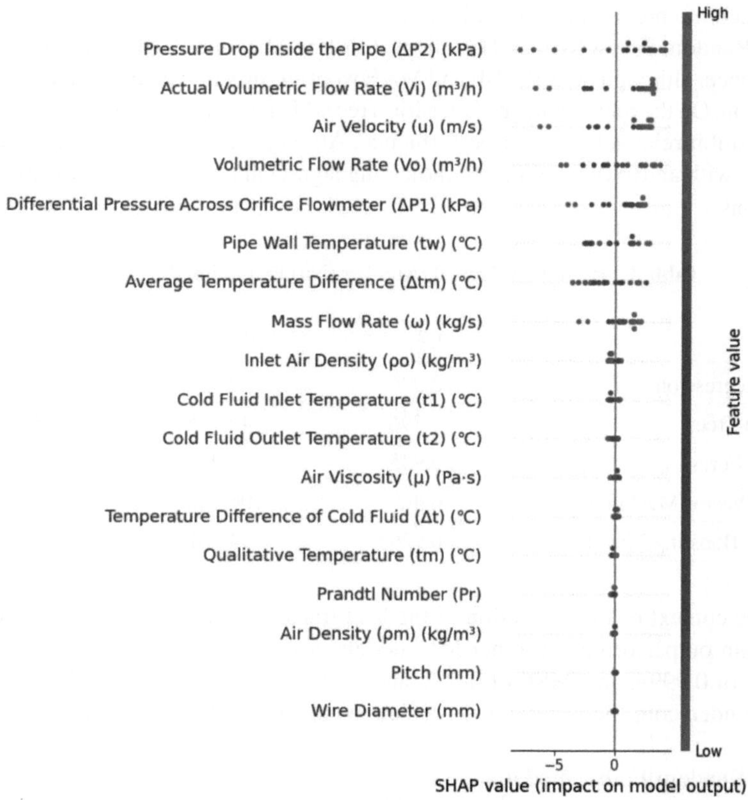

Fig. 2. SHAP Summary Plot for Random Forest Model, depicting feature importance and contribution to predictions for Nusselt Number.

3 Results and Discussion

For the dataset in the first phase of the study, the performance of the models was evaluated using an 80–20 training split and 5-fold cross-validation. The results were assessed based on key regression metrics, including the Coefficient of Determination (R^2), the Root Mean Squared Error (RMSE), and the Mean Absolute Error (MAE). Separate evaluations were conducted to predict the Nusselt Number (Nu) and the Heat Transfer Rate (Q). Based on all the models, the linear factors seemed to overpower the nonlinearity existing in the relationship between the features. To observe the feature importance (see Fig. 2 above), the SHAP summary is provided for the Random Forest model that shows how the most influential features are the pressure drop inside the pipe (ΔP_2), actual volumetric flow rate (V_i), and air velocity (u) in the prediction of Nusselt number.

3.1 Results on 80–20 Training Split

The results for the prediction of the Nusselt Number (Nu) using the 80–20 training split are summarized in Table 1 below. The Linear Regression (LR) has demonstrated the best performance among all the models, achieving an R^2 value of 0.996, an RMSE of 0.798, and an MAE of 0.662. These metrics indicate that Linear Regression has provided highly accurate predictions with minimal error.

The Random Forest (RF) and Gradient Boosting Machines (GBM) also deliver good performance, although their RMSE and MAE were slightly higher as compared to Linear Regression. On the other hand, the Decision Tree (DT) and the Support Vector Regression (SVR) exhibit relatively weaker performance. Among these, the SVR exhibits the least accuracy, with an R^2 value of 0.469, indicating significant room for improvements in its predictions.

Table 1. Results for Nusselt Number (Nu) Prediction (80–20 Split)

Model	R^2	RMSE	MAE
Linear Regression	0.996	0.798	0.662
Decision Tree	0.796	40.716	5.637
Random Forest	0.935	12.939	3.138
Support Vector Machines	0.469	106.180	8.322
Gradient Boosting Machines	0.9295	14.101	3.229

In the context of the prediction of the heat transfer rate (Q), the Linear Regression once again outperformed all other ML models, achieving near-perfect results with an R^2 value of 0.9999, an RMSE of 0.002, and an MAE of 0.035 (refer to Table 2). These metrics underscore the exceptional predictive capability of Linear Regression for this task.

The Random Forest model and Gradient Boosting Machines also produced competitive results with a good R^2 value, although their RMSE and MAE were slightly higher than those of Linear Regression. In contrast, the Decision Tree and Support Vector Regression exhibited significantly lower accuracy. The SVR, in particular, yielded the lowest R^2 value of 0.131, indicating poor predictive performance for this dataset.

Table 2. Results for Heat Transfer Rate (Q) Prediction (80–20 Split)

Model	R^2	RMSE	MAE
Linear Regression	0.9999	0.002	0.035
Decision Tree	0.857	461.901	16.811
Random Forest	0.925	243.847	10.658
Support Vector Machines	0.131	2811.377	40.223
Gradient Boosting Machines	0.9659	110.2436	8.7924

3.2 Results on 5-Fold Cross-Validation

The robustness of the models was further validated using 5-fold cross-validation. For Nu prediction (Table 3; see below), LR maintained exceptional performance, achieving a perfect mean $R^2 = 1.0$, while RF delivered consistent results with a mean $R^2 = 0.89$. SVR's performance declined further in this scenario ($R^2 = 0.26$), highlighting its limitations in capturing the relationships in the dataset.

Table 3. Fold Cross-Validation Results for Nu Prediction

Model	Mean R^2	Mean RMSE	Mean MAE
Linear Regression	1.0	1.03	0.76
Decision Tree	0.83	6.68	5.43
Random Forest	0.89	5.64	4.28
Support Vector Machines	0.26	14.53	11.88
Gradient Boosting Machines	0.92	4.79	3.76

For Q prediction (refer to Table 4), Linear Regression again excelled with mean $R^2 = 1.0$ and low errors (RMSE: 0.06, MAE: 0.04), as shown below. Random Forest demonstrated reliable performance with mean $R^2 = 0.93$, while Support Vector Machines showed significant deviations ($R^2 = -0.11$), indicating poor generalization.

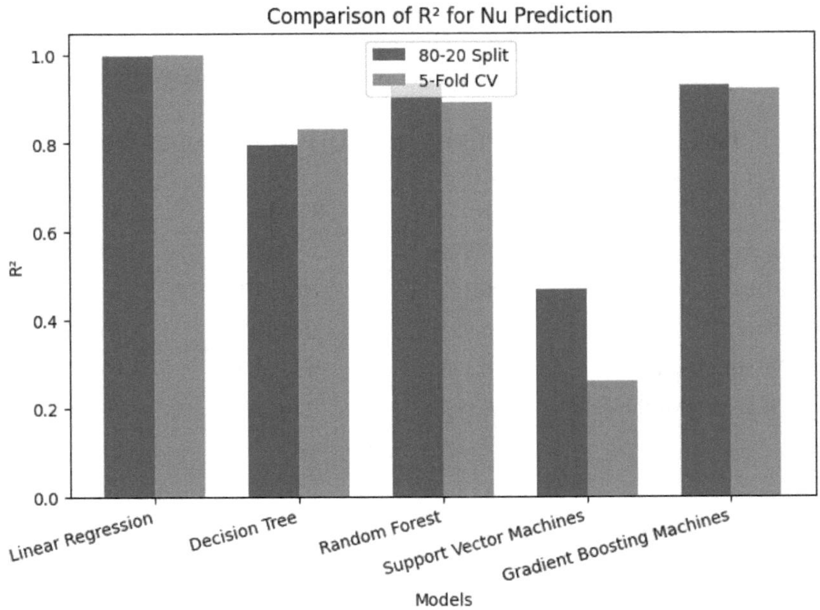

Table 4. Fold Cross-Validation Results for Q Prediction

Model	Mean R^2	Mean RMSE	Mean MAE
Linear Regression	1.0	0.06	0.04
Decision Tree	0.89	21.51	17.35
Random Forest	0.93	16.99	12.52
Support Vector Machines	−0.11	69.83	55.32
Gradient Boosting Machines	0.96	13.55	10.79

Linear Regression (LR) and Random Forest (RF) outperformed Decision Trees (DT) and Support Vector Regression (SVR) in predicting forced convection heat transfer rates, due to their ability to capture the dataset's underlying relationships effectively. LR achieved near-perfect accuracy ($R^2 \approx 1.0$), indicating a strong linear correlation between features aligning with fundamental heat transfer equations. RF, while slightly less accurate than LR, handled localized non-linearities (e.g., turbulence effects) better through ensemble learning, feature importance ranking, and noise reduction. In contrast, DT suffered from overfitting, leading to high variance, while SVR performed the worst ($R^2 \approx 0.13$ for heat transfer rate prediction), likely due to its reliance on complex kernel transformations, which were unnecessary given the dataset's strong linear trends. Future work could explore deep learning approaches for more complex heat transfer systems with strong non-linearity, provided a comprehensive dataset is aggregated.

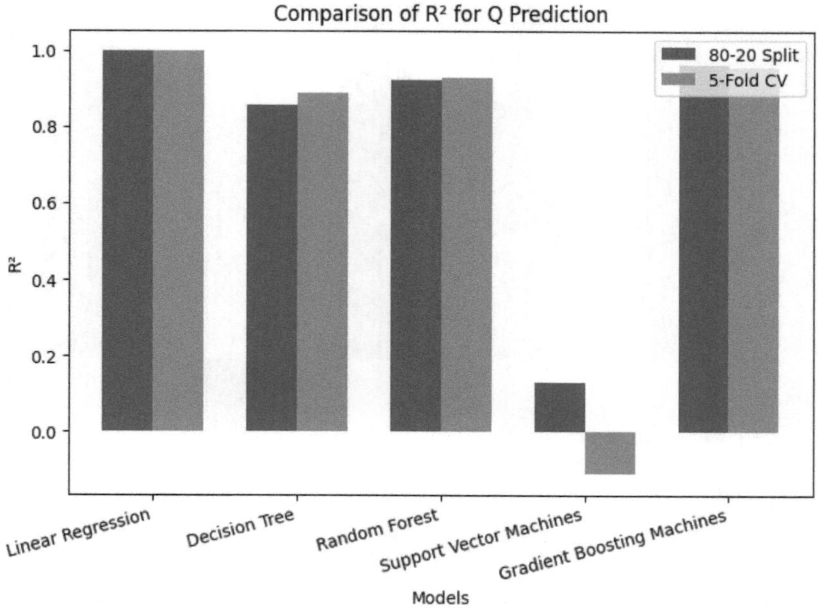

When the above models were applied to the second dataset (assessing the effectiveness of plate-fin heat exchanger), some results were consistent with the previous ones, while others showed some deviations.

Table 5. Results for Cold fluid outlet temperature (T2) (80–20 Split)

Model	R^2	RMSE	MAE
Linear Regression	0.9967	0.4913	0.4002
Decision Tree	0.9937	0.6823	0.5640
Random Forest	0.9875	0.9596	0.8918
Support Vector Machines	−0.2555	9.6203	8.2579
Gradient Boosting Machines	0.9970	0.4732	0.4085

The performance of the models in predicting the cold fluid outlet temperature (T2) followed a similar trend observed in the previous experiments, with Linear Regression (LR) and Gradient Boosting Machines (GBM) achieving the highest accuracy.

Table 5 summarizes the results for T2 prediction using the 80–20 training split. GBM demonstrated the best performance, achieving the highest R^2 value of 0.9970 and the lowest RMSE (0.4732) and MAE (0.4085), indicating superior predictive capability. LR also performed exceptionally well, with an R^2 of 0.9967, an RMSE of 0.4913, and an MAE of 0.4002, making it a strong competitor to GBM.

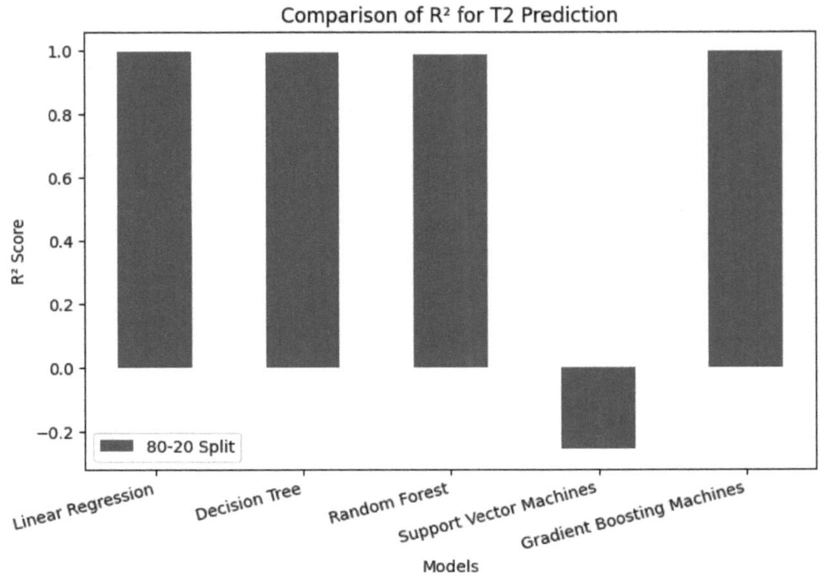

The Decision Tree (DT) model showed slightly lower accuracy than LR and GBM, with an R^2 of 0.9937, but its RMSE (0.6823) and MAE (0.5640) were higher, indicating that it introduced more errors in prediction. Meanwhile, Random Forest (RF) also tried to maintain good performance.

In contrast, Support Vector Regression (SVR) exhibited significantly poor performance, with an R^2 value of -0.2555, an RMSE of 9.6203, and an MAE of 8.2579. These results suggest that SVR struggled to generalize well for T2 prediction, likely due to the dataset's inherent linear relationships, which made the complex kernel transformations ineffective.

The noticeable errors observed in the Support Vector Regression (SVR) model align with findings from previous studies, such as those utilizing Artificial Neural Networks (ANN) for plate-fin heat exchanger performance prediction [8]. In [8], despite ANN's capability to capture complex nonlinear relationships, significant errors were reported in certain cases, particularly when the dataset exhibited strong linear correlations. This supports the observation (of this study) that SVR, which relies on kernel transformations to map data into higher-dimensional spaces, struggles to generalize effectively and leads to poor predictive performance.

SVR's high RMSE and MAE values suggest that its kernel-based approach was not well-suited for this dataset, likely due to the dominance of linear dependencies in heat exchanger parameters. Similar trends were noted in the reference study, where ANN models exhibited noticeable deviations, reinforcing the idea that complex models may not always outperform simpler regression-based approaches when dealing with datasets with inherent linearity.

Overall, the findings reaffirm that LR and GBM are the most reliable models for predicting the cold fluid outlet temperature (T2) due to their ability to capture the dominant patterns in the dataset effectively. Future research could further explore hybrid models or deep learning approaches to enhance predictive accuracy for more complex heat exchanger systems.

The performance evaluation for predicting the hot fluid outlet temperature (T4) using an 80–20 training split reveals notable differences across the models, as seen in Table 6. Linear Regression (LR) outperformed other models, achieving the highest R^2 value of 0.9336 and the lowest RMSE (0.3917) and MAE (0.2907), indicating its superior ability to capture the linear relationships in the dataset.

Gradient Boosting Machines (GBM) demonstrated moderate performance, with an R^2 value of 0.4191 and an RMSE of 1.1580, but its MAE (0.8613) was significantly higher than LR. The Decision Tree (DT) model showed limited accuracy, with an R^2 value of 0.5498 and higher RMSE (0.8760) and MAE (1.0195), indicating potential overfitting and reduced generalization.

Random Forest (RF) and Support Vector Machines (SVR) exhibited the weakest performance. RF achieved an R^2 of only 0.3011 and reported higher error metrics (RMSE: 1.2703, MAE: 1.6136), while SVR performed the worst, with a negative R^2 value (−0.2380), highlighting its inability to generalize for this task effectively. The error analysis across different models highlights critical factors influencing prediction performance for the given tasks.

Table 6. Results for Hot fluid outlet temperature (T4) (80–20 Split)

Model	R^2	RMSE	MAE
Linear Regression	0.9336	0.3917	0.2907
Decision Tree	0.5498	0.8760	1.0195
Random Forest	0.3011	1.2703	1.6136
Support Vector Machines	−0.2380	1.4689	1.6906
Gradient Boosting Machines	0.4191	1.1580	0.8613

Linear Regression (LR) consistently achieved near-perfect results ($R^2 \approx 1.0$) with minimal RMSE and MAE across predictions, emphasizing the dataset's strong linear correlations and the suitability of LR for capturing these relationships. Random Forest (RF) and Gradient Boosting Machines (GBM) delivered competitive performance, particularly for tasks requiring moderate non-linearity handling, such as turbulence effects in heat transfer. However, their slightly higher errors compared to LR indicate potential overfitting to localized patterns, reducing generalization, particularly evident in predictions like the hot fluid outlet temperature (T4). Decision Tree (DT), despite offering interpretable results, exhibited significant overfitting, resulting in elevated error metrics and reduced accuracy. Support Vector Regression (SVR), on the other hand, consistently underperformed, with high RMSE and even negative R^2 values in some cases. This can be attributed to its reliance on complex kernel transformations, which proved unnecessary for the dataset's inherent linear trends, leading to poor generalization. These findings reaffirm the dominance of LR for datasets with strong linear dependencies, while ensemble methods like RF and GBM remain viable alternatives for scenarios involving moderate non-linearity or noise.

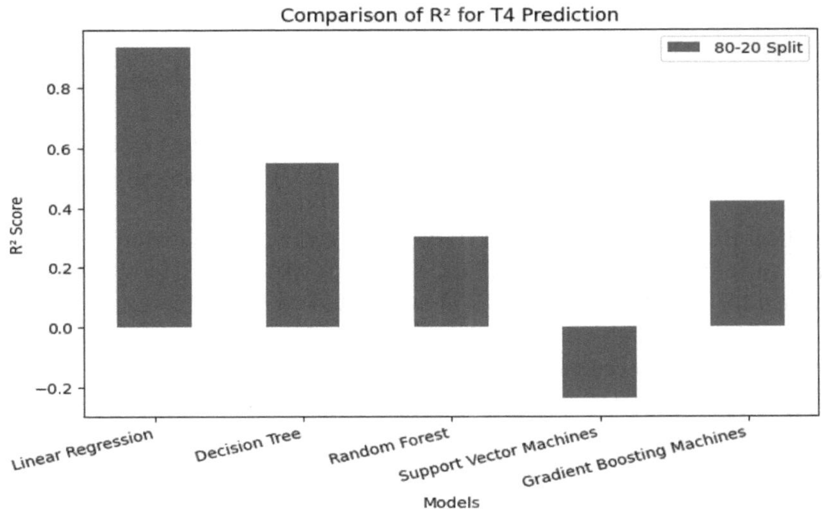

4 Conclusions and Future Scope

The results amply highlight the superior performance of the Linear Regression method in the prediction of forced convection heat transfer rate $(\dot{Q})$,, when the relationships between the 'features' and the 'targets' are nearly linear. The Random Forest technique also performs well in accurately predicting both the Nusselt number (Nu) and heat transfer rate $(\dot{Q})$, demonstrating its capability to model non-linear interactions. In contrast, the techniques of Decision Trees and Support Vector Regression show limitations in handling the complexity of the dataset, with the SVR performing poorly across all metrics.

The above findings underline the importance of model selection in achieving accurate predictions for fluid flow and heat transfer in an experimental study on convection. While linear regression is effective for simpler relationships, ensemble methods like the Random Forest offer a more generic solution for capturing non-linear dependencies, making them suitable for complex applications in thermal engineering.

As can be concluded from the present study, the application of machine learning in a practical heat exchanger configuration promoting forced convection heat transfer opens up promising avenues for innovation and practical benefits. By expanding the dataset (from experimental and/or CFD studies) to include diverse exchanger geometries, thermal boundary conditions, and material properties, deep learning models can achieve higher accuracy and robustness. For instance, these models can optimize the design and operation of heat exchangers, reduce energy losses, and ensure flow and thermal stability in critical processes. Moreover, this approach can accelerate the exploration of using novel materials and configurations for augmenting heat transfer rates, supporting advancements in fields such as renewable energy, microelectronics cooling, and aerospace engineering.

As computational power increases and access to high-fidelity experimental and/or simulation data improves, these models have the potential to bridge the gap between empirical methods and theoretical frameworks, promoting data-driven thermal management solutions. By leveraging these innovations, industries can achieve sustainable and cost-effective solutions for state-of-the-art thermal engineering challenges.

References

1. Pacio, J.C., Dorao, C.A.: A review on heat exchanger thermal hydraulic models for cryogenic applications. Cryogenics **51**(7), 366–379 (2011). https://doi.org/10.1016/j.cryogenics.2011.04.005
2. Cavallini, A., et al.: Condensation inside and outside smooth and enhanced tubes-a review of recent research. Int. J. Refrig **26**(4), 373–392 (2003). https://doi.org/10.1016/S0140-7007(02)00150-0
3. Rifert, V. G., & Sereda, V. V.: Condensation inside smooth horizontal tubes: Part 1. Survey of the methods of heat-exchange prediction. Thermal Science, 19(5), pp. 1769–1789 (2015). https://doi.org/10.2298/TSCI140522036R
4. Omidi, M., Farhadi, M., Jafari, M.: A comprehensive review on double pipe heat exchangers. Appl. Therm. Eng. **110**, 1075–1090 (2017)https://doi.org/10.1016/j.applthermaleng.2016.09.027
5. Hosseini, S.H., Moradkhani, M.A., Valizadeh, M., Zendehboudi, A., Olazar, M.: A general heat transfer correlation for flow condensation in single port mini and macro channels using genetic programming. Int. J. Refrig **119**, 376–389 (2020). https://doi.org/10.1016/j.ijrefrig.2020.06.021
6. Zhou, L., Garg, D., Qiu, Y., Kim, S.M., Mudawar, I., Kharangate, C.R.: Machine learning algorithms to predict flow condensation heat transfer coefficient in mini/micro-channel utilizing universal data. Int. J. Heat Mass Transf. **162**, 120351 (2020)
7. Wang, Y.Z., He, X.J., Hua, Y., Chen, Z.H., Wu, W.T., Zhou, Z.F.: Closed-loop forced heat convection control using deep reinforcement learning. Int. J. Heat Mass Transf. **202**, 123655 (2023) https://doi.org/10.1016/j.ijheatmasstransfer.2022.123655
8. Acikgoz, O., Çolak, A.B., Camci, M., Karakoyun, Y., Dalkilic, A.S.: Machine learning approach to predict the heat transfer coefficients pertaining to a radiant cooling system coupled with mixed and forced convection. Int. J. Therm. Sci. **178**, 107624 (2022). https://doi.org/10.1016/j.ijthermalsci.2022.107624
9. Gupta, A.K., Kumar, P., Sahoo, R.K., Sahu, A.K., Sarangi, S.K.: Performance measurement of plate fin heat exchanger by exploration: ANN, ANFIS, GA, and SA. Journal of Computational Design and Engineering **4**(1), 60–68 (2017). https://doi.org/10.1016/j.jcde.2016.07.002
10. Ghalandari, M., Irandoost Shahrestani, M., Maleki, A., Safdari Shadloo, M., & El Haj Assad, M.: Applications of intelligent methods in various types of heat exchangers: a review. Journal of Thermal Analysis and Calorimetry, pp. 1–12 (2021) https://doi.org/10.1007/s10973-020-10425-3
11. Çorumlu, V., Altıntaş, V., Abuşka, M.: Evaluation of prediction and modeling performance using machine learning methods for thermal parameters of heat sinks under forced convection: The case of external validation. Int. Commun. Heat Mass Transfer **151**, 107228 (2024). https://doi.org/10.1016/j.icheatmasstransfer.2023.107228
12. Huang, M.: Theory and Implementation of linear regression. In: 2020 International Conference on Computer Vision, Image and Deep Learning (CVIDL), pp. 210–217 (2020). https://doi.org/10.1109/CVIDL51233.2020.00-99

13. Sedkaoui, S., Khelfaoui, M.: Linear regression. In: Sharing Economy and Big Data Analytics, pp. 141–170 (2020). https://doi.org/10.1002/9781119695035.ch9
14. Bhattacharya, S., Verma, M.K., Bhattacharya, A.: Predictions of Reynolds and Nusselt numbers in turbulent convection using machine-learning models. Phys. Fluids **34**, 025102 (2022). https://doi.org/10.1063/5.0083943

Simulated Longitudinal Dynamic Decision Networks (SLDDN)

Swayambhu Dutta, Himadri Sikhar Pramanik, Soumya Banerjee(✉), and Manish Kirtania

Corporate Marketing Research, Tata Consultancy Services Limited, Kolkata, India
{swayambhu.dutta,himadri.pramanik,Soumya13.b,manish.k}@tcs.com

Abstract. Many decisions are derivatives of cognitive illusions, biases, contextual influences and are taken on limited data. Structured, knowledge-based, methodical decision-making enables some improvements. Linear models like decision trees, multi-variate methods or hierarchical models provide enablement in the context of structured decision-making. However, such models are constrained, to factor in changing future scenarios, state changes and adaptive contexts – thus impacting decision utility and future consequences. To factor in complexities like non-linearity, decentralization, interconnectedness, interdependences and multiplicity application of Decision Dynamic Networks (DDN) are observed – specifically in scenarios, that dynamically adapt themselves over time to changing and uncertain systemic triggers and state changes. This paper proposes extension to existing capabilities in decision networks by incorporating probabilistic modelling of adaptive contexts across future time slices. Proposed in this paper is a conceptualized simulated longitudinal dynamic decision network model to factor interactions among elements of decision networks and identify newer elements for consideration. Newer considerations including decisions, actions, results and the influence and interactions are moderated to optimize utility. It is premised that decision networks need to evolve over simulated time slices alongside future state changes. Maximizing the utility of decisions therefore needs to be understood over valid time slices in the future. Measure of utility over evolving longitudinal decision networks leads to discovery of a valid time which is likely to yield maximum decision utility. Time association with decision executions is imperative as proposed.

CCS (Computing Classification System) CONCEPTS • Applied Computing → Artificial Intelligence (AI) → Dynamic Decision Networks (DDN) → Longitudinal analysis.

Keywords: Dynamic Decision Network (DDN) · Generative Simulation · Utility Maximization (U-Max) · Longitudinal

1 Introduction

Decision-making is a cognitive process, which helps in selecting a specific course or process or action among the available alternatives. Decision-making could be at an individual level or at an institutional level. Temporal static decisions could be regarded as

S. Dhar et al. (Eds.): AGC 2025, CCIS 2686, pp. 159–171, 2026.
https://doi.org/10.1007/978-3-032-09153-6_10

a culmination of beliefs, culture, environmental and contextual factors in the present. All such factors have 'moderating' and 'mediating' effects on decision-making. Flawed decisions come with negative consequences, and these may become further amplified through multipliers in the future. Systems which enable decision makers to gain insights or predict probabilistic scenarios are gaining ground, owing to the consideration of incorporating varied conditions and to mitigate future uncertainties. This paper proposes how dynamic decision networks may be applied to understand the utility of decisions into future time slices. Longitudinal generation of decision networks through probabilistic simulations help navigate the future. The paper proposes a Simulated Longitudinal Decision Network Framework/Model enabled through data, analytics and relevant artificial intelligence and has application across multiple decision scenarios. The framework/model is operationalized through systems and integration of AI-based probabilistic simulations/modelling in adaptive contexts using dynamic feedback loops, across future time slices/state changes (Longitudinal).

2 Literature Review

2.1 Background and Methodology

Prevailing research proposes several ways, processes, guidance and frameworks through which decision-making could be effective, institutionalized or streamlined [1]. Researchers have proposed that incentives for accuracy coupled with effective monitoring procedures could help in increasing the quality and volume of efficient decision-making [2]. Others have augmented decision-making with the concept of contextualization [3]. Metacognitive confidence along with the related calibration are used by researchers to validate proposed decision frameworks.

This novel theoretical and systemic proposal for SLDDN is grounded in existing prior research in this domain. Research methodology involves qualitative analysis of the varied decision-making frameworks as available across secondary data sources, high impact journal, conference publications and public disclosures. The prior knowledge explorations include studying probabilistic simulations, case study appraisals, coupled with theoretical modelling for maximizing utility of the decisions. The propositions made in this paper lead to extending the body of knowledge. Explored are related areas including – considerations in decision-making, role of probabilistic simulations, tensors and vectors to understand directionality in decisions to consider future utility alongside Markov Chains [4],to understand state change scenarios. Further we explore how artificial intelligence (AI) can be deployed [5, 6] for effective decisions aligned to this novel proposition.

2.2 Considerations in Decision-Making

Prevailing research indicates that apart from satisfying normative goals, spontaneous affective reactions, conscious deliberations, expert opinions, technological help, institutions could play an important role in influencing the meta-choice for solving problems [7]. These problems could be strategic, operational and even tactical like the level of algorithmic controls [8]. In recent times, researchers have argued that despite different meta-factors and influences, institutional decision-making tends to be based more on logical explanations and hybridization techniques [9]. For large organizations or institutions, decision-making could also be model based on linear programming [10], differential equations, Bayesian approach and many others with fair degree of overlap and distinctions.

These relate to data driven decision-making and attempts to reduce biasedness and errors and thereby increase effectiveness of the decisions [11]. Existing research indicates application of Lens Model in analysis – used by experts for effective strategies in decision-making [12]. Studies indicate non-linear decision models including system thinking and cybernetics are likely to improve wider decision considerations and thereby influence accuracy in decision-making process [13]. There are methodological issues with over dependence on human dynamics [14] in decision-making at institutional level. These play out evidently in scenarios of multivariate decision processes to generate a composite model towards optimization of objectives [15] in interaction among various procedures, model learning in sequential or parallel decision-making. Non-linearity affects proportionality or confidence in decision-making, decentralization adds to increased time and complexity of options. Interconnectedness calls for a systematic or ecosystem view and interdependencies among the contextual factors leading to diversity of options. While we see approaches, especially in mitigating complexities, including parallel, multiplicity of decisions, interconnectedness in the decisions and the need for agile and effective decision-making remains increasingly imperative.

2.3 Role of Simulations, Tensors and Vectors to Attribute Directionality in Decision-Making Alongside Markov Chain for Probabilistic State Changes

Simulations have been deployed mostly in conceptualizing scenarios relevant to the decision context. As an extension it can help incorporate multiple considerations of decisions, actions and results as well. Probabilistic simulations could act as an important enabler for the researchers in development of several decision-making instruments [16]. Notwithstanding possibilities of simulation with advanced computation capabilities there are mixed views on the limitations of simulation techniques. Nevertheless, simulations are capable in generation of indicative future processes and operations, scenarios increasingly considering behavioral and normative aspects of decisions including uncertainties [17] of the future.

Like simulations, fuzzy tensors and vectors play a significant role in capturing and explaining directionality in real-life complexities especially in cases of multiplicity [18]. Applications of Markov's Chain highlights how future state changes are independent of prior transient or absorption state scenarios [19]. However, feasibility and efficiency achieved through modelling with the concepts remain complex and probabilistic, due to the various built-in assumptions [20]. Hence, its approach for solving complex models and to decide on several aspects remain potential but less explored in research. Option generation regarding considerations and associated directionality are key applications in decision-making [21].

2.4 Longitudinal Considerations to Assess Future Utility of Decisions

Decisions and associated time are important and require studying together. Some research indicates time pressure determines effectiveness for decision makers due to inherent or trained capabilities among the individuals or groups [22]. While research stresses both intellectual and cognitive ability to be the major forces of defining decisions but limits itself in terms of explaining the time validity of the decisions or time till which utility realized from the decision is maximum. Decision utility variance over time while is of great significance is somewhat unexplored concept in research. This is partly because of constraints in design of experiment to effectively study this. It is premised that decisions in the present time would generate varied utility in future state changes attributable to upcoming scenarios and past decision networks. The solution here proposes generation of dynamic decision networks into future time slices to optimize utility through continuous learning.

Past research in decision-making frameworks or guidelines have tried to weave the varied constructs of human beings and society to provide a meaningful logical explanation of the dynamic influence across the different variables. While some frameworks have talked about contextuality, some have highlighted exhaustiveness, while others have embraced reality or modelling techniques of different variables in the processes. Dynamic belief networks [23] indicate normative paradigms of decisions.

Simulations in decision-making are explored by researchers. However, when it comes to applications, limitations exist with respect to specific industry mentions and considerations. There are few studies in terms of developing an overarching framework where simulation applications generate future conditions and results in a probabilistic manner, interpreting how present conditions or decisions could be altered to get the expected outcomes for future. Decision utility maximization also involves choosing from alternatives to achieve aggregate goals. Time variation effects are a key consideration in such simulated future scenarios as Net Present Value (NPV) [24] in financial management where decisions utility is deterministic over likely future scenarios.

2.5 Role of AI and Neural Networks in Decision-Making: Importance of Feedback

Considering varied techniques for Decision-Making, AI plays an important role; for example, in rule-based systems, the pre-determined rules and actions enable decision-making as per the specific instances or the considerations being set in the system. Expert systems enable decision-making through human interventions. Machine Learning (ML) utilizes algorithms data, patterns, processes and operational methods are being taught to the system, which acts as an improved version of rule-based system [25]. AI and ML systems may provide benefits of scalability or exhaustiveness, and even improved accuracy. Transparency, interpretability and accountability remain as concerns in this technology proliferation. Systems of AI, human interventions are limited to number of inputs [26]. Generative AI provides new, unique and even innovative content and even simulative scenarios. However, the quality of the output may need necessary validations [27].

Extending on the concept of Neural Networks, one of the major issues has always been finding the optimal solution which would help in either minimizing or maximizing utility function. This is mainly one of the major requirements for multi-objective optimization problems, where there are complex interactions across the multi-level variables. However, research by Yu (2023), suggests the use of AI and NN (Neural Networks) for complex problems related to system optimization. This proposed solution also takes into consideration the uncertainty in the nature of the complexity involved in the problems and thereby suggests a model which has advantages of both the AI and NN. With the help of data-preprocessing techniques, this model also takes into consideration the time complexity. Another research by Fei (2024) talks about achieving optimization of algorithms in AI network, for management of network security [28].

With increase in technology capability some AI systems are adaptive. There have been efforts to design AI systems to complement or overcome the cognitive limitations [29]. However, the models often fail to understand the human mental models [30, 31]. Through experimental setups, effectiveness of human-AI collaborations is being explored with ambivalent results [32]. While a time-strategy allocation is being considered, when it comes to the definite past data considerations for effective influence on decisions to be more accurate is still non-conclusive and limited to specific cases [33, 34]. Varied ML models for different types of problems in machine learning algorithms, real-world applications and research [35], including longitudinal approaches are being considered [36].

3 Proposition: Longitudinal Utility Maximizing Model Considering Evolving Contexts in Dynamic Decision Networks

3.1 Novelty in the Model

Proposing a novel model capable of incorporating probabilistic future information, and at the same time remain dynamic and agile so that inter-influence can be captured in an effective way. The Longitudinal Utility Maximization model focuses on integrating the concept of past information capture, along with incorporating the AI techniques and evolving contextual information at a real time. The recognition of the complex interconnectedness being proposed through this model makes it even more relevant from a decision-making perspective. The function being proposed could be seen as a continuous function of optimizing the process of decision-making. The decisions are evaluated over future time slices through generation of simulated probabilistic scenarios.

SLDDN incorporates dynamic simulations, adaptability adjustments, probabilistic modelling and accuracy. The traditional models being static in nature often struggles when it comes to the changing of contexts with associated state changes. However, due to the dynamic nature, learning capabilities of SLDDN it is adaptive and evolves, thereby allowing improved utility maximization. Figure 1. Illustrates the process of how SLDDN is operationalized for institutional decision making. The capabilities are based on diverse multi-modal data inputs across databases leading to generation of adaptive contexts leading to simulations indicative of future state changes deriving optimal decisions alongside longitudinal utility.

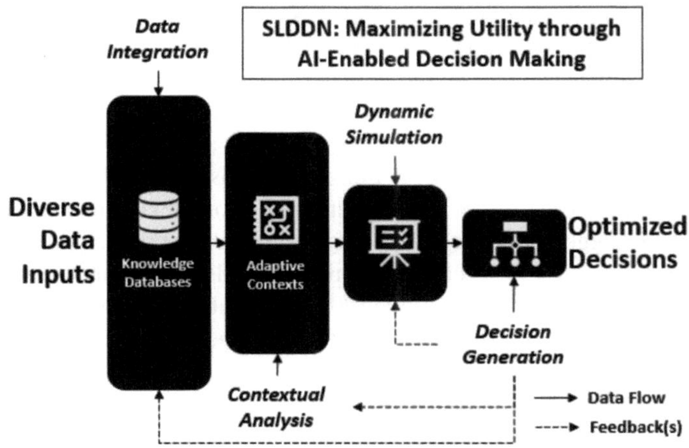

Fig. 1. SLDDN: Maximizing utility through AI-enabled decision making

Artificial Intelligence (AI) to Operationalize Proposed SLDDN Framework/Model Iterative learning capability by leveraging AI, helps mature the accuracy of probabilistic simulations, with more contextual information as obtained continuously through dynamic feedback loops. Machine Learning will help to analyse the outcomes of the past and incorporate variations of the decisions, to help maximize the utility of the decisions in real-time. The evolutionary approach of making the knowledge base robust to help in enhancing decision resilience, accuracy and adaptability, over effective outcomes and decision-making enhances performance in complex environments.

3.2 Decision Utility Maximization Across Evolving Contexts and Dynamic Networks

A decision renders utility over a period in the future. This may be considered as the decision validity time. Over this period multiple consequences of the decision play out. The utility of any decision can therefore be extrapolated into such valid future time slides. Utility can be viewed as a function of decisions, actions, results and inter-influence, as shown in Fig. 2. And be captured through a simple mathematical expression.

$$Decision\ Utility = Function\ (Decisions, Actions, Results, Inter - Influence)$$

Figure 2. Shows progression of time is marked with emergence of newer decisions, associated actions, derived results – evolution of future states. SLDDN proposes to generate these future states with machine learning and probabilistic generation of future states. The new scenarios may be generated through AI-driven probabilistic simulations. Each of these is influenced by time-dependent historic and emerging decisions. Over a period, the relative core-context of a decision is further expanded to include an extended context. The variables (decision, action, result and interactions) impacting utility are transient over the period. The emergence of dynamic decision networks needs to be factored into paradigms of decision-making. Numerous decision planes over valid time slices render variable such decision networks. The magnitude of utility over time continuum will vary with a definite maximum at some point in time. This is critical from a time modulation perspective. While some decisions may benefit from early phasing out, some may realize potential utility maximization over extended time. This is key, from the perspective of decision execution. Clearly decision utility maximization is a derivate of the utility function over the time continuum of decision onset to decision validity closure time.

$$Decision\ Utility\ Maximization = Function\ (Decisions, Actions, Results, Inter - Influence)$$

Time considerations from decision onset to decision validity closure

The evolution of decision networks is generative and simulated with AI, through contextual and related extended context over time. The evolution of decision networks is influenced by proliferation of its key components including newer decisions, associated actions, derived results and inter-influences. Such proliferations while intuitive and probabilistic in early stages of this decision-making approach can become a system learned accurate practice premised on some normative guidelines over system learning cycles. Variations and commonalities in decision networks of multiple decisions can be used to derive key insights for predicting associated context behaviors and avoid biases [38]. The ability to project the proliferation of a decision network into future time and state changes alongside its associated consequences helps gain future wisdom. Ability to measure utility over multiple simulated variations will help guide future steps and make more effective and accurate choices at decision onset time.

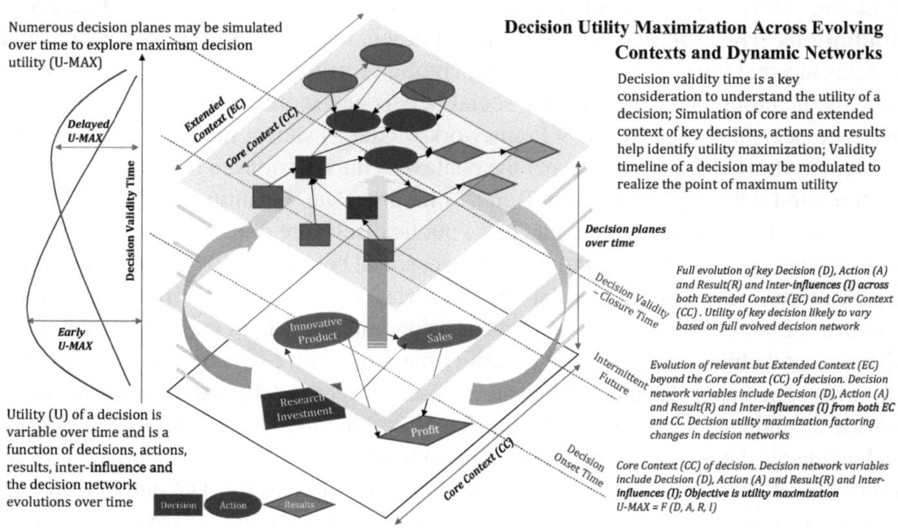

Fig. 2. Decision utility maximization across evolving contexts and dynamic networks

This proposed solution is premised on multiple theories relevant to institutional and complex decision-making bounded by practice experience. Focus on time, future scenario planning has been related to decision-making. However, such decisions are mostly viewed singularly and not within a decision network context. Beyond the conceptual proposition we suggest system interventions for enabling this at institutions.

As shown in Fig. 2., Delayed U-MAX (Delayed Utility Maximization over a period), Early U-MAX (Early Utility Maximization at the onset of the time), Decision Validity Time (Time Period for validity), Core Context (CC), and Extended Context (EC) are obtained as the thresholds, to be established from contextual and historical data analysis of the use cases under consideration. The thresholds are obtained by using Statistical Process Control (SPC) techniques and will vary across decision and related contexts. Arriving at context specific thresholds are statistical, simple leads to context-aware decisions systems. This ensures consistency, as SPC identifies the Control Limits and thereby enable dynamic adjustments facilitating Utility Maximization of decisions as per the contextual, dynamic and longitudinal data points.

Here, Core Context (CC) of decision. Decision network variables include Decision (D), Action (A) and Result(R) and Inter-influences (I); Objective is utility maximization.

U-MAX = F (D, A, R, I); Evolution of relevant but Extended Context (EC) beyond the Core Context (CC) of decision. Decision network variables include Decision (D), Action (A) and Result(R) and Inter-influences (I) from both EC and CC. Decision utility maximization factoring changes in decision networks across multiple longitudinal state changes.

3.3 System Enablement

Dynamic Decision Networks with Artificial Intelligence and Longitudinal Simulations The proposed solution is an AI based model that considers a set of input contexts in a real time environment, for generating decisions, considering a set of human contexts including language, culture, social beliefs and scores of other factors that can influence decisions. The SLDDN Framework/Model as shown in Fig. 1., incorporates different multimodal data sets, as per the context of the decisions to be made. Inputs can be structured data (e.g., Historical Trends, timeline Data, Quantitative data or metrics etc.) or can also be unstructured data (e.g., public disclosure or advocate website narratives or journal narratives or culture, language, social beliefs etc.). Contextual attributes (e.g., culture, social beliefs, regulations, laws etc.) and dynamic attributes (e.g., Geographical climate, triggers of environment, economic indicators etc.), all enabling the AI-based probabilistic simulations for utility maximization of the decisions in dynamic scenarios. The paper discusses application cases in Sect. 4.1 **Practice Implications** to provide an understanding about the data elements as inputs. The system leverages existing knowledge databases that get updated periodically for evolving new sets of decisions, actions, results and interactions for a known set of input contexts. The output decisions from the system for a set of input contexts, may be evaluated with alternative human in the loop decisions for the same set of input contexts, under the same set of conditions, for matching and calculating deviation, and to enable learning with back propagation, for adjusting the set of considerations within the system through updating knowledge base accordingly, during the machine learning phase. The projection of the decision into longitudinal future time slices through generative scenarios and other evolving transitions help validate the decision and assess utility over time.

Utility Measurement and Modulating Decisions to Maximize Utility Measuring the utility in relevance to the decisions taken would further help in measuring the time validity consideration. Decisions around large programs and initiatives span over a period. It could be further measured for similar problem statements to derive the optimal utility benefits. Utility maximization for the decisions could be achieved by interpreting the results, understanding the actions, analyzing the inter-influence and further modifying the decisions with time validity considerations. This will help in arriving at the optimal set of probabilistic scores to take the right decision for utility maximization. Utility maximization functions may be used and deployed over generative time slices in the future to evaluate optimal time validity of present decisions. Statistical controls are administered to operationalize quantitative measurement of utility functions. For comprehensive validation of utility maximization multiple techniques are adopted including - Compared Performance (t-tests of SLDDN outputs); Sensitivity Analysis (Measurement of Changes in the Utility, with variability in changes); Scenario Testing (AI-Based Simulations to get probable outcomes and adjustment of utility maximization). The analysis of Utility Maximization Score (U-Max) to be done through aggregation of different Performance Indicators a) Utility Gain (UG), b) Temporal Stability (TS), c) Decision Accuracy (DA), d) Impact from Inter-Influence (II), e) Adaptability Index (AN) and f) Efficiency of Computation (EC). U-Max = W1*UG + W2*TS + W3*DA + W4*II + W5*AN + W6*CE, where W1-W6 are weights assigned to the Performance Indicators as per the context and the environment requirements. The weights are context specific, and variable based on data. While the hypothesized function is additive, this may be productive depending on the association of variables.

Here the Performance Indicators are defined as below:

- UG = Enhanced Utility Delivered by SLDDN
- TS = Consistency of Utility values over time = 1 – (Standard Deviation of Utility Values across Time Slices/ Mean of Utilities)
- DA = Indicating Reliability of SLDDN Model = | (Predicted Outcome – Optimized Outcome)/Optimized Outcome|*100
- II = Measure of interdependencies in between actions, results and decisions = (Sum of Interactions of Utilities/Total Decisions of Utility) ….. Here the utilities of the interactions are derived from the joint outcomes
- AN = Assessing the adaptability to changes = (Total Number of decisions for adjustments/ Total number of decisions)
- CE = Time required for SLDDN for processing/ Time required by the traditional models

Longitudinal Simulation Analysis SLDDN model includes AI-Enabled Probabilistic Simulation modelling, occurring across multiple state changes. First the As-Is State, where the current Decision Dynamics and Contexts are evaluated. This is followed by multiple Future State generations where capabilities of generating probabilistic potential scenarios are demonstrated. The SLDDN passes through Transition States. What-If Analysis is deployed to explore exhaustiveness in state generations.

4 Implications

4.1 Practice Implications

The system renders multiple practice implications including preemptive decision-making based on situational inputs, to initiate real time actions for target achievement. Temporal assessment and evaluation of decision sequences to preempt future decisions can also be accomplished by leveraging this system. This helps in incorporating a new dimension of time in dynamic decision network. While dynamic decision networks help in incorporating uncertainty through feature-based action representation [23], the time validity incorporated by this proposed model provides a rational boundary to quasi-static decisions.

The AI generative scenarios into future time slices help understand the future consequences of present decisions – thereby validations are available beforehand. This improves further through human in loop interventions and machine learning. This is particularly relevant for long-term projects and large program executions. The solution simulates a set of alternatives and models disruptive influences that can affect standard decision-making approaches in a dynamic environment. Finally, evolving continuous modelling through incremental adaptations for enabling business continuity and resilience can be achieved through this system with the objective of utility maximization. The utility evaluation across future time slices helps derive optimal time validity of decisions.

Instance of SLDDN Application Practice applications of SLDDN range across varied domains – (i) In Healthcare Sector, it could be used to optimize the plans for treatment of the patients, by understanding their context and probable responses to treatment and various drugs over time. This will enable enhanced patient recovery; (ii) Allocation of limited resources during disaster recovery management, to have maximum possible impact as per the available response strategies; (iii) For any sector, supply chain decisions could be optimized to adjust demand fluctuations as per dynamic inventory situations [37]; (iv) Minimizing carbon emission scenarios as per simulated environments of the future; (v) Modelling predictions for investments in the financial sector for maximizing returns. These use cases are not exhaustive but indicative of the wide applicability of SLDDN across industries and (vi) Existing research explains the role of simulations in IT (Information Technology) Service strategy by stating how the simulations on the service capacity and customer needs coupled with previously agreed SLAs (Service Level Agreements) for the customers could help to identify the limit of abandonment of service requests [38].

4.2 Academic Implications

Longitudinal utility-maximizing decision-making models can be leveraged for enhancing existing self-adaptive models, taking into consideration the parameters and contexts pertaining to language, culture, social beliefs, environmental constraints, technological advancements, futuristic probabilities of varied results and dynamic interrelations between the constructs. Chaos theory [37] or Dynamic Decision Networks [23] or Bayesian Approaches [11] touch upon uncertainty, feature based actions or data enablement, respectively. However, the conceptualization here complements integrating existing uncertainty views with longitudinal utility maximization and time validity of decisions. The proliferations in artificial intelligence and capabilities in generating simulated future time slices will render rigor and robustness to Simulated Longitudinal Dynamic Decision Networks (SLDDN). Back propagation and feedback looping incorporated in intelligent SLDDN solutions can lead to newer directions in self-learning for decision-making approaches. Temporal assessment and evaluation of decision sequence characteristics of this model can be leveraged by future research for time-based structured decision-making and in contextualizing simulated evaluations. The modular nature of the proposed solution, with evolving knowledge base, can be plugged in and utilized in existing rule based decision-making solutions.

4.3 Limitations an Mitigation Strategies

Limitations of this framework/model could arise out of computational complexity, quality of data, issues related to scalability and issues of security vulnerabilities. However, as per the situation and the availability of the contexts for application, several strategies could be deployed like that of algorithm optimization, robust data pre-processing techniques, implementing parallel processing, using distributed computing to avoid scalability issues and integrating security protocols for data integrity. AI driven generative simulations may be costly, complex and require processing capabilities.

5 Conclusion

The model is capable of guiding for embracing disruptive forces in decision-making process, for taking necessary actions for sustaining and thriving amidst adaptive future environments. The proposed SLDDN solution can be utilized in multiple contexts, including business, social, education, healthcare, drug development among others. Time bound utility maximization is factored alongside constraints and uncertainties through continuous learning mechanisms. Understanding decision consequences in future time slices in an unbiased manner will help decision-makers globally.

References

1. Keestra M.: Understanding human action: Integrating meanings, mechanisms, causes, and contexts. Case studies in interdisciplinary research, pp. 225–258. (2012)

2. Koriat, A., Goldsmith, M.: Monitoring and control processes in the strategic regulation of memory accuracy. Psychol. Rev. **103**(3), 490–517 (1996)
3. Jackson, S.A., Kleitman, S.: Individual differences in decision-making and confidence: Capturing decision tendencies in a fictitious medical test. Metacogn. Learn. **9**(1), 25–49 (2014)
4. Chan, K.C., Mills, T.M.: An introduction to Markov Chains. in The Mathematical Association of Victoria For The Forty-Nineth Annual Conference, Brunswick (2012)
5. Alasmri, N., Basahel, S.: Linking artificial intelligence use to improved decision-making, individual and organizational outcomes. Int. Bus. Res. **15**(10), 1–15 (2022)
6. Niu, S., Ma, J., Yin, Q. Wang, Z., Bai L. Yang, X.: Modelling patient longitudinal data for clinical decision support: A case study on Emerging AI healthcare technologies. Information Systems Frontiers (2024)
7. Engel, C., Weber, E.U.: The impact of institutions on the decision how to decide. J. Inst. Econ. **3**(3), 323–349 (2007)
8. Dutta, S. Pramanik, H.S. Rajan, S., Rajan, R.G., Satapathy, S.: "Proliferations in Algorithmic Control: Review of the Phenomenon and Its Implications," in Transfer, Diffusion and Adoption of Next-Generation Digital Technologies. , S. D. Y. M. B. L. B. E. A. Sharma, Ed., Springer, Cham. p. 44–54. (2024) https://doi.org/10.1007/978-3-031-50188-3_5
9. Teixeira, M.G., Roglio, M.G., Marcon, B.R.: Institutional logics and the decision-making process of adopting corporate governance at a cooperative organization. J. Manage. Governance (2016)
10. Vance, C.M., Groves, K.S., Paik, Y., Kindler, H.: Understanding and measuring linear-nonlinear thinking style for enhanced management education and professional practice. Academy of Management Learning & Education **6**(2), 167–185 (2007)
11. Angria, L., Sari, Y.D., Zarlis, M., Tulus, M.: Data-driven Modelling for decision making under uncertainty. in IOP Conference Series: Materials Science and Engineering (2018)
12. Karelaia, N., Hogarth, R.M.: Determinants of linear judgment: A meta-analysis of lens model studies. Psychological Bulleti **134**(3), 404–426 (2008)
13. Kim, C.N., McLeod, J.R.: Expert, linear models, and nonlinear models of expert decision making in Bankruptcy prediction: A lens model analysis. J. Manag. Inf. Syst. **16**(1), 189–206 (1999)
14. Schmidt, C.F.: Understanding Human Action. aclweb.org (1975)
15. McCann, J.M., Miller, J.G., Moskowitz, H.: Modelling and testing dynamic multivariate decision processes. Organ. Behav. Hum. Perform. **14**(2), 281–303 (1975)
16. Morgan, G.B., Moore, C.A., Floyd, H.S.: On using simulations to inform decision making during instrument development. J. Psychoeduc. Assess. **36**(1), 82–94 (2017)
17. A. O. Victor, "General Simulation Techniques for Decision-Making in Real Situation, A proficiency of STEM Research," in A Proceeding of the 56th Annual Conference of the Science Teachers Association of Nigeria , 2015
18. Deng, S., Liu, J., Wang, X.: The properties of fuzzy tensor and its application in multiple attribute group decision making. IEEE **27**(3), 589–597 (2019)
19. Yang, X.: Markov Chain and Its Applications. SSRN. (2020)
20. Wang, C.-N., Nguyen, V. T., Chyou, J.-T., Lin, T.-F., Nguyen, T. N.: Fuzzy Multicriteria Decision-Making Model (MCDM) for Raw Materials Supplier Selection in Plastics Industry. MDPI, vol. 7, no. 10. (2019)
21. Kalis, A., Kaiser S., Mojzisch, A.: Why we should talk about option generation in decision-making research. Frontiers in Psychology (2013)
22. Kocher, M.G., Schindler, D., Trautmann, S.T., Xu, Y.: Risk, time pressure, and selection effects. Exp. Econ. **22**, 216–246 (2019)

23. Bencomo, N., Belaggoun, A., Issarny, V.: Dynamic decision networks for decision-making in self-adaptive systems: A case study. in 2013 8th International Symposium on Software Engineering for Adaptive and Self-Managing Systems (SEAMS) (2013)
24. Chen, X.: Theoretical analysis of net present value. BCP Business & Management **30**, 683–687 (2022)
25. Balbaa, M.E., Abdurashidova, M.S.: The impact of artificial intelligence in decision making: A comprehensive review," EPRA International Journal of Economics. Business and Management Studies (EBMS) **11**(2), 27–38 (2024)
26. Shalev-Shwartz, S., Ben-David, S.: Understanding machine learning: From theory to algorithms. Cambridge University Press, Cambridge (2014)
27. Chan, C.K.Y., Hu W.: "Students' voices on generative AI: perceptions, benefits, and challenges in higher education," International Journal of Educational Technology in Higher Education. (2023)
28. Fei, W.: Research on optimization algorithms for artificial intelligence network security management based on All IP Internet of Things fusion technology. Computers and Electrical Engineering, vol. 11, (2024)
29. Cummings, M.: Automation Bias in Intelligent Time Critical Decision Support Systems. Decision making in aviation, pp. 289–294. (2017)
30. Russell, S.: Human Compatible: Artificial Intelligence and the Problem of Control, Viking (2019)
31. Lai, V., Chen, C., LIAO, Q.V., SMITH-RENNER A., TAN, C.: Towards a Science of Human-AI Decision Making: A Survey of Empirical Studies. (2023)
32. Rastogi, C., Zhang, Y., Wei, D., Varshney, K.R., Dhurandhar, A., Tomsett, R.: Deciding Fast and Slow: The Role of Cognitive Biases in AI-assisted Decision-making. (2022)
33. Kim, H.-S.: Decision-Making in artificial intelligence: Is it always correct? J Korean Med Sci. **35**(1), (2020)
34. Franke, F., Franke, S., Riedel, R.: AI-based improvement of decision-makers' knowledge in production planning and control. IFAC-PapersOnLine **55**(10), 2240–2245 (2022)
35. ISarker, I: Machine Learning: Algorithms, Real-World Applications and Research Directions. SN Computer Science, vol. 2, no. 160 (2021)
36. Chacon, A. Kausel, E.E., Reyes T.: A longitudinal approach for understanding algorithm use. J. Behav. Decis. Mak. (2022)
37. Kumara, S., Ranjan, P., Surana, A., Narayanan, V.: Decision making in logistics: A chaos theory based analysis. CIRP Ann. **52**(1), 381–384 (2003)
38. Orta, E., Ruiz, M.: A simulation approach to decision making in IT service strategy. Sci. World J. 1–9. (2014)

The Effects on Bug Resolution Time: A Causal Discovery Perspective

Reshmi Maulik[1(✉)], Subhajit Datta[2], and Subhashis Majumdar[2]

[1] Meghnad Saha Institute of Technology, Kolkata, India
reshmi.maulik@msit.edu.in
[2] Heritage Institute of Technology, Kolkata, India
subhajit.datta@acm.org, subhashis.majumder@heritageit.edu

Abstract. Rapid resolution of bugs is a key software engineering outcome. Past studies have shown that different aspects of developer interaction are associated with bug resolution times. However, such studies are predominantly correlational in nature; they do not establish causal connections between variables of interest. Software intensive systems play critical roles in nearly all aspects of our lives today. Thus understanding what causes software bugs to be resolved faster is of significant importance to the state of art and practice of software engineering. In this paper, we apply the Greedy Equivalence Search (GES) algorithm on real-world data from a large software development ecosystem to identify the causal influences on bug resolution times. Our results indicate that the amount of time needed to resolve a bug is affected by different types of interactions among developers, in addition to workload and experience of the developer owning the bug, and the number of developers working on the bug. The conclusions from our results can offer insights for researchers and practitioners on further causal investigations of software engineering outcomes.

Keywords: Causal discovery · Greedy Equivalence Search (GES) · Software development · Bug resolution

1 Introduction and Motivation

The quest for causation lies at the heart of all scientific endeavors. With the dominant role software plays in our lives today, understanding causal influences on software quality is of utmost importance. Bug resolution time is a key determinant of software quality; thus isolating factors that cause bugs to be resolved faster has wide implications for various stakeholders. Randomized controlled trials (RCT) – though effective for establishing causation – are often impractical to execute in real-world software development ecosystems. Naturally, in view of

Supplementary Information The online version contains supplementary material available at https://doi.org/10.1007/978-3-032-09153-6_11.

S. Dhar et al. (Eds.): AGC 2025, CCIS 2686, pp. 172–183, 2026.
https://doi.org/10.1007/978-3-032-09153-6_11

end-user interests bugs can not be randomly segregated into control and treatment groups and resolution effort ("treatment") selectively withheld from those in the control group.

With the increasing availability of software development data from open source development ecosystems, techniques that seek to establish causation by analyzing historical records – without the need for active intervention in an ongoing experiment – can be useful. *Causal inference* and *causal discovery* are some such techniques.

The goal of causal inference is to investigate potential outcomes related to altering a particular system. The process of developing models that identify the causal connections present in data is known as causal discovery. Causal discovery uses a dataset's inherent structure to distinguish between causal and correlational links, as correlation may not necessarily imply causation [1].

Causal discovery techniques use constructs such as Directed Acyclic Graphs (DAGs) and the concept of *d-separation*; which is used to assess if a set of variables is conditionally independent of another set, given a third set. Paths between variable sets are investigated in this procedure to determine if they are blocked or unblocked, indicating independence or dependence respectively. The process entails creating an ancestral graph, connecting parents to promote it, rearranging the graph, and finally eliminating the specified variables and their edges. The variables are regarded as independent if they are shown to be disconnected after this procedure. This process finds wide applications in the field of epidemiology, genetics, economics, machine learning and biology [2,3].

In this paper, we use a causal discovery based approach to analyze data from the Eclipse[1] development ecosystem. In an earlier regression based study on this dataset, it was reported that higher levels of repeat developer interactions relate to more time taken for bug resolution, after controlling for other peripheral influences [6]. We seek to investigate whether a causal perspective leads to similar observations in the current study. With that view, we examine the following research question in this paper: *What are the causal influences on bug resolution times in a large software development ecosystem?*.

2 Causal Discovery

The process of determining the causal connections between the variables in a dataset is known "causal discovery". In contrast to conventional statistical techniques, which are usually correlation-based, causal discovery seeks to identify the mechanisms underpin the relationship between two variables. Beyond mere correlation, underlying processes that produce a cause-and-effect relationships between variables is referred to as the core mechanism in causal discovery. It involves figuring out how one variable might directly affect another to produce measurable results. While correlation only indicates the existence of a relationship between two variables, causal discovery aims to explain how and why two variables are related [7,11,23].

[1] https://www.eclipse.org/.

A *network* or a *graph* is frequently used to visually represent a causal structure that specifies how one predictor influences other predictor(s) and/or the response of interest. A tuple made up of two sets is represented mathematically as a graph $(G) = (V, E)$. The variables describing the phenomenon of interest are represented by the first set (V), which is a finite set of vertices, and the second set (E), which is a finite set of edges connecting the vertices [30]. We will now describe the algorithm used in this paper for causal discovery.

GES is a score-based algorithm that seek to identify the causal structure by optimizing a suitably defined score function [4,7,10]. GES makes the following assumptions:

1. *Causal Sufficiency*: A *confounder* is any variable that influences both the independent and dependent variables and produces a confusing association. As a result, figuring out the actual cause and effect relationship between the variables becomes difficult. This assumption specifies that every pertinent variable is present in the dataset and that there are no unobserved confounders influencing the correlations between the variables that are being observed.
2. *Markov Condition*: According to this assumption, every variable in the causal graph is conditionally independent of its non-descendants, given its parents. Two variables that are d-separated in a causal network are assumed to be probabilistically independent by the causal Markov condition. Variables which are d-separated are those for which there is no path (taking into account the direction of arrows and conditioning on specific variables) in the graph.
3. *Acyclicity*: This assumption states that causal model can be represented by a directed acyclic graphs (DAG); that is the graph can not have cycles, which are paths that begin and terminate at the same vertex.
4. *Faithfulness (or Stability)*: According to this assumption, the causal graph's structure is responsible for every conditional independency that has been found in the data.
5. *Linear relationship with Gaussian noise terms*: This assumption states that When there is a linear relationship between variables and Gaussian noise terms, the relationship can still be explained by a linear equation but with the addition of random noise that has a Gaussian (normal) distribution. In this context, Gaussian noise refers to random departures from the ideal linear relationship, which are modelled as normally distributed errors. This enables the linear model to account for real-world errors and variability in the data.
6. *Score Equivalence*: To assign the same score to graphs that represent the same set of conditional independencies, a scoring function such as *Bayesian Information Criterion* (BIC), which is employed in GES, must be score comparable. BIC is a criterion for choosing a model from a limited number of options. Models with more parameters are penalized, which helps to strike a balance between model fit and complexity. Equation 1 is used for calculating the BIC [5]:

$$BIC = -2.ln(L) + k.ln(n) \tag{1}$$

In Eq. 1, L denotes the model's likelihood, k the number of parameters, and n is the sample size. Based on the data, the likelihood of each model is computed, which indicates the extent to which the observed data can be explained by the model. The goodness of fit is represented by the first term $-2.ln(L)$, and the model's complexity is penalized by the second term $k.ln(n)$. A better model is indicated by a lower BIC value.

7. *Large Sample Size*: There must be a sufficiently large sample to precisely estimate the conditional independencies and the scores of various graph configurations. The complexity of the causal model, the magnitude of the causal effects, and the degree of measurement noise all play a role in determining the appropriate sample size for causation discovery in GES.

Causal Bayesian Networks (CBNs) are probabilistic graphical models that show causal relationships between a set of variables. They are an extension of Bayesian networks that are explicitly intended to model and infer causality rather than merely correlations [29]. GES retrieves the model with the greatest Bayesian score it discovers by heuristically searching the space of causal Bayesian networks. The Bayesian score is a metric used to assess the performance of a Bayesian network. It assesses how effectively a network structure describes observable data. The score combines the likelihood of the data presented to the network (how likely the data is) with a prior probability on the network topology itself [24].

GES uses an empty graph to begin its search. The next step is a forward search, where the goal is to raise the Bayesian score by adding edges between the nodes. Until no single edge addition raises the score, this process is repeated [4,13]. Lastly, it executes a backward search, eliminating edges until the score cannot be raised by removing any single edge. For this approach, multiprocessing is not supported [8]. We present a simple example illustrating the application of GES in the Appendix (available online[2] and shared as a supplementary material).

3 Methodology

3.1 Collecting Data

As mentioned earlier, we have analyzed data from the Eclipse development ecosystem for this study. With an extensible plug-in architecture, Eclipse is widely used as an open source integrated development environment (IDE) for desktop environments and is increasingly positioned as a platform to support collaboration. We used the Eclipse dataset that was collected, curated, and shared for research by the author [12].

[2] https://bit.ly/AGC2025-GES-Appendix.

3.2 Organizing Data

Bugs in the dataset were divided into various groups. To ensure that there was closure for every study work unit, we limited the selection of defects for our analysis to those that were fixed and made public. Only fixed and published bugs that have received feedback from three or more different developers were chosen, as we are studying developer interaction with regard to bugs. By using this filtering criterion, we can make sure that the bugs we are looking at had meaningful developer engagement. After preparing the dataset, we obtained 13,248 bugs and 227,718 comments from 11,326 Eclipse contributors. All potential outliers have been removed from the dataset through cleaning.

3.3 Identifying Developer Types

If a developer has never commented on a bug prior to the one we are examining, they are deemed novices. A developer who has commented on at least one problem prior to the one we are currently examining is considered experienced. Based on these concepts, we now go on to explain various types of developer interactions as follows [6]:

1. *NewInteraction (NI):* Two novice developers co-commenting on a problem for the first time is known as a NewInteraction in a development ecosystem.
2. *FreshInteraction (FI):* A FreshInteraction is characterized as an event in which two developers, one inexperienced and the other experienced, co-comment on an issue in an ecosystem for development.
3. *SeasonedInteraction (SI):* A "seasoned interaction" is defined in a development ecosystem as the first time two seasoned developers co-comment on a problem.
4. *RepeatInteraction (RI):* In the development ecosystem, a RepeatInteraction occurs when two seasoned developers who have previously co-commented on a bug together comment on another problem together once more.

3.4 Model Variables

We have considered the following model variables along with the different developer types defined in the previous section for our study.

1. *Contributors(CC):* How quickly a problem is resolved is largely determined by the overall number of developers participating in the bug-resolution process. [14]. One of the most important ways to assist in fixing a bug in the context of our study is to comment on it. Therefore, this amount is derived from the total number of unique developers who have remarked on a bug.
2. *CommunityInterest(CI):* While each bug has its owner, challenging bugs need greater involvement from the development community to be resolved as soon as possible. The extent to which the development community goes beyond the bug's immediate owner has been found to be a crucial component in bug resolution. [15]. To quantify this variable, we measure the number of comments from developers who are not the owner of a particular bug.

3. *OwnerEngagement(OE):* The result of each developer's specific assignment depends on how involved they are in the development ecosystem as a whole [16–19]. The variable being examined is calculated by adding up the owner's comments for every other problem in the development ecosystem. This is meant to represent the level of involvement a bug owner has outside of their official responsibilities.
4. *OwnerWorkload(OW):* "Owner workload" refers to the responsibilities and tasks assigned to the person responsible for monitoring and resolving issues in software development. Identifying and properly documenting issues found during testing or from user feedback is another aspect of this effort.
5. *Priority(Pr):* Software development in particular and collaborative enterprises in general are known to be impacted by the task's relevance [17,19]. Each problem in our Eclipse dataset has a priority assigned to it. The seven priority levels in Eclipse range from "trivial" to "blocker". We assigned a numerical value to each priority level, 1 being the lowest and each subsequent level indicating a higher level of importance assigned to a certain bug.
6. *ElapsedTime(ET):* The resolution time of a specific problem in our study setup is determined by calculating the difference in time between the bug's initial and last recorded activity [20]. Our datasets are filtered to contain only bugs that have been released and resolved, therefore, the time difference is taken to be the interval between the opening and closing of a bug.

3.5 Exploratory Data Analysis

To begin our research, we have first analyzed the data using a basic statistical technique to determine the distribution of the data. Table 1 gives a summary of all the 10 variables in the dataset.

Table 1. Summary statistics for the Eclipse dataset

	ET	CC	Pr	CI	OE	OW	NI	SI	RI	FI
count	13248.00	13248.00	13248.00	13248.00	13248.00	13248.00	13248.00	13248.00	13248.00	13248.00
mean	18426.11	6.20	3.61	16.78	689.41	2723.44	3.21	6.07	14.73	8.79
std	21559.51	5.78	1.04	11.89	1372.97	4185.36	57.87	47.55	138.79	93.36
min	0.26	3.00	1.00	2.00	0.00	1.00	0.00	0.00	0.00	0.00
25%	2167.03	4.00	3.00	10.00	0.00	270.00	0.00	0.00	3.00	0.00
50%	9359.10	5.00	3.00	14.00	109.00	955.00	0.00	0.00	6.00	0.00
75%	27559.47	7.00	4.00	20.00	732.00	2882.00	0.00	3.00	11.00	4.00
max	11801.06	162.00	7.00	241.00	9379.00	15529.00	5050.00	2198.00	13041.00	6161.00

The mean values of the variables vary significantly, with the smallest mean being 3.21 for NI and the largest mean being 18426.11 for ET. The standard deviation (std) values also show considerable variation among the variables,

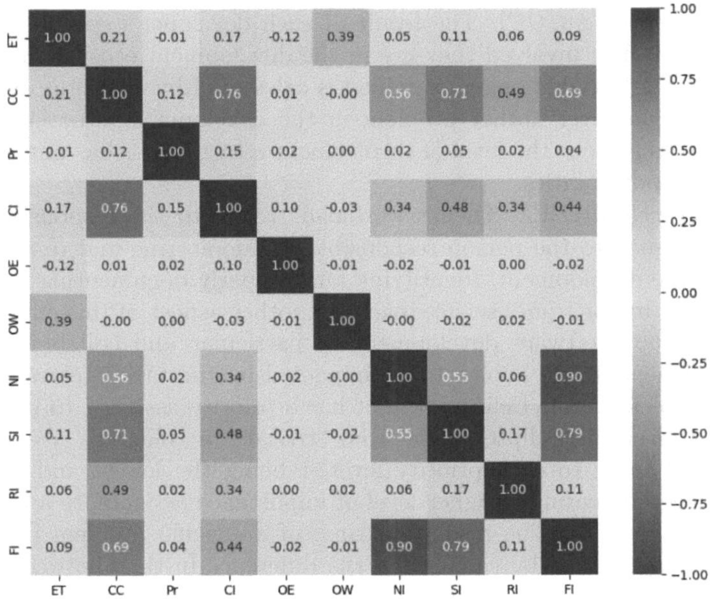

Fig. 1. Correlation plot for the Eclipse dataset

indicating differing levels of dispersion around the mean for each variable. The smallest standard deviation is 1.04 for Pr, while the largest is 4185.36 for OW. The 25th percentile (25%), 50th percentile (median or 50%), and 75th percentile (75%) values for each variable also show a wide range of values. According to the research, the variable distributions are probably going to differ and might be skew to varying degrees. Any relationships that may exist can be shown with the help of a diagram. The Correlation matrix Fig. 1 displays the correlations, such as those between CI and CC, ET and NI, CC and OW, etc., that require more analysis to determine their causal linkages.

4 Results and Discussion

4.1 Running GES on the Eclipse Dataset

GES was run on the Eclipse dataset, resulting in the causal network shown in Fig. 2. We used Python's *DoWhy*[3] and *causal-learn*[4] libraries to develop and execute our causal discovery model. A directed graph without any closed loops is called a directed acyclic graph (DAG), and a causal network is represented as a DAG [31]. Given two vertices X and Y, a causal network has the following three components, with specific connotations: (1) There is an arc (directed link) from X to Y if and only if X causes Y; (2) There is an edge (undirected link)

[3] https://www.pywhy.org/dowhy/.

[4] https://github.com/py-why/causal-learn.

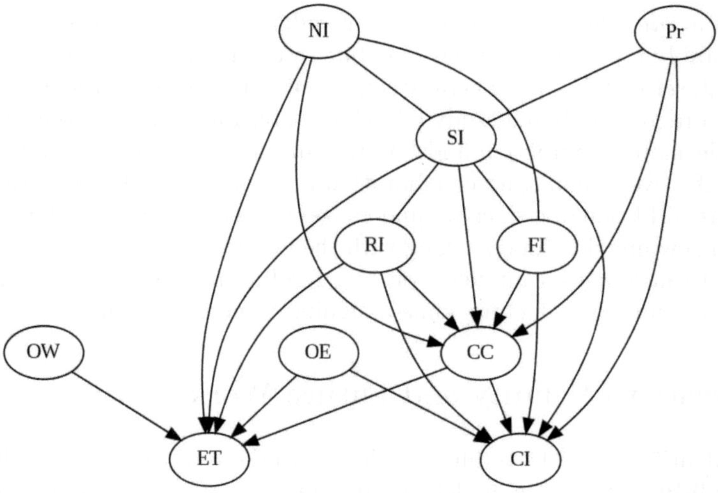

Fig. 2. Causal structure derived from the Eclipse dataset by the GES algorithm

from X to Y if and only if X causes Y or Y causes X; and (3) There is no edge from X to Y if and only if X and Y do not directly cause one another [9].

With reference to this notation explained above, we interpret Fig. 2. From the arrows pointing to the vertex representing bug resolution time (ET), we observe that it is affected by repeat interactions (*RI*), along with new interactions (*NI*) and seasoned interactions (*SI*). We also find evidence that bug owning developers' workload (OW), bug owning developers' experience (OE) and the number of contributors in the bug resolution process (CC), also affect bug resolution times. Additionally, it is observed that fresh interactions (*FI*) and priority (Pr) does not have any causal impact on bug resolution time (ET).

It is further manifested in the causal network (Fig. 2) that the number of contributors is affected by the numbers of repeat, new, fresh, and seasoned interactions, as well as priority (CC has arrows pointing from NI, *RI*, SI, FI, and Pr). Additionally, we see that community interest is affected by owner experience, repeat interactions, number of contributors, fresh interactions, seasoned interactions, and priority (CI has arrows pointing from OE, *RI*, CC, SI, and Pr).

However, the direction of causation cannot be determined between new interaction and seasoned interaction (*NI* and *SI* have an edge, that is an undirected link between them), repeat and seasoned interaction (edge between *RI* and *SI*), seasoned and fresh interaction (edge between *SI* and *FI*), and seasoned interaction and priority (edge between *SI* and *Pr*).

4.2 Implications and Utility

Let us address the implications and utility of the above results in light of our research question examining the causal influences on bug resolution times. As presented above, the causal influences of different types of developer interactions – notably repeat interactions – on bug resolution times as derived from the Eclipse dataset by GES are generally congruent with the associations established earlier in a regression based study [6]. The establishment of a causal connection between developer interaction and bug resolution times underscores the importance of effectively managing collaboration in large scale software development ecosystems. Additionally, since we find causal influence of workload, and experience of bug owning developers, along with the number of developers commenting on a bug to have causal influence on bug resolution times, it is imperative to take into account these factors while assigning ownership of bugs.

5 Threats to Validity and Future Work

We now identify and address different threats to the validity of our results. **Construct validity**: All of the model variables in this study are precisely defined in Sect. 3, derived from data from reliable sources, and their methods of collection and curation have undergone peer review [28]. Thus threats to construct validity are not notably present in our study. **Internal validity**: The performance of GES can be negatively impacted by noise and outliers in the data, which can result in inaccurate causal conclusions [10]. The Gaussian distribution of the data and the absence of any hidden confounders are the two underlying assumptions of GES. Consequently, the algorithm's usefulness may be limited in situations where these assumptions are not true in real life. In cases where faithfulness is compromised, GES may yield less-than-ideal outcomes. Comparative analysis of different causal discovery algorithms can be found in the Appendix (available online[5] and shared as a supplementary material) which helps address the threats to internal validity. **External validity**: This is concerned with the degree to which the outcomes can be generalized. Since we have considered only one dataset, we do not claim our results to be generalizable as yet, even as we are able to derive interesting insights from them. **Reliability**: Given access to the data, our results are reproducible and hence there are no threats to reliability. While GES is able to discover causal relationships between variables from data, it does not specify the directionality of the discovered causal influence, that is whether the increase in magnitude of a variable causes another variable to increase, or vice-versa. In our **future work**, we plan to investigate the directionality of causal influences on bug resolution times discovered in this study using techniques that can augment GES.

6 Related Work

We briefly outline some of the related work in this area. Huang et al. propose an approach to causal discovery that is resilient to different kinds of data and

[5] https://bit.ly/AGC2025-GES-Appendix.

distributions [7]. Without assuming any particular functional forms or distribution parameters, the authors presented generalized scoring functions that efficiently determine causal links. Chickering et al. presents new findings regarding the intricacy of techniques for learning discrete-variable Bayesian networks from data [22]. They establish that even if the learning algorithm applies a uniform scoring criterion throughout a huge sample, the task would still be computationally difficult. Hasan and Gani propose KGS: a novel method for causal discovery that combines the Greedy Equivalency Search (GES) with previous information. KGS guides the search for a causal graph with the use of structural priors, which are constraints in the form of established causal edges, in addition to observational data [8]. Ramsey et al. suggest improvements to GES through parallelization and caching, resulting in an enhanced approach, called Fast Greedy Equivalence Search (fGES) [21]. This was tested on a functional magnetic resonance imaging (fMRI) dataset of one million Gaussian variables, to establish its effectiveness in high-dimensional scenarios. Lu et al. highlight the drawbacks of GES, which frequently yield less-than-ideal results when dealing with finite data or when the causal faithfulness requirement is broken [23]. The authors suggest an ideal score-based exhaustive search strategy that works better. Siebert had made a thorough analysis of the applications of statistical causal inference techniques in software engineering; 32 papers published between 2010 and 2022 demonstrate the growing use of these techniques in the field [25]. Furia et al. investigate causal correlations, specifically the impact of programming languages on the results of coding competitions [26]. Peralta et al. examine factors that influence bug fixing time using causal inference models [27]. Results from this study complement the existing literature in this area.

7 Conclusions

In this paper, we have applied the Greedy Equivalence Search (GES) algorithm to data from the Eclipse software development ecosystem to derive the causal influences on bug resolution times. We found evidence that repeated interaction, as well as new interactions and seasoned interactions between developers, along with developer workload, experience of developers owning bugs, and the number of developers participating in the bug resolution process, are causal factors influencing bug resolution times. Identifying relations that are causal – vis-a-vis those merely correlational – in software development ecosystems can inform vital decisions at individual and organizational levels. Given the nature of the GES algorithm, these causal relations are inferred directly from the data, without any predefined model. This ensures that our results are free from biases that assumptions underlying a predefined model may introduce. Our results can inform team assembly and governance in complex software development scenarios.

References

1. Pearl, J.: The seven tools of causal inference, with reflections on machine learning. Commun. ACM **62**(3), 54–60 (2019)
2. Spirtes, P., Glymour, C., and Scheines, R.: Constructing Bayesian networks models of gene expression networks from microarray data. In: Proceedings of the Atlantic Symposium on Computational Biology, North Carolina(2000)
3. Spirtes, P., Glymour, C., Scheines, R.: Causation, Prediction, and Search. MIT press (2001)
4. Chickering, D.M.: Optimal structure identification with greedy search. J. Mach. Learn. Rese. 3(Nov), 507–554 (2002)
5. Schwarz, G.: Estimating the dimension of a model. Ann. Stat., 461–464 (1978)
6. Datta, S., Roychoudhuri, R., Majumder, S.: Understanding the relation between repeat developer interactions and bug resolution times in large open source ecosystems: a multisystem study. J. Softw.: Evol. Process **33**(4), e2317 (2020)
7. Huang, B., Zhang, K., Lin, Y., Schölkopf, B., Glymour, C.: Generalized score functions for causal discovery. In: Proceedings of the 24th ACM SIGKDD International Conference on Knowledge Discovery and Data Mining, pp. 1551–1560 (2018)
8. Hasan, U. and Gani, M.O.: KGS: causal discovery using knowledge-guided greedy equivalence search. arXiv preprint: arXiv:2304.05493 (2023)
9. Neapolitan, R.E.: Learning Bayesian Networks, vol. 38. Upper Saddle River: Pearson Prentice Hall (2004)
10. Wang, Y., Squires, C., Belyaeva, A., Uhler, C.: Direct estimation of differences in causal graphs. In: Advances in Neural Information Processing Systems, vol. 31 (2018)
11. Zhalama, Zhang, J., Mayer, W.: Weakening faithfulness: some heuristic causal discovery algorithms. Int. J. Data Sci. Analytics **3**, 93–104 (2017)
12. Gonzalez-Barahona, J.M., Robles, G., Izquierdo-Cortazar, D.: The MetricsGrimoire database collection. In 2015 IEEE/ACM 12th Working Conference on Mining Software Repositories, pp. 478–481 (2015)
13. Afrianto, N., et al.: Applying PC algorithm and GES to three clinical data sets: heart disease, diabetes, and hepatitis. In: IOP Conference Series: Materials Science and Engineering, vol. 1077, no. 1, p. 012067. IOP Publishing, February (2021)
14. Datta, S., Sarkar, P., Das, S., Sreshtha, S., Lade, P., Majumder, S.: How many eyeballs does a bug need? An empirical validation of Linus' Law. In: Agile Processes in Software Engineering and Extreme Programming: 15th International Conference, XP 2014, Rome, Italy, May 26-30, Proceedings 15, pp. 242–250. Springer International Publishing (2014)
15. Datta, S., Sindhgatta, R., Sengupta, B.: Talk versus work: characteristics of developer collaboration on the jazz platform. In: Proceedings of the ACM International Conference on Object Oriented Programming Systems Languages and Applications, pp. 655–668 (2012)
16. Fong Boh, W., Slaughter, S.A., Espinosa, J.A.: Learning from experience in software development- a multilevel analysis. Manage. Sci. **53**(8), 1315–1331 (2007)
17. Wagstrom, P., Herbsleb, J.D., Carley, K.M., :Communication, team performance, and the individual- bridging technical dependencies. In: Academy of Management Proceedings, vol. 2010, no. 1, pp. 1–7. Academy of Management, Briarcliff Manor (2010)
18. Cataldo, M., Wagstrom, P.A., Herbsleb, J.D., Carley, K.M.: Identification of coordination requirements: Implications for the design of collaboration and awareness

tools. In: Proceedings of the 2006 20th Anniversary Conference on Computer Supported Cooperative Work, pp. 353–362 (2006)
19. Cataldo, M., Herbsleb, J.D., Carley, K.M.: Socio-technical congruence: a framework for assessing the impact of technical and work dependencies on software development productivity. In: Proceedings of the Second ACM-IEEE International Symposium on Empirical Software Engineering and Measurement, pp. 2–11 (2008)
20. Datta, S. and Lade, P.: Will this be quick? A case study of bug resolution times across industrial projects. In: Proceedings of the 8th India Software Engineering Conference, pp. 20–29 (2015)
21. Ramsey, J., Glymour, M., Sanchez-Romero, R., Glymour, C.: A million variables and more: the fast greedy equivalence search algorithm for learning high-dimensional graphical causal models, with an application to functional magnetic resonance images. Int. J. Data Sci. Analytics, 121–129 (2017)
22. Chickering, M., Heckerman, D., Meek, C.: Large-sample learning of Bayesian networks is NP-hard. J. Mach. Learn. Res. **5**, 1287–1330 (2004)
23. Lu, N.Y., Zhang, K., Yuan, C.: Improving causal discovery by optimal Bayesian network learning. In: Proceedings of the AAAI Conference on Artificial Intelligence, vol. 35, no. 10, pp. 8741–8748 (2021)
24. Kitson, N.K., Constantinou, A.C., Guo, Z., Liu, Y., Chobtham, K.: A survey of Bayesian network structure learning. Artif. Intell. Rev. **56**(8), 8721–8814 (2023)
25. Siebert, J.: Applications of statistical causal inference in software engineering. Inf. Softw. Technol. **159**, 107198 (2023)
26. Furia, C.A., Torkar, R., Feldt, R.: Towards causal analysis of empirical software engineering data: The impact of programming languages on coding competitions. ACM Trans. Softw. Eng. Methodol. **33**(1), 1–35 (2023)
27. Peralta, S.R.O., Washizaki, H., Fukazawa, Y., Noyori, Y., Nojiri, S., Kanuka, H.: Unraveling the influences on bug fixing time: a comparative analysis of causal inference model. In: Proceedings of the 28th International Conference on Evaluation and Assessment in Software Engineering, pp. 393–398 (2024)
28. Eclipse Testing tool. https://www.eclipse.org/. Accessed 26 Aug 2024
29. Acharya, J., Bhattacharyya, A., Daskalakis, C., Kandasamy, S.: Learning and testing causal models with interventions. In: Advances in Neural Information Processing Systems, vol. 31
30. Naser, M.Z.: Discovering causal models for structural, construction and defense-related engineering phenomena. Defence Technol. (2024)
31. Glymour, C., Zhang, K., Spirtes, P.: Review of causal discovery methods based on graphical models. Front. Genet. **10**, 524 (2019)

Vulgar Language Detection Using Web-Scraped X Data: A Machine Learning Approach

Diya Patra(✉), Sk. Shihab, and Nisha Banerjee

NSHM Knowledge Campus, Kolkata, India
diyapatra2nd@gmail.com

Abstract. The exponential increase in the use of social media jargons and trend of fast paced conversations has led to an increase in the practice of slangs and vulgar language among netizens, for various different contexts. These use of vulgar words and over sexualization of expression in conversation causes a loss of depth in them, which also negatively affect our cognitive conditioning and Linguistic Intelligence, and majorly affect children as they lack the Emotional, Cognitive and Integral Intelligence, which is basically required to differentiate between right and wrong. Machine learning can help us detect vulgar language in texts on digital platforms, for further understanding and study with the intention of creating a healthier cyber world in the realm of digital communication and expression. This research work focuses on detecting vulgar language in texts, on social media. The study has evaluated four different machine learning classifiers by training them on the same web scraped X (formerly Twitter) dataset using supervised machine learning approach. The evaluation of the proposed approach on X dataset shows that Logistic Regression and Naïve Bayes achieves 94% accuracy, whereas Decision Classifier and Random Forest achieves 96% accuracy, on the same test dataset.

Keywords: Vulgar Language Detection · Web Scraping · Sentiment Analysis · Natural Language Processing · Machine Learning · Text Classification · Social Media Texts

1 Introduction

The usage of social media is increasing rapidly over the past few years, and majorly after 2020 the screen-time of people on an average has increased subsequently. As of July 2024, there were 5.17 billion social media users globally [1]. Social media provides us a platform where people can express themselves in a much casual manner, but it also brings its own share of cons, the algorithm of these platforms are designed in such a way that it can often cause behavioural addiction [2], as it triggers a dopamine release each time an interesting post or text is encountered, which in turn creates an urge for further reinforcements [3]. The average time spent on social media by a single user is around 2 and a half hours [4]. The vast availability of high-speed internet has also majorly contributed to its growth. As social media has become an integral part of our lives, it has

S. Dhar et al. (Eds.): AGC 2025, CCIS 2686, pp. 184–200, 2026.
https://doi.org/10.1007/978-3-032-09153-6_12

conditioned this generation to adopt a more casual form of expression over the internet [2], which often include the practice of various jargons, slangs and vulgar language. This paper considers vulgar language as any form of expression that includes the use of informal words that can be regarded as disrespectful, offensive, inappropriate or sexually abusive in nature. Vulgar language is often used to express negative sentiments or intensify the various sentiments present in texts (in the context of social media platforms) [5], it is also used to emphasize a point or build rapport in certain cases [6]. The usage of swear words can be further categorized into five different functions i.e., Abusive (offensive or intention to psychological harm), Cathartic (emotion released in response to pain or frustration), Dysphemistic (expression of negative emotion), Emphatic (intention to emphasize or draw attention) and Idiomatic (intended informally with little to no meaning) [7]. Neuroscientific research has found that expressing strong emotions through swearing or using vulgar language might be located in different parts of the brain compared to other speech regions. This might activate parts of 'Limbic systems' (which includes features– basal ganglia and amygdala). These complex structures are responsible for instinctive processes like memory and emotion processing [8]. In simple terms, it gradually desensitizes our basic emotions which are required in order to function as a healthy member of society and can negatively affect one's personal, cultural and professional lives in various circumstances. The purpose of this study is to detect vulgar language in social media texts using machine learning classifiers, with the intention of taking it into account, monitoring and reducing the same, creating a healthier and more productive cyber space for the netizens. However, typical deep learning techniques were not used as the primary goal of the analysis deals with short sentences of up to 280 characters (as per allowed by X platform). The text classification approaches rely on two types of techniques, i.e., a) Lexicon based and b) Machine Learning based, and three extraction levels, which are, a) Feature or Aspect level, b) Document level and c) Sentence level [9]. Our paper has adopted the supervised machine learning approach with document level extraction method to train our classification models for detecting vulgar words using two labels: a) True (1), b) False (0). The training dataset is generated using posts web scraped from X (formerly Twitter) as it is a microblogging platform which creates an easy access to vast amounts of diverse text data. The four selected classifiers are trained using the same training dataset to avoid domain bias and determine the best machine learning model to detect vulgar language in text. A test dataset is generated, and the models are tested using the same. The observations are noted for analysis and potential scopes of future study.

2 Literary Review

During the literature review stage of this paper, a lack of sufficient existing literatures on the study of vulgar language detection using machine learning classifiers and predictive algorithms were identified. However, several existing research articles have explored text classification, pattern detection and sentiment analysis, with various papers also addressing abusive or offensive language detection and a few on vulgar language detection, using various machine learning algorithms. This review primarily examines researches employing classifiers and prediction-based algorithms, as well as select works utilizing lexicon-based algorithms, neural networks and transformers, to gain a comprehensive understanding of the field. The reviewed papers are noted as follows:

Eshan et al. [10] focuses on detecting abusive text specially in Bengali language content using various machine learning classifiers including Multinomial Naïve Bayes, Support Vector Machine (Linear, RBF, Polynomial, Sigmoid) and Random Forest. The dataset is generated using Facebook comments. The highest accuracies were achieved by Multinomial Naïve Bayes and Support Vector Machines (SVM) with a linear kernel especially when using CountVectorizer and TF-IDF vectorizations. Kwok et al. [11] focuses on detecting anti-black racist tweets using a Binary Naïve Bayes classifier. The dataset is generated using tweets collected from X (formerly Twitter). The classification model has achieved an accuracy of 76%. Sharma et al. [12] focuses on detecting harmful speech on Twitter, classifying them into different degrees of hateful intent. For the task they have used Naïve Bayes Classifier and Support Vector Machine and Logistic Regression Classifier, with Support Vector Machine achieving the best accuracy of 87%. Waseem et al. [13] focuses on the classification of different types of hate speech on X (formerly TWITTER), they have also discussed the impact of extra-linguistic features and character n-grams in detecting hateful language. The dataset is generated using publicly available corpus consisting of more than 16,000 tweets. The corpus is fed into the Logistic Regression Classifier and achieved an accuracy of 73%. Islam et al. [14] focuses on detecting abusive language in Bengali texts using various machine learning classifiers including Naive Bayes, Support Vector Machine, Logistic Regression, Random Forest, and Decision Tree. The dataset is generated using texts collected from Facebook and YouTube. SVM outperformed other classifiers with an accuracy of 88%. Chakraborty et al. [15] focuses on detecting threats and abusive language on social media by employing Multinomial Naïve Bayes, Support Vector Machine, and a Convolutional Neural Network combined with Long Short-Term Memory (LSTM) for classification. The dataset is generated primarily by collecting the text from interactions and comments on Facebook. SVM outperformed other models with an accuracy of 78%. Nobata et al. [16] focuses on detecting language in online content using a combination of supervised machine learning models including Logistic Regression and Support Vector Machine. The dataset is generated using data collected primarily from Yahoo Finance and News Forums. The best accuracy was achieved by Logistic Regression by combining it with syntactic and semantic features, achieving an accuracy of 77%. Malmasi et al. [17] focuses on detecting hate speech on social media using linear Support Vector Machine classifier with several lexical features including character n-grams (from bigrams to 8-g) and word n-grams. The dataset is generated using a collection of English-language social media posts. The highest accuracy of 78% was achieved by the character 4-g model. Köffer et al. [18] focuses on automatic detection of hate speech in user comments by employing a Word2Vec model with Extended 2-g. The dataset is generated using user comments on news articles discussing the Refugee crisis in Germany during 2015 to 2016. The model achieved an accuracy of 70%. Abro et al. [19] focuses on automatic detection of hate speech in social media content (primarily X/Twitter) using various machine learning classifiers including Naïve Bayes, Support Vector Machine, Logistic Regression, Decision Tree, and Random Forest. They used a publicly available dataset containing a total of 14,509 tweets. The Support Vector Machine outperformed other classifiers achieving an accuracy of 89%. Essa et al. [20] focuses on detecting hateful comments on social media employing Naïve Bayes Classifier, and also utilizing a neural algorithm to improve performance. The dataset for this study was sourced from Kaggle, comprising 30,000 tweets. The Naïve Bayes classifier achieved an accuracy of 62.75% which was deemed unsatisfactory, however the neural algorithm significantly improved the performance upto 87%. Tarmizi et al. [21] focuses on the study of usage of vulgar words in cyberbullying activities employing a lexicon-based machine learning model using machine learning. The dataset was generated by collecting X (formerly Twitter) posts of 50 different users, targeting tweets containing vulgar language. The model achieved an accuracy of approximately 69%. Singh et al. [22] focuses on analysing social media texts by classifying them into positive and negative emotions, using Naïve Bayes, J48, BFTree & OneR. The dataset is generated by collecting reviews from Amazon e-commerce site and IMDb (Internet Movie Database). OneR outperformed other models with an accuracy of 91.3%. Jemai et al. [23] focuses on analysing sentiment using Naïve Bayes, Multinomial Naïve Bayes, Bernoulli Naïve Bayes, Logistic Regression, Support Vector Machine. The dataset is collected from NLTK library. All the models performed almost similarly with Naïve Bayes slightly outperforming other classifiers with an accuracy of 99.73%. Bhatt et al. [24] focuses on analysing sentiment in customer reviews by classifying into positive, negative and neutral, by employing Support Vector Machine, Naïve Bayes and Random Forest. It has also highlighted the utilization of various feature extraction methods like TF-IDF and Unigrams with SentiWordNet for enhancing the task. The corpus for generating the dataset is sourced from multiple social media platforms like Twitter, Facebook, etc. Random Forest combined with Unigram and SentiWordNet including negation handling provide the highest accuracy of 95.6%. Jiang et al. [25] focuses on analyzing sentiment on Yelp reviews data by classifying them into positive and negative, using Logistic Regression, Random Forest, K-Nearest Neighbours and Naïve Bayes. The Kaggle Yelp Challenge dataset was used to train the models, as it contains extensive reviews and business information across multiple metropolitan areas. The Logistic Regression with TF-IDF features achieved the highest accuracy of 84%. Raza et al. [26] focuses on detection of offensive terms in resource-poor language utilizing Decision Tree, Logistic Regression, Support Vector machine, Naïve Bayes and Random Forest. Two cross-platform datasets were gathered for the task, one from Twitter platform (collected from a previous work by Amjad et al. [27]) and the other from YouTube, and the models were alternatively trained and tested on both the datasets. Decision Tree achieved the highest accuracy of 83.72%. Razzak et al. [28] focuses on classification of vulgar comments utilizing Convolutional Neural Network, XGBoost and Support Vector Machine. The dataset is collected from a previous work of Beniwal et al.[29]. The CNN outperformed XGBoost and SVM with 98% precision score. Ashraf et al. [30] focuses on detecting abusive language in YouTube comments and replies using machine learning algorithms. The dataset is generated by collecting comments and replies from 29 YouTube videos, based on politics, religion and other such genres. For the task, they employed Logistic Regression, Multilayer Perception, Adaboost, Random Forest, Support Vector Machine, Naïve Bayes, Decision Tree, and VotingClassifier for abusive language detection(task 1) and topic classification (task 2). For the dataset containing only comments, Decision Tree achieved the highest accuracy of 87.38%, and for the dataset containing both comments & replies, Adaboost outperformed other models with an accuracy of 92.32%. They also used two Neural Network models viz. 1-Dimensional Convolutional Neural Network (1D-CNN) and Long Short-Term Memory Networks with additional max-pooling and attention layers for abusive language detection (task 1) and topic classification (task 2). In this approach, for the dataset containing only comments, 1D-CNN with max pooling, achieved the highest accuracy of 90.28%, and for the dataset containing both comments & replies, the 1D-CNN with an attention layer feature, achieved the highest accuracy of 91.76%. Salminen et al. [31] focuses on detecting hateful comments across multiple social media platforms employing various machine learning algorithms including Logistic Regression, Naïve Bayes, Support Vector Machines, XGBoost, and Neural Networks. The dataset is generated using 197,566 comments collected from YouTube, Reddit, Wikipedia, and Twitter. The models were trained using a combination of feature representations including BERT features, TF-IDF, Word2Vec, and Bag-of-Words. The BERT features were found most impactful for predictions with XGBoost outperforming other models with an accuracy of 92%. Emon et al. [32] focuses on detecting abusive Bengali text by employing Linear SVC with CountVectorizer, Artificial Neural Network, Recurrent Neural Network and Linear SVC with TF-IDF Vectorizer. The dataset was generated using public comments collected from YouTube, ProthomAlo and different Facebook pages. The Recurrent Neural Network outperformed other models with an accuracy of 82.20%. Aurpa et al. [33] focuses on abusive Bengali comments detection by employing Bidirectional Encoder Representations from Transformers (BERT) and Efficiently Learning an Encoder that Classifies Token Replacements Accurately (ELECTRA), which are transformer-based pre-trained language models. The models are fine-tuned using a Bengali abusive content dataset consisting of 44,001 comments collected from various Facebook posts. Both models performed almost similarly with an accuracy of 85% for BERT and 84.92% for ELECTRA, which shows that BERT was slightly more effective for this task.

3 Methodology

In this section, we will outline the experimental procedures employed in this analysis, detailing the methods and approaches applied, and the machine learning classifiers implemented. A two-phase approach has been adopted for the model development and evaluation: The Training Phase and the Testing Phase. This systematic approach is designed to ensure both accuracy and optimal performance of the models. For a better understanding of the model training and evaluation process, a visual representation of the architecture of the proposed methodology has been outlined in Fig. 1.

3.1 Data Collection

The corpus for generating the training dataset is collected through web scraping X (Formerly Twitter) posts, of two different users. X is a mini-blogging site which encourages spontaneous expression and frequent use of informal language, providing a constant stream of user generated content with minimal filtering, making it an ideal medium for capturing everyday language and trends in profanity [34]. The tweets are webscraped utilizing a scraper from Python library—'Ntscraper'.

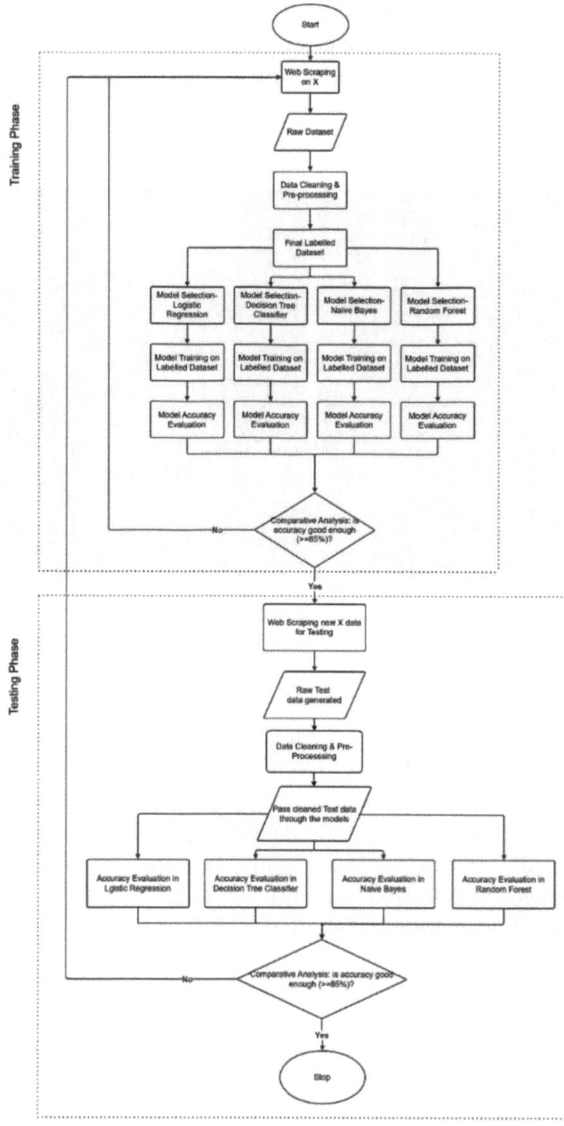

Fig. 1. Architecture of the proposed Methodology

- The corpus is collected through web scraping posts of the following X handles: i) @uniuser_01 & ii) @oneuser_02
- A total of 208 tweets were fetched on the first attempt and a preliminary dataset was generated. But a major class imbalance was encountered.
- On the second attempt, a total of 425 tweets have been fetched, still the accuracies were too low, with huge amounts of false positives.
- On the third attempt, a total of 529 tweets were fetched, which reduced the class imbalance, but still the models were giving a moderate amount of false positives.

- On the fourth attempt, a total of 790 tweets were fetched, which further reduced the class imbalance with very few false positives to be taken into account.
- On the last attempt, a total of 844 tweets were fetched and the final dataset is generated for the observation and analysis (Fig. 2).

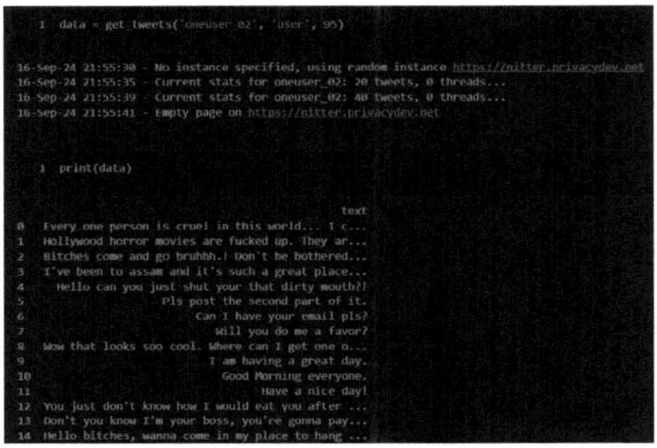

Fig. 2. Web scraping tweets using Ntscraper

3.2 Data Annotation

As the corpus has been already collected, which contains a total of 844 tweets or texts, the texts must be manually classified into different classes for labelling, and a structure must be employed for the dataset with all the necessary columns, to prepare it for training the algorithm. There are three columns in our dataset: 'Handle' (containing the usernames for the tweets fetched), 'Tweets' (containing the text data) and 'Vulgar' (containing the labels for vulgar language detection training).The Vulgar labels have been divided into two different classes: i) Non-Vulgar (denoted by 0) & ii) Vulgar (denoted by 1) (Fig. 3).

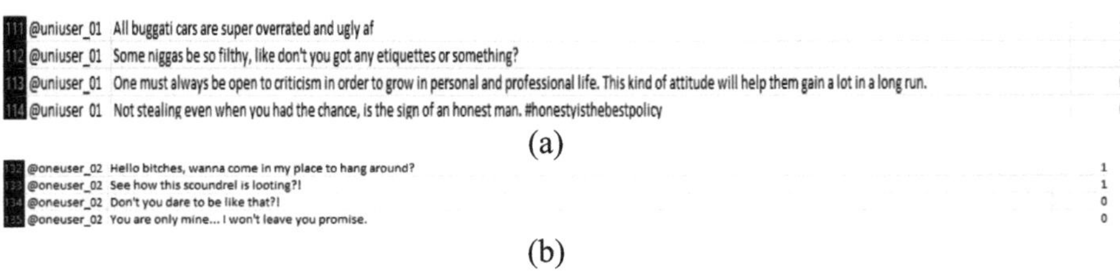

Fig. 3. Structure of the Dataset (a) Tweets from @uniuser_01 (b) Tweets from @oneuser_02

3.3 Data Pre-Processing

As the data collection and annotation has been done, data pre-processing is the next crucial part. Data pre-processing is done to remove noise from the data and prepare it into a format that can be fed to the algorithm. There are several SNS comments, special characters and emoji or emoticons that are used by people in posts. So, we must remove those unnecessary things first to reduce computational complexities. We have performed the following steps for the pre-processing of our data:

3.3.1 Removing Punctuations

Punctuation are those characters which aren't essential to the text analysis process and must be removed. The punctuations targeted for exclusion are:'"!#()* +,-./:; < = > ?@[]'|, etc. According to the research of Mahmud et al. [35] removing punctuations can improve the text classification by traditional ML models. Following this approach, we have also removed the punctuations to enhance the accuracy of our text classification as traditional ML models have been employed for this analysis.

3.3.2 Stemming

Stemming is the process of reducing the dimensionality of a word by consolidating its different forms into the basic root word, which increases the likelihood of words with similar meanings to be treated as the same word, enhancing the model's ability to detect patterns. For example, the word "running", "runner", "runs" will be "run" after applying stemming. The "Snowball Stemmer", an improved version of Porter Stemmer, is utilized for stemming our dataset for its efficiency and enhanced abilities of handling linguistic nuances (Fig. 4).

	Handle	Tweets	Vulgar	stemmed_tweets
0	@oneuser_02	I haven't met any teacher like her before. She...	0	i haven t met ani teacher like her befor she s...
1	@oneuser_02	She's the most beautiful girl I've ever seen...	0	she s the most beauti girl i ve ever seen
2	@oneuser_02	I wish you starve to death like thaah...	0	i wish you starv to death like thaah
3	@oneuser_02	Women are retarted. I don't think they are cap...	1	women are retart i don t think they are capabl...
4	@oneuser_02	Such a queer...	1	such a queer
5	@oneuser_02	I think he's a fag. Otherwise why he would do ...	1	i think he s a fag otherwis whi he would do th...
6	@oneuser_02	That tramp is just made for my dick...	1	that tramp is just made for my dick
7	@oneuser_02	Those crackers are looking soo good with their...	1	those cracker are look soo good with their tor...

Fig. 4. Final Dataset after stemming

3.3.3 Feature Engineering

Feature Engineering or Vectorization can be defined as the process of converting text data to its numerical representation, a format which is interpretable for the machine learning algorithms. Traditionally, TF-IDF Vectorizer, Count-Vectorizer (CV) and N-gram are used for this task. Algorithms like Naive Bayes and SVM often perform well on sparse matrices, making CV a suitable choice for tasks involving these models, as it produces high sparsity, enhancing model compatibility. Thus, Count-Vectorizer is utilized for its efficiency over small datasets and compatibility with sparse data.

3.4 Model Training

The analysis is done by employing four selected ML classifiers: i) Logistic Regression, ii) Decision Tree Classifier, iii) Naïve Bayes, and iv) Random Forest.

3.4.1 Logistic Regression (LG)

Logistic Regression has been selected for this analysis as it requires a small amount of data to train and its efficiency over prediction on small texts. The implementation of this model is rather simple than other ML models, which makes it an ideal choice for data analysis tasks.

The underlying mathematical model of Logistics Regression calculates the predictions using the logistic or sigmoid function. The following equation explains how this function models the relationship between input features and the binary outcome, for calculating the probability:

$$P(y = 1|X) = \frac{1}{1 + \varrho - (\beta 1X1 + \beta 2X2 + ... + \beta nXn)1} \tag{1}$$

where,

1. $P(y = 1 \mid X)$ represents the probability of the outcome being 1, given the input features X
2. β_0 represents the intercept term
3. β_i represents the coefficients for each feature X_i

3.4.2 Decision Tree Classifier (DT)

The most obvious reason for selecting Decision Tree Classifier is its ability to perform analysing tasks with raw data or mildly cleaned data, thus it can compensate for any errors or oversights that went unnoticed during the data pre-processing stage. It is also a fast, flexible, and easy to interpret algorithm.

The underlying mathematical model of Decision Tree Classifier evaluates and splits data based on feature values to increase class purity. The following equation explains how it is achieved by selecting splits, such as Gini Impurity and Entropy, which minimizes impurity measures:

Gini Impurity for a node:

$$G = 1 - \sum_{i=1}^{c} p_i^2 \tag{2}$$

Entropy for a node:

$$H = -\sum_{i=1}^{c} p_i log_2(p_i) \tag{3}$$

where,

1. p_i represents the probability of a sample belonging to class i within the node
2. C represents the total number of classes.

3.4.3 Naïve Bayes (NB)

The reason for selecting this classifier is its scalability and efficiency in handling discrete values. The model classifies the data points based on the likelihood of features in each class. Moreover, it is fast and highly accurate for classifying texts and has low computational costs.

The underlying mathematical model of Naive Bayes Classifier is based on Bayes' theorem, assuming that features are conditionally independent given the class label. The following equation explains how the model calculates the probability of a class y given the input features X:

$$P(y|X) = \frac{P(y)\Pi_{i=1}^{n} P(X_i|y)}{p(X)} \tag{4}$$

Since $P(X)$ remains constant across classes, the classifier simplifies to:

$$arg\ max_y P(y)\prod_{i=1}^{n} P(X_i|y) \tag{5}$$

where,

1. $p(y)$ represents the prior probability of class y
2. $P(X_i|y)$ represents the probability of feature X_i given class y
3. n is the total number of features

For continuous data, Gaussian Naive Bayes is often used, assuming each feature follows a Gaussian (normal) distribution. In this case:

$$P(X_i|y) = \frac{1}{\sqrt{2\pi\sigma_y^2}} exp(-\frac{(X_i - \mu_y)^2}{2\sigma_y^2} \tag{6}$$

where,

1. μ_y represents the mean of feature X_i for class y
2. σ_y represents the standard deviation of feature X_i for class y

3.4.4 Random Forest (RF)

This model has been selected as it generally provides higher accuracy, reduces overfitting, and handles missing data. And it can also handle categorical data.

The underlying mathematical model of the Random Forest Classifier is built upon an ensemble of individual decision trees, where each tree is trained on random subsets of data and features, enhancing the prediction accuracy and robustness. The following equation shows how Random Forest achieves a more stable and accurate prediction by aggregating the results of multiple trees, through majority voting:

$$\hat{y} = mode(\{\widehat{y_1}, \widehat{y_2}, ..., \widehat{y_T}\}) \tag{7}$$

where,

1. $\hat{y}$ represents the final predicted class for a given input
2. $\hat{y}_t$ represents the predicted class from each tree t in the ensemble
3. T represents the total number of trees in the forest

3.5 Performance Evaluation Metrics

Now that we have discussed how the data is processed and which classifiers are employed for the analysis, the models have been evaluated based on their performance on two datasets. For the first evaluation, the training dataset consisting of 844 tweets was used for measuring the accuracy of its prediction on known data. After the training and evaluation of the model on the training dataset; a new test dataset has been generated by web-scraping 200 tweets for the second evaluation, to measure the model's accuracy on unknown text data.

For the evaluation, the following accuracy (ACC) metrics has been taken into consideration:

$$Accuracy(ACC) = \frac{Number\ of\ Correct\ Predictions}{Total\ Number\ of\ Predictions} \tag{8}$$

4 Result & Analysis

In this section, the result and analysis have been presented for vulgar language detection on social media texts, using the above-mentioned methods. Firstly, texts are taken as input, stemmed, and vectorized, then fed into all four models to detect vulgar language in the texts, the data is visualized, and the accuracies are measured based on their performance on both testing and training data.

4.1 Data Visualization

Data visualization is done to gain an insight of how the model is going to interpret the labeled text data. Word Cloud is a common method of visualizing the data, it provides a graphical representation of the common theme in the corpus, allowing a better conceptual exploration of the same. For this task, we have utilized the WordCloud class from the wordcloud library in Python. Figure 5 represents the wordclouds for Vulgar and Non vulgar words as visualized by the algorithm.

(a)

(b)

Fig. 5. Word Cloud for (a) Non-Vulgar Words (b) Vulgar Words

4.2 Vulgar Language Detection

For vulgar detection, the same training dataset containing 844 tweets has been used to train all the models, then the models are tested using another test dataset which contains overall 200 tweets. After the evaluation on the train data, Logistic Regression achieved 94.80% Accuracy, Decision Tree Classifier achieved 100.00% Accuracy, Naïve Bayes achieved 95.60% Accuracy and Random Forest achieved 100.00% Accuracy. Figure 6 shows the performance of the four models on the training dataset in a graphical representation. After the evaluation on the test data, Logistic Regression achieved 94.00% Accuracy, Decision Tree Classifier achieved 96.00% Accuracy, Naïve Bayes achieved 94.00% Accuracy and Random Forest achieved 96.00% Accuracy. The Fig. 7 shows the performance of the four models on the test dataset in a graphical representation.

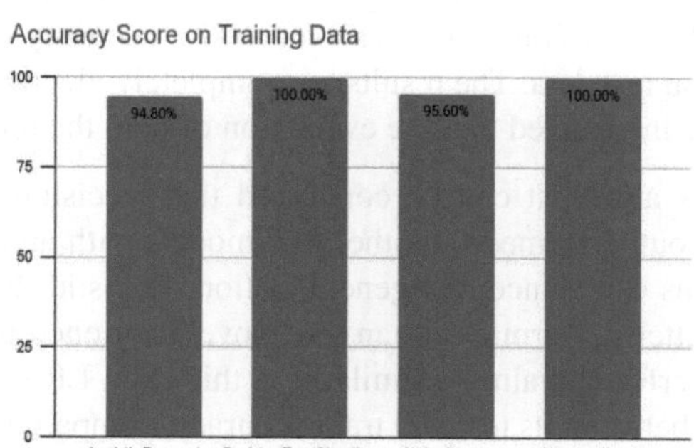

Fig. 6. Accuracy results on Training Data

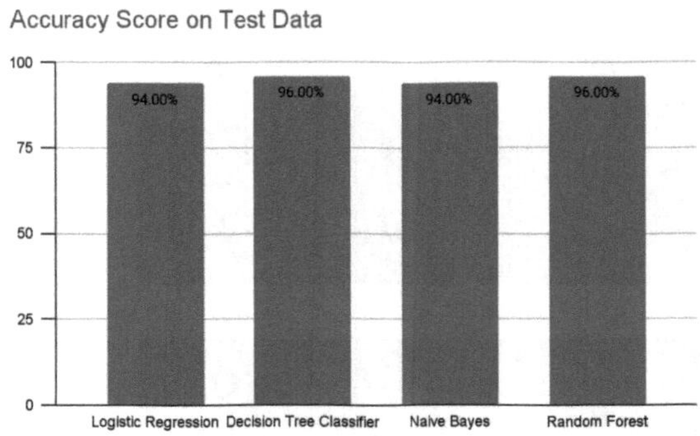

Fig. 7. Accuracy results on Test Data

4.2.1 Comparison Between Training and Test Data Accuracy

In terms of accuracy, it can be considered that the models showed a pretty standard (more than or equals to 85%) performance. Now, let's compare the accuracy results of train and test data for further analysis of the models:

- Logistic Regression: Logistic Regression has achieved 94.80% accuracy on training data and 94.00% accuracy on test data, which is almost identical. This performance shows that this model has learned generalized patterns well without overfitting or underfitting and can mostly predict new data accurately, which makes this model highly reliable for the task.
- Decision Tree Classifier: Decision Tree Classifier has achieved 100.00% accuracy on training data and 96.00% accuracy on test data. The 100.00% accuracy on the training data shows it has recognized patterns in known data with utmost precision while also achieving a high test-accuracy. This shows that it is a quite balanced model which can recognize known data efficiently while generalizing new data as well.
- Naïve Bayes: Naïve Bayes has achieved 95.60% accuracy on training data and 94.00% accuracy on test data. This shows that the model has performed extremely well with only 1.60% of gap, which indicates minimal overfitting and a strong generalization.
- Random Forest: Random Forest has achieved 100% accuracy on training data and 96.00% accuracy on test data. The results are completely identical with the Decision Tree, thus it can be interpreted that the evaluation of both the models are the same.

From the analysis above, it can be concluded that Decision Tree Classifier and Random Forest have outperformed the other two models with an accuracy of 96% on test data. But, In terms of balance and generalization, Logistic Regression and Naïve Bayes have shown better performance than the above mentioned models, even though the LR and NB has performed almost similarly in this case, LR slightly outperformed with only 0.80% gap between its test and train accuracy, compared to NB which shows a gap of 1.60%.

5 Discussion

6 Contribution

The major contributions of this paper include, creating a labelled dataset from scratch through web scraping X(Twitter) posts, for vulgar language detection training. A detailed literature review of existing research papers in similar fields of work has been conducted. Selected machine learning classifiers are trained on the created dataset for Vulgar Language Detection, the models are evaluated based on their accuracies and a constructive analysis has been performed. The result and observations are noted for future reference, contributing valuable insights to further research in the field.

6.1 Practice in Industry

This paper can be studied to gain insights on 'Vulgar Language Detection' and can serve as a basis for further advancements in the same. Analysing the strengths and limitations of the specific models, they can be refined for real-time applications in detecting vulgar language within social-media texts. The implementation of these models can aid towards the vision of a healthier, cleaner, more productive cyber space with less digital pollution.

6.2 Limitation/Drawbacks

Every study has its own limitations and drawbacks. Few limitations and drawbacks have also been noted in this study, which we are going to discuss in this section:

- Firstly, classification-based models were selected for vulgar language detection as they perform keyword-based analysis, however it was observed that these models struggle to detect patterns in lengthy texts.
- Also, it was noted that because of a small dataset, Decision Tree and Random Forest are overfitting on the training dataset with an accuracy of 100%.
- The study uses a single dataset with texts from only one social media site; the models also lack multilingual and contextual understanding of texts, these limitations can raise potential biases such as selection bias and moderation bias.
- Pruning is generally done in models like Decision Tree Classifier to handle overfitting and increase generalization, which was not taken into account in this analysis.

6.3 Ethical Considerations

The topic of this research paper is concerned with the human population (social media users/netizens) and the use of vulgar language used by netizens on the internet. So, certain ethical guidelines are crucial to be taken into consideration: The article "X / Twitter Web Scraping: Legality, Methods & Use Cases in'24" states that posts that are publicly visible and accessible without logging in, are permissible to be web scraped [36]. However, ethical guidelines and data protection laws suggest that data collection even in public platforms should adhere to the privacy policies of the platform, and use of copyright content without permissions must be avoided. If personal data is collected for research, the user's information must be anonymized or pseudonymized to protect individual privacy. And, if copyright data is used, fair use compliance must be ensured through obtaining permission from the copyright holders. Hence, in order to avoid the violation of privacy policies and issues for access permissions, the data is collected from accounts that were created specifically for this purpose (users handles mentioned in the 'Data Collection' section). Posts from real users were taken into account with attention to the language practiced, while writing these posts, capturing the authentic linguistic dynamics of language of these platforms and minimizing biases.

7 Conclusion & Future Scope

The objective of this research has been successfully fulfilled. The Machine Learning Classifiers have been trained and analysed for Vulgar Language Detection, the observations are noted, and the results are analysed and discussed in the paper in detail. Limitations and scopes of betterment are also taken into consideration. Decision Tree Classifier and Random Forest showed almost identical performance, achieving the highest accuracy of 96%. While Logistic Regression and Naïve Bayes proved to be more balanced models, performing almost similarly in this criteria, Logistic Regression slightly outperformed with just 0.8% margin between its testing and training data.

This analysis is definitely a stepping stone in the field of 'Vulgar Language Detection', adding value as a detailed investigation of machine learning classifiers for this task, and can be taken into consideration for future study. However, there are always scopes of betterment in every study. The models can be trained with more amounts of data to capture a vast range of real-life language for the analysis. Also, Neural Network and Transformer based features or models can be utilized to capture the context of the data more precisely and detect vulgar language in lengthier texts more efficiently. With certain improvements and modifications in the ML models, they can be prepared for real-life applications and real-time vulgar language detection.

References

1. Social media usage in India - statistics & facts: https://www.statista.com/topics/5113/social-media-usage-in-india/#topicOverview
2. Social Media Addiction: https://www.addictioncenter.com/behavioral-addictions/social-media-addiction/
3. Mujica, A.L., Crowell, C.R., Villano, M.A., Uddin, K.M.: Addiction by design: Some dimensions and challenges of excessive social media use. Med. Res. Arch. **10**(2) (2022).
4. Social Media Usage & Growth Statistics: https://backlinko.com/social-media-users
5. Wang, N.: An analysis of the pragmatic functions of "swearing" in interpersonal talk. En: Griffith Working Papers in Pragmatics and Intercultural Communication, 6, 71–79, (2013)
6. Holgate, E., Cachola, I., Preoţiuc-Pietro, D., Li, J.J.: Why swear? Analyzing and inferring the intentions of vulgar expressions. In Proceedings of the 2018 Conference on Empirical Methods in Natural Language Processing, pp. 4405–4414, (2018)
7. Pinker, S. The Stuff of Thought: Language as a Window into Human Nature. Penguin Books (2007)

8. The power of swearing: How obscene words influence your mind, body and relationships https://www.deccanherald.com/science/the-power-of-swearing-how-obscene-words-influence-your-mind-body-and-relationships-1154918.html
9. Singh, J., Singh, G., Singh, R.: A review of sentiment analysis techniques for opinionated web text. CSI Trans. ICT **4**, 241–247 (2016)
10. Eshan, S.C., Hasan, M.S.: An application of machine learning to detect abusive Bengali text. In 2017 20th International Conference of Computer and Information Technology (ICCIT), pp. 1–6. IEEE (2017, December)
11. Kwok, I., Wang, Y.: Locate the hate: Detecting tweets against blacks. In Proceedings of the AAAI Conference on Artificial Intelligence, Vol. 27, No. 1, pp. 1621–1622, (2013, June)
12. Sharma, S., Agrawal, S., Shrivastava, M.: Degree based classification of harmful speech using twitter data. arXiv preprint arXiv:1806.04197 (2018)
13. Waseem, Z., Hovy, D.: Hateful symbols or hateful people? predictive features for hate speech detection on twitter. In Proceedings of the NAACL Student Research Workshop, pp. 88–93 (2016, June)
14. Islam, T., Ahmed, N., Latif, S.: An evolutionary approach to comparative analysis of detecting Bangla abusive text. Bull. Electr. Eng. Inform. **10**(4), 2163–2169 (2021)
15. Chakraborty, P., Seddiqui, M.H.: Threat and abusive language detection on social media in Bengali language. In 2019 1st International Conference on Advances in Science, Engineering and Robotics Technology (ICASERT), pp. 1–6. IEEE (2019, May)
16. Nobata, C., Tetreault, J., Thomas, A., Mehdad, Y., Chang, Y.: Abusive language detection in online user content. In Proceedings of the 25th International Conference on World Wide Web, pp. 145–153 (2016, April)
17. Malmasi, S., Zampieri, M.: Detecting hate speech in social media. arXiv preprint arXiv:1712.06427 (2017)
18. Köffer, S., Riehle, D.M., Höhenberger, S., Becker, J.: Discussing the Value of Automatic Hate Speech Detection in Online Debates. In MKWI (pp. 83–94) (2018, October)
19. Abro, S., Shaikh, S., Khand, Z.H., Zafar, A., Khan, S., Mujtaba, G.: Automatic hate speech detection using machine learning: A comparative study. Int. J. Adv. Comput. Sci. Appl. **11**(8) (2020)
20. AlZarouni, E.: Detection of Hateful Comments on Social Media (2022)
21. Tarmizi, N., Saee, S., Ibrahim, D.H.A.: Detecting the usage of vulgar words in cyberbully activities from Twitter. Int. J. Adv. Sci. Eng. Inf. Technol. **10**(3), 1117–1122 (2020)
22. Singh, J., Singh, G., Singh, R.: Optimization of sentiment analysis using machine learning classifiers. HCIS **7**, 1–12 (2017)
23. Jemai, F., Hayouni, M., Baccar, S.: Sentiment analysis using machine learning algorithms. In 2021 International Wireless Communications and Mobile Computing (IWCMC), pp. 775–779. IEEE (2021, June)
24. Bhatt, N.T., Swarndeep, S.J.: Sentiment analysis using machine learning technique: a literature survey. Int. Res. J. Eng. Technol. (IRJET) **7**, 12 (2020)
25. Jiang, S., Kumar, E. Sentiment analysis based on machine learning algorithms: A comprehensive study. ScienceOpen Preprints
26. Raza, M.O., Mahoto, N.A., Hamdi, M., Al Reshan, M.S., Rajab, A., Shaikh, A.: Detection of offensive terms in resource-poor language using machine learning algorithms. PeerJ Comput. Sci. **9**, e1524 (2023)
27. Amjad, M., Zhila, A., Sidorov, G., Labunets, A., Butt, S., Amjad, H. I., ... Gelbukh, A.: Urduthreat@ fire2021: Shared track on abusive threat identification in Urdu. In Proceedings of the 13th Annual Meeting of the Forum for Information Retrieval Evaluation (pp. 9–11) (2021, December)
28. Razzak, R., Taki, S.T.O., Mim, M.R., Patwary, M.S.H., Pavel, M.I.: Vulgar Comments Classification: Comparison between CNN, XGBoost and SVM (2023)

29. Beniwal, R., Maurya, A.: Toxic comment classification using hybrid deep learning model. In Sustainable Communication Networks and Application: Proceedings of ICSCN 2020, pp. 461–473. Springer Singapore (2021)
30. Ashraf, N., Zubiaga, A., Gelbukh, A.: Abusive language detection in Youtube comments leveraging replies as conversational context. PeerJ Comput. Sci. **7**, e742 (2021)
31. Salminen, J., Hopf, M., Chowdhury, S.A., Jung, S.G., Almerekhi, H., Jansen, B.J.: Developing an online hate classifier for multiple social media platforms. HCIS **10**, 1–34 (2020)
32. Emon, E. A., Rahman, S., Banarjee, J., Das, A. K., Mittra, T.: A deep learning approach to detect abusive Bengali text. In 2019 7th International Conference on Smart Computing & Communications (ICSCC) (pp. 1–5). IEEE (2019, June)
33. Aurpa, T.T., Sadik, R., Ahmed, M.S.: Abusive Bangla comments detection on Facebook using transformer-based deep learning models. Soc. Netw. Anal. Min. **12**(1), 24 (2022)
34. Wang, W., Chen, L., Thirunarayan, K., Sheth, A.P.: Cursing in English on twitter. In Proceedings of the 17th ACM Conference on Computer Supported Cooperative Work & Social Computing, pp. 415–425 (2014, February)
35. Mahmud, T., Ptaszynski, M., Eronen, J., Masui, F.: Cyberbullying detection for low-resource languages and dialects: Review of the state of the art. Inf. Process. Manage. **60**(5), 103454 (2023)
36. X/Twitter Web Scraping: Legality, Methods & Use Cases in '24 https://research.aimultiple.com/twitter-web-scraping/

Financial Security

ATM Fraud Detection Using Streaming Data Analytics

Yelleti Vivek[1], Vadlamani Ravi[1](✉), Abhay Anand Mane[2], and Laveti Ramesh Naidu[2]

[1] Centre for Artificial Intelligence and Machine Learning, Institute for Development and Research in Banking Technology, Castle Hills Road #1, Masab Tank, Hyderabad 500076, India
{yvivek,vravi}@idrbt.ac.in

[2] Center for Development of Advanced Computing, #1 Old Madras Road, Bangalore 560038, Karnataka, India
{abhaym,rameshl}@cdac.in

Abstract. Gaining the trust and confidence of customers is the essence of the growth and success of financial institutions and organizations. Of late, numerous instances of fraudulent activities impacted the financial industry. Furthermore, owing to the generation of large voluminous datasets, it is highly essential that the underlying framework is scalable and meets real-time needs. To address this issue, in the study, we proposed ATM fraud detection in static and streaming contexts respectively. In the static context, we investigated parallel and scalable machine learning algorithms for ATM fraud detection that are built on Spark and trained with various machine learning (ML) models, including Naive Bayes (NB), Logistic Regression (LR), Support Vector Machine (SVM), Decision Tree (DT), Random Forest (RF), Gradient Boosting Tree (GBT), and Multi-layer perceptron (MLP). We also employed several balancing techniques such as Synthetic Minority Over-sampling Technique (SMOTE) and its variants, Generative Adversarial Networks (GAN), to address the rarity in the dataset. Additionally, we proposed a streaming-based ATM fraud detection in the streaming context. Our sliding window-based method collects ATM transactions that are performed within a specified time interval and then utilizes them to train several ML models, including NB, RF, DT, and K-Nearest Neighbour (KNN). We selected these models based on their less model complexity and quicker response time. In both contexts, RF turned out to be the best model. RF obtained the best mean AUC of 0.975 in the static context and a mean AUC of 0.910 in the streaming context. RF is also empirically proven to be statistically significant than the next-best performing models.

Keywords: Fraud Detection · Spark · Streaming · Imbalance · GAN · DStreams

1 Introduction

Customers ubiquitously utilize automated teller machines (ATM) as one of the widely accepted forms of payment on a "*day-to-day*" basis. Over the past two decades, the demand for ATMs has surged as financial institutions strive to fulfill client demand and

---

An earlier version is submitted to arxiv: https://arxiv.org/abs/2303.04946

S. Dhar et al. (Eds.): AGC 2025, CCIS 2686, pp. 203–222, 2026.
https://doi.org/10.1007/978-3-032-09153-6_13

provide sophisticated service management. However, this immense growth attracted the attention of scammers, who have targeted and caused significant financial loss. This inflicts a severe loss to the customers and further affects the '*bank-customer*' relationship. This could be avoidable by taking proper measures, which indeed gain the trust and builds the confidence of the customer over the financial institutions and organizations.

Financial fraud is a broad term that encompasses various types of deceptive activities aimed at obtaining money or other assets through fraudulent ways. Among these fraudulent activities, the following are the most prevalent, credit card fraud, money laundering, debit card fraud, etc. When any of these financial crimes occur, financial organizations suffer significant losses. Therefore, it is imperative for any financial institution or organization to handle them carefully. The negative repercussions of various financial fraud have captivated the consciousness of stakeholders.

Organizations base their policies and critical decisions on the analyses of the vast volumes of data, or "Big Data". These financial fraud datasets in particular fall within the big data paradigm. The five Vs – Volume, Velocity, Variety, Veracity, and Value are the primary characteristics of big data. Hadoop and Spark are two popular big data processing tools, which enable organizations to collect and manage huge data. Recently, Spark [4] gained popularity owing to its unified data analytics engine which can handle massive datasets. Spark offers a variety of abstractions, including data frames (DF) and resilient distributed data frames (RDD), which are immutable, cacheable, and support persistence. They are partitioned into multiple data partitions, thereby supporting parallel and distributed operations. Additionally, Spark offers extensions for various programming languages such as Scala, Python, Java, and R. Furthermore, Spark extends its support to SQL with SparkSQL, developing scalable machine learning algorithms with SparkMLlib, and streaming analytics with Discretized streams (DStreams) [39] as the abstraction. DStream is a collection of RDDs collected over time that facilitates batch stream processing. Spark can easily integrate with other ingestion tools such as Apache Kafka, Kinesis, or TCP sockets and other big data processing tools such as Hadoop, Hive, etc. making it to manage the continuous stream of data.

Among the types of financial fraud discussed earlier, ATM fraud detection has captivated our interest. The ATM transaction history can be examined to detect fraudulent transactions. We determined that the following scenarios might assist us in identifying fraudulent transactions: (i) stolen or lost debit/credit card, (ii) unusual transaction history, (iii) transactions initiated from an unusual location, and (iv) transactions involving bulk amounts, which could increase the probability of becoming fraudulent. It is quite difficult to conduct such analysis in real-time within a limited time interval. Therefore, there is a need to automate this process by employing various machine learning techniques. Detecting ATM fraud detection is indeed a binary classification task, where fraudulent transactions are regarded as a positive class and non-fraudulent transactions as a negative class.

Owing to the significance of identifying fraud detection under the big data paradigm. Through this study, we addressed ATM fraud detection in the big data paradigm using scalable machine learning algorithms and efficient solution methodology in both static and streaming contexts. To achieve this, we conducted the study as follows:

- Firstly, in the static context, we developed machine learning models on the historic ATM transaction data collected from India. To create scalable ML models, we used Spark MLlib.
- Secondly, in the streaming context, we used DStreams to capture data at regular intervals and employed a sliding window-based fraud detection approach.
- Further, the imbalance in the data is handled in both streaming and static contexts.

In the current study, we analyzed India-sourced ATM transaction data for fraud detection. As discussed earlier, this ATM fraud detection problem has a significant impact on daily living and has motivated us to work toward providing an effective and efficient solution. In this study, we performed ATM fraud detection using binary classification tasks. To accomplish this, we employed several machine learning techniques: NB, LR, SVM, DT, RF, GBT, and MLP. The collected ATM transaction data is highly imbalanced in nature. This data imbalance is handled by employing the following techniques such as SMOTE and its variants (SMOTE-ENN, SMOTE-Tomek), and ADASYN. Furthermore, we employed the GAN variants (i.e., V-GAN and W-GAN), too, owing to its recent advancements and achievements.

The major contributions of the proposed work are as follows:

- Proposed the resilient ATM fraud detection framework under Spark.
- Explored and compared ML algorithms for ATM fraud detection in the static context.
- Investigated the use of GAN variants (V-GAN and W-GAN) for handling data imbalance in ATM fraud detection.
- Proposed sliding window approach-based ATM fraud detection using streaming analytics, thereby handling the latency and time-to-time decisions.
- Handled the imbalance by employing various data-balancing techniques.

The remainder of the paper is arranged as follows: Sect. 2 presents the literature review. Section 3 discusses the proposed framework for ATM transactional data using binary modeling under static and streaming contexts. Section 4 describes the dataset. Section 5 discusses the results of various techniques. Finally, Sect. 6 concludes the paper.

2 Literature Review

Data mining algorithms play a quintessential role in detecting fraud post-facto [5]. These consist of LR, MLP, DT, etc. When cutting-edge methods, such as semi-supervised and unsupervised techniques, are used to mine the data that arrives in a stream, it may also be discovered in almost real-time [1, 6]. Online classification, online clustering, outlier detection, etc., are some of the most sophisticated approaches. A reasonable amount of research is reported concerning credit card fraud detection [5, 7, 8], however, ATM/Debit card fraud detection is seldom the researched topic. To the best of our knowledge, one research article appeared dealing with ATM transactional fraud, where a specialized language is proposed for proactive fraud management in financial data streams [6]. Depending, on how an ATM card is used – at an ATM, a point of sale (POS) device, or online banking, there are three alternative options. Other frauds can be perpetrated, and as the resultant data vary from structured to semi-structured to unstructured, so does the method they can be detected. Wei et al. [9] observed various characteristics and challenges involved in credit card fraud detection. The development of a scalable system for credit card fraud detection utilizing distributed settings was pioneered by Chan et al. [10]. The authors demonstrated the significance of considering scalability into the consideration when developing the solution.

Chen et al. [11] proposed a binary support vector machine, a hybrid of SVM, and a genetic algorithm (GA). Here, the optimal mix of support vectors is chosen using a genetic algorithm. To estimate the distribution of the input data, a self-organizing map (SOM) is also included. In order to fulfill real-time analytics, Dorronsoro et al. [12] proposed an online system for credit card fraud detection based on a neural classifier. To detect fraud, the authors proposed a non-linear variation of Fisher's discriminant analysis. This strategy could separate a significant amount of fraudulent activities from legitimate traffic. A real-time fraud detection system was presented by Quah et al. [13] that uses the SOM to identify fraudulent credit card transactions and consumer behaviour online. Sanchez et al. [14] applied association rule mining to find the pattern of anomalous behaviour that would assist them to identify a fraudulent transaction. On the transaction database, the authors showed how well the suggested technique worked. Hidden markov model (HMM), Srivastava et al. [3] devised a method that comprises the following for credit card fraud detection: (i) Initially, the set of customers with identical spending patterns are clustered together using a clustering technique. (ii) After that, the HMM is executed to identify the critical patterns and sequences, which capture the fraudulent transactions. Zaslavsky and Strizhak [15] applied the SOMs for credit card fraud detection. IT-based paradigm was presented by Wang et al. [2] to address various frauds: (i) system attacks, and (ii) non-system attacks. The authors studied the effectiveness of their framework on the Taiwan financial fraud dataset.

Credit card transaction data frequently concern scalability as one of the primary concerns that must be resolved in real-time. Syeda et al. [16] developed a parallel granular neural network with shared memory architecture to address the above-mentioned issue. It is applied to Visa card transactions to identify fraudulent transactions. Due to the considerable increase in online activities, several fraudsters are attacking the most often used websites. Several studies have been proposed to identify such attacks in the literature [17]. Arya et al. [18] applied a spiking neural network is inspired by neuroscience, and applied it to several classification problems.

Ravisankar et al. [19] proposed a fraud detection framework to identify the critical financial ratios obtained by analyzing the financial statements. With this, it would be possible to assess whether the company is conducting too much debt or inventory. The presence of missing values hampers the performance of the models in the financial data. Farquad et al. [20] proposed a modified active learning framework, that addresses churn prediction and insurance fraud detection, and uses the SVM to extract the if-then rules. It is a three-phase approach where: (i) SVM is used for the recursive feature elimination, (ii) synthetic data is generated by using active learning, and (iii) rules are generated using DT and NB. Vasu and Ravi [21] proposed an undersampling technique to address the imbalance challenges in many banking and finance applications. K-Reverse nearest neighbourhood is used to remove the outliers from the majority class. Then, the classifiers are used to predict fraudulent transactions. Sundarkumar and Ravi [22] analyzed the dataset available in the CSMINING group. The authors invoked mutual information as the feature selection criteria are applied after extracting the features from the API calls. The dataset is imbalanced in nature. Hence, the authors employed oversampling techniques over several ML techniques.

Customers sometimes wonder if they are conducting transactions at the rightful website. Determining if the website is a target of a cyberattack is therefore critical. A strategy to recognize phishing websites was presented [23]. To determine the validity of a website, their methodology examines the server's IP address, domain, and internet domains to identify website legitimacy. Lakshmi and Vijaya [24] proposed a supervised learning mechanism to accomplish phishing detection. The extraction of the features from 200 website URLs by the authors served as evidence of the success of their methodology. Tayal and Ravi [26] developed a fuzzy rule-based association rule mining by formulating the above as a combinatorial optimization problem. This fuzzy-based rule miner is trained using PSO and applied to a transactional database to extract the rule, which will help to detect phishing attacks. Further, PSO-trained classification association rule mining (PSOCARM) [25], a rule-based classifier indicates whether or not phishing is detected. Pradeep et al. [27] also proposed a rule-based classifier based on Firefly and threshold-accepting algorithms to classify fraudulent financial statements.

Malware detection poses a number of difficulties and needs a robust, advanced security system. Application program interface (API) calls are typically used by malware attackers to steal credit card details, important personal data, etc. To retrieve the information, Sundarkumar et al. [28] employed this API call information and topic modelling approach. Their findings demonstrate that DT and SVM outperformed the competition, with DT securing first place in the interpretability category. Another study by Sheen et al. [29], uses features retrieved from executable files and an ensemble of classifiers to identify malware. Their main goal was to employ the harmony search algorithm with the fewest possible classifiers. On the other hand, fraud detection can also be solved using One class classification (OCC) methods. The rarity problem in fraud detection is handled by, Sundarkumar and Ravi [31], who proposed an undersampling method where the K-Reverse nearest neighbourhood and OCSVM are applied in tandem and remove the unnecessary samples in the majority class. The authors demonstrated the effectiveness of their approach in the automobile insurance fraud detection dataset and customer credit card churn prediction dataset. Kamaruddin and Ravi [30] implemented the above hybrid architecture, i.e., PSO-AANN architecture, under the Spark environment to make it suitable for big datasets. The proposed architecture outperformed the extant approaches, such as OC-SVM, etc.,

Now, we will discuss the studies related to fraud detection solving using real-time analytics. Thennakoon et al. [32] utilized the ML model and application processing interface (API) mechanism to flag fraudulent transactions. Arya and Sastry [33] proposed a credit card fraud detection based on predictive modelling and named it deep ensemble learning for detecting fraudulent transactions in the real-time data stream. The authors tried addressing the overfitting issues, which is one of the major concerns of the previous studies. They employed an ensemble of deep learning networks to identify credit card fraud. Mittal and Tyagi [34] discussed the pros and cons of various computational techniques for real-time credit card fraud detection. Abakiram et al. [35] proposed a real-time model for credit card fraud detection based on deep learning. The authors employed a deep autoencoder to classify fraudulent transactions. Rajeswari and Sathish Babu [36] proposed a credit card fraud detection system in real-time. They proposed a framework where the models are developed and relied on historical data with which the fraudulent transactions are identified in real-time. Carcillo et al. [37] proposed a scalable framework for credit card fraud detection using Spark. The authors integrated the ingestion tools such as Kafka and Cassandra into their framework and simulated the credit card fraud transactions coming in the form of data streams. They employed three different mechanisms to identify fraudulent transactions. Wahid et al. [40] proposed an online fraud detection technique that employs a Neural factorization autoencoder (NFA) to analyze the calling patterns of the customers to detect fraudulent calls. NFA integrates both neural factorization machines (NFM) and autoencoders (AE) to capture customer usage patterns. This model is evaluated on a real-world dataset and turned out to be better than several SOTA methods. Abbasi et al. [41] proposed a real-time architecture to identify digital transaction fraud by employing Isolation Forest (IForest), which integrates spark streaming, Kafka, and PostgreSQL. The efficacy of this approach is studied on two real datasets. Khodabandehlou et al. [42] proposed an unsupervised scalable approach and named it FiFraud to depict the manipulator's behavior in the transaction stream. In their approach, they utilized a stream of graphs to detect fraudulent traders based on exploiting the density signals. Nguyen et al. [43] proposed a generic fraud detection framework by utilizing the sub-graphs and query-by-example approach to index and extract the k most similar diverse fraud subgraphs by utilizing the previous frauds.

3 Proposed Methodology

This section starts with the proposed methodology in the static context and then continues with a discussion on the streaming context.

3.1 Binary Classification Methodology Under Spark

The proposed methodology is depicted in Fig. 1, and the block diagram is depicted in Fig. 2, and can be used in a static environment, where the model is trained on historical data. The process starts with the data collection step. In the current work, we collected ATM transaction data from India. In India, all banks maintain software to monitor the daily transactions that occur in an ATM. The collected dataset contains transactions that occurred in three different modes viz., (i) ATM transaction data, (ii) at the point of sale (POS), and (iii) internet transaction. It includes critical private information related to the customers. To maintain the data integrity and privacy of the customers, we masked the feature names.

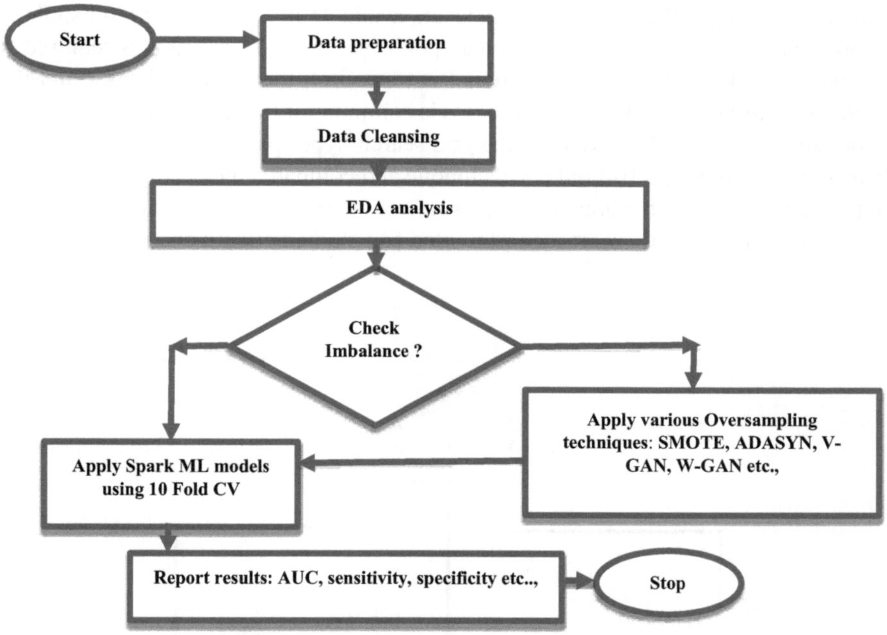

Fig. 1. Flowchart of the proposed methodology in the static environment

Since, '*garbage data in results in garbage analysis out*' is very suitable in machine learning. Therefore, to increase the richness of the data, we include the following data cleansing steps: (i) removing duplicate transactions because the same information is possessed twice, which indeed does not add any value, (ii) removing incomplete data within a dataset, which includes corrupted features either due to human error or some other, and (iii) filtering out the features having almost null values. As we know, the missing data need to be dealt with through imputation techniques, but if the feature has more than 90% null values, these imputation techniques will fail miserably.

Hence, we dropped off the features having more null values. Thereof, the data can be analyzed to obtain vital insights and check assumptions with the help of generated statistical summary. All the features are normalized. Now, the dataset is observed to be of imbalance type. It is handled by applying various balancing techniques such as ADASYN, SMOTE, etc. The dataset is partitioned into training and test sets in the ratio 80:20. While doing so, stratified random sampling is adopted, where the same proportion of both the positive and negative classes is maintained in both the training and test sets. All of these balancing techniques are discussed in detail in the subsequent sections. On a note, while performing oversampling, the test dataset must be kept intact. Once, the above operations are performed, the dataset is now stored in RDD [38].

Now, the modelling phase begins, where we build the machine learning model such as NB, LR, DT, etc. All of these models are built under the Spark environment, and we utilized Spark MLlib. This is achieved by constructing a pipeline, which consists of a group of stages. Each stage utilizes the data produced by the previous stage. The pipelines in Spark follow Directed Acyclic Graph (DAG), where the stages are specified in topological order. We used feature indexers available in Spark MLlib, to handle the features and then transform the entire data point into a single list. This is required for the Spark MLlib models. Then the model is trained on the training dataset and then tested on the test dataset. The above process is common for the rest of the spark models. While building the models, we used the grid search hyper-parameter tuning technique. On a note, we employed 10-Fold CV technique and compared the performance thereof. All the metrics such as the area under the receiver operator characteristic curve (AUC), sensitivity, and specificity scores are presented and studied effectively.

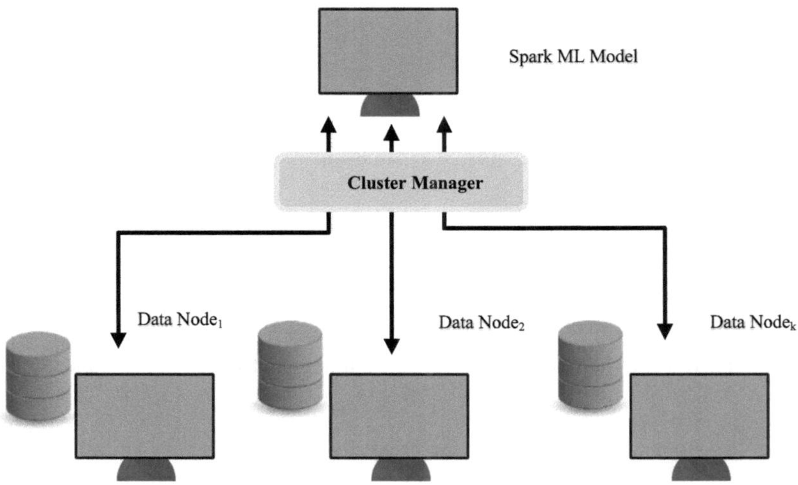

Fig. 2. Block diagram for ATM fraud detection in static context

3.2 Sliding Window for Streaming Data

The methodology discussed in Sect. 3.1 is very much suitable in the static environment. However, when it comes to processing the stream of data that needs to be handled within the specified interval of time demands a new mechanism. To meet these challenges and demands, we proposed a sliding window-based streaming approach for ATM fraud data detection. The algorithm is presented in Algorithm 1.

Spark provides DStream, an array of RDDs where each RDD comprises the transaction information collected at a certain interval of time. All the operations, which are to be handled by DStream are processed in a streaming context. Hence, one needs to initialize the streaming context at the beginning of the process and must close it once the operations are finished. All the variables, models, broadcasting variables, etc. are confined to this streaming context. On a note, Spark provides batch stream processing. Hence, the operations are performed on the data, which is collected in the form of batches. By using the ingestion tools, streaming data is processed into the HDFS cluster. In our context, we did not use any ingestion tool. However, we simulated the data ingestion by using QStreams, which is provided by Spark. We created an unbounded data frame, where the streams of data are coming in multiple batches.

Algorithm 1: Sliding Window approach for ATM Fraud detection under the streaming context

```
Collect the dataset
Initialize ws ← window size
Initialize sl ← sliding interval
Create an Unbounded Table ← ϕ
Start StreamingContext(Time)
while IncomingStreams == True do
    UnboundedTable ← Append the IncomingStream
    If Window == Full then
        TrainData ← Window[:ws-1]
        TestData ← Window[ws]
        Train the Model on TrainData
        Validate the Model on TestData
        AUC ← Compute AUC score on TestData
        Sensitivity ← Compute Sensitivity on TestData
        Specificity ← Compute Specificity on TestData
        Report AUC, Sensitivity and Sensitivity
    end if
    Slide window by according to sl across UnboundedTable
end while
Stop StreamingContext
```

In our approach, we have utilized the sliding window approach. The sliding window approach is designed as follows: transaction data streams are collected in the form of batches within the specified time interval. The data collected thus far collected over a particular interval of time is sitting in a single RDD. The number of RDDs collected is dependent on the window size. Too much window size increases the latency. Let the sliding window length be *ws*, and then the training is done on the first '*ws-1*' batches and testing on the latter batch of the sliding window. If the required number of batches has not yet been collected, it will wait until it is full. As discussed earlier, we know that a large window size demands more previous data to be processed. It further increases the latency too. By considering these criteria, we fixed the window size as 2. By doing so, we are restricting ourselves to the most recent data. Hence, the underlying machine learning model is trained on the first RDD of the window and then tested on the second RDD of the window. We kept the stride, the moving speed of the window to be 1. Hence, now the second RDD of the second window becomes the first RDD of the next window. This process is continued until the new data is available and ingested into the cluster.

4 Overview of the Dataset

We collected India-sourced ATM transactional data for the current study. This dataset consists of a few days worth of transactions totalling one million transactions over three different channels, including ATM transaction data, point of sale (POS) transaction data, and internet transaction data. It includes vital facts on this kind of transaction, payment method, client information, transaction amount, etc. We renamed these features to preserve data privacy and integrity while adhering to bank regulations. We employed all the data pre-processing techniques for both static and streaming contexts to clean up the data. After the dataset was refined, it was seen that there were 7,46,724 transactions. The number of features was reduced to 10 as a consequence of the removal of several unique and null-valued characteristics. The data proportion of non-fraud and fraud classes is 87.80:12.20. Hence, it is an imbalanced dataset. All the following experiments were performed in 10-Fold Cross Validation (CV).

4.1 Evaluation Measures

We considered AUC to measure the robustness of the employed classifier. In addition to AUC, we considered sensitivity and specificity as the other measures for the binary classification.

4.1.1 Auc

It is proved to be a robust measure while handling imbalanced datasets and is an average of specificity and sensitivity. The mathematical representation of AUC is given in Eq. (1).

$$AUC = \frac{(Sensitivity + Specificity)}{2} \tag{1}$$

4.1.2 Sensitivity

It is also well known as True Positive Rate (TPR). It is the ratio of the positive samples that are truly predicted to be positive to all the positive samples. The mathematical notation is given in Eq. (2).

$$Sensitivity = \frac{TP}{TP + FN} \tag{2}$$

where TP is a true positive, and FN is a false negative.

4.1.3 Specificity

It is also well known as True Negative Rate (TNR). It is the ratio of the negative samples that are truly predicted to be negative to all the negative samples. The mathematical notation is given in Eq. (3).

$$Specificity = \frac{TN}{TN + FP} \tag{3}$$

where TN is a true negative, and FP is a false positive.

Table 1. Hyperparameters for all the techniques

Model	Hyperparameters
LR	'regularization parameter': [0.1, 0.01, 0.001] 'maximum iterations': [10,20,30,40,50,100,500]
SVM	'regularization parameter': [0.1, 0.01] 'maximum iterations': [30,40,50,100]
DT	'criterion': {'gini','entropy'}; 'maxdepth': [5,10, 15, 20]
RF	'maxdepth':[5, 10, 15, 20]; 'estimators':[10,20,30,40,50] 'criterion': {'gini','entropy'}; 'maxdepth': [5,10, 15, 20]
GBT	'maxdepth':[5, 10, 15, 20]; 'estimators':[10,20,30,40,50] 'criterion': {'gini','entropy'}; 'maxdepth': [5,10, 15, 20]
MLP	'solver': {'l-bfgs', 'gd} 'maximum iterations': [50, 100, 200, 300]
V-GAN	**Discriminator:** Hidden layer 1: 128 neurons, activation function:'leaky ReLU' Hidden layer 2: 64 neurons, activation function:'leaky ReLU' Hidden layer 3: 32 neurons, activation function:'leaky ReLU' Hidden layer 4: 8 neurons, activation function:'leaky ReLU' Epochs = 10,000
W-GAN	**Discriminator:** Hidden layer 1: 256 neurons, activation function:'leaky ReLU' Hidden layer 2: 128 neurons, activation function:'leaky ReLU' Hidden layer 3: 64 neurons, activation function:'leaky ReLU' Hidden layer 4: 32 neurons, activation function:'leaky ReLU' Epochs = 10,000

5 Results and Discussion

In this section, we presented the results in two different contexts, viz., static and streaming contexts, respectively. Firstly, we will discuss static context results followed by streaming context results.

5.1 Static Context Results

In the current research study, we utilized the hyperparameters presented in Table 1. We presented all the hyperparameters after performing rigorous experiments. As discussed in Sect. 3, all the experiments followed the same identical data cleansing, and data pre-processing technique. We observed that the ATM transaction dataset collected from India is highly imbalanced, i.e., 87.80: 12.20. Hence, we incorporated various balancing techniques viz., SMOTE [47], SMOTE-ENN / SMOTE-Tomek [48], ADASYN [49], V-GAN [50], W-GAN [51]. It is to be noted that the experiments were performed in 10-Fold Cross Validation (CV). In the imbalanced classification datasets, AUC is proven to be a robust measure, which is quite evident in previous case studies such as credit card fraud detection. Hence, while choosing the best model, AUC is given more preference. The corresponding results were presented in Table 2. Further, we also discussed the sensitivity and specificity scores in Table 3 and 4 respectively.

Table 2. Mean AUC obtained by various ML models under 10-Fold CV

Model	Balancing Technique						
	Imbalanced	SMOTE	SMOTE-Tomek	SMOTE-ENN	ADASYN	V-GAN	W-GAN
NB[40]	0.721	0.775	0.775	0.777	0.692	0.897	0.798
LR[41]	0.709	0.803	0.803	0.805	0.690	0.893	0.869
SVM[42]	0.807	0.806	0.805	0.807	0.694	0.890	0.853
DT[43]	0.755	0.823	0.823	0.824	0.775	0.953	0.904
RF[44]	0.759	0.837	0.837	0.838	0.791	**0.975**	0.920
GBT[45]	0.759	0.837	0.837	0.838	0.791	0.975	0.920
MLP[46]	0.757	0.837	0.831	0.831	0.721	**0.977**	0.904

Out of all ML models, MLP turned out to be the best model in terms of AUC (refer to Table 2) after balancing the dataset with the V-GAN balancing technique. The second and third places are secured by GBT and RF respectively. It is interesting to note that both RF and GBT obtained similar AUC. The same is also reflected in the sensitivity and specificity scores. As a decision-maker, while choosing the best model, the complexity of the model should also need to be considered along with the numerical AUC. In such a case, RF turned out to be the best model because of its less number of learnable parameters than MLP and low complexity than GBT. Further, it is observed to be having a very minor difference (AUC of 0.2%) in terms of mean AUC compared to the other two best models. Overall, all the tree-based models performed relatively well when compared to other ML models. In terms of balancing techniques, V-GAN clearly outperformed the other balancing techniques in obtaining the best AUC.

Table 3. Mean Sensitivity scores obtained by various ML techniques on 10-Fold CV

Model	Balancing Technique						
	Imbalanced	SMOTE	SMOTE-Tomek	SMOTE-ENN	ADASYN	V-GAN	W-GAN
NB	0.546	0.779	0.778	0.780	0.699	0.871	0.674
LR	0.540	0.782	0.781	0.783	0.660	0.787	0.745
SVM	0.736	0.775	0.774	0.775	0.632	0.781	0.707
DT	0.607	0.789	0.787	0.789	0.738	0.930	0.824
RF	0.618	0.773	0.773	0.776	0.769	0.961	0.850
GBT	0.618	0.773	0.773	0.776	0.769	0.961	0.850
MLP	0.613	0.781	0.781	0.781	0.675	**0.969**	0.825

Table 4. Mean Specificity scores obtained by various ML techniques on 10-Fold CV

Model	Balancing Technique						
	Imbalanced	SMOTE	SMOTE-Tomek	SMOTE-ENN	ADASYN	V-GAN	W-GAN
NB	0.895	0.772	0.773	0.774	0.685	0.924	0.922
LR	0.878	0.825	0.825	0.826	0.720	0.999	0.992
SVM	0.877	0.837	0.836	0.838	0.757	**1.0**	0.999
DT	0.902	0.857	0.859	0.859	0.813	0.975	0.991
RF	0.901	0.901	0.902	0.901	0.812	0.989	0.991
GBT	0.901	0.901	0.902	0.901	0.812	0.989	0.991
MLP	0.902	0.882	0.882	0.882	0.767	0.986	0.982

Yet another tie-breaker that can be considered while choosing the best model. It is a well-known fact that DT is the explainable model in the tree-based models. The rules of DT obtained after balancing with V-GAN are presented in Table 5 showed that X_{14} is the most important feature followed by X_0 in deciding whether a transaction is fraudulent or not. As we know, both the RF and GBT are ensemble models. Hence, complex models stand second in terms of explainable aspects. But, if one closely observed that the AUC is 2% less than the RF and GBT, which is quite unacceptable in critical applications such as fraud detection. Hence, despite having lesser interpretability, RF and GBT are ranked over DT. Because of its numerical superiority and less model complexity, RF ranked as the best model.

Similar observations were noticed with respect to the sensitivity scores (refer to Table 3). As we know that sensitivity is another critical parameter that artefacts the number of fraudulent transactions truly identified to be as fraudulent ones. MLP outperformed the rest of the model by obtaining higher mean sensitivity scores. However, when considering the model complexity and numerical superiority as tie-breakers, RF turned out to be the best model in terms of mean sensitivity.

On the other hand, a similar observation is made on the specificity score (refer to Table 4), which artifacts the number of non-fraudulent transactions truly identified to be non-fraudulent ones. Specificity is another critical factor because no loyal customer's transaction should be tagged as a fraudulent transaction. It is observed that the SVM obtained a sensitivity score of 1.0 yet the specificity score is remarkably inferior to tree-based models. After considering tie-breakers, we concluded that RF turned out to be the best model in the static context.

Table 5: Rules obtained by Decision Tree in static context after balancing with V-GAN

Rule No.	**RULE**
1	IF $X_{14} \leq 0.50$ AND $X_0 \leq 0.083$ AND $X_8 \leq 0.50$ THEN **Negative Class**
2	IF $X_{14} \leq 0.50$ AND $X_0 \leq 0.083$ AND $X_8 > 0.50$ AND $X_3 \leq 0.001$ THEN **Positive Class**
3	IF $X_{14} \leq 0.50$ AND $X_0 \leq 0.083$ AND $X_8 > 0.50$ AND $X_3 \leq 0.001$ THEN **Negative Class**
4	IF $X_{14} \leq 0.50$ AND $X_0 \leq 0.083$ AND $X_8 \leq 0.50$ AND $X_0 \leq 0.083$ THEN **Positive Class**
5	IF $X_{14} > 0.50$ THEN **Negative Class**

5.2 Statistical Testing of the Results.

The two-tailed t-test analysis is conducted on the AUC obtained by models over 10-Fold cross-validation (from Table 2) and is reported in Table 6.

The null hypothesis is, H_0: *Both the algorithms are statistically equal.*

The alternate hypothesis is, H_1: *Both the algorithms are statistically not equal.*

The significance level is chosen to be 5%, and the degree of freedom is 18 (10 + 10−2). Hence, the p-value should be less than 5% to reject the null hypothesis. Table 6 infers that when the p-values are less than 0.05; the null hypothesis is rejected, and the alternate hypothesis is accepted in the t-test conducted between RF and DT. This concludes that in the static context, RF is more statistically significant than DT in terms of AUC. However, RF and GBT are proved to be statistically equal. However, in terms of complexity, RF is less than GBT, which makes RF to be considered significant over DT.

Table 6. Paired t-test results

Model	Parameter	
	t-statistic	p-value
RF vs GBT	0.0	1.0
RF vs DT	33.39	1.21×10^{-17}

5.3 Streaming Context Results

Now, we will discuss the performance of the models in the streaming context. As we already discussed, the ATM transaction dataset is highly imbalanced. Hence, the transactions collected over a window might not have any fraudulent transactions. Further, the binary classification models cannot be built when only one single class-related data point is given. Hence, in our approach, one assumption is made in the streaming context with respect to the proportion of the negative and positive class transactions collected over a window. Hence, in every data stream, we will include the fraudulent transactions (positive class samples) with the collected ATM transaction data. Hence, the balanced datasets are only considered for being evaluated in the streaming context. Furthermore, latency is another critical parameter while evaluating the model in the streaming context. Hence, the models should be less complex. Therefore, we considered the simple and tree-based models and compared their performance. It is important to note that all the models are evaluated in a sliding window fashion, and the metrics such as AUC, sensitivity, and specificity are calculated and discussed in this section. After employing the balancing technique, the dataset has now become 10 lakh transactions. Each window comprises 1000 transactions, thereby making the total number of windows are 1000.

Further, here also, we considered AUC as the suitable metric to choose the best model, and the results are presented in Table 7. We presented the AUC obtained in Figs. 3, 4, 5, 6, 7 and 8 to illustrate the outlier behaviour of the employed ML models. Overall, the RF model performed the best irrespective of the balancing technique. DT stands second on the list. The rules obtained by DT where it achieved the highest AUC by employing a sliding window are given in Table 10. The rules show that feature X_9 is the most important feature followed by feature X_3 which also plays a key role in the identification of fraudulent class. Further, Fig. 7 shows that the RF model has very fewer outlier behaviour, which makes it more stable while identifying the positive class. However, DT showed higher variability that ranks it second despite having an interpretability aspect. The same is observed in the sensitivity and specificity scores presented in Table 8 and 9, respectively. Interestingly, RF not only showed superior performance in mean AUC, sensitivity, and specificity scores but also showed lesser variability that supports its robustness. The same observation is noticed with respect to other balancing techniques (refer to Fig. 6 to Fig. 8). V-GAN is empirically obtained to be an efficient balancing technique in this setup, which is followed by SMOTE-ENN.

Table 7. Mean AUC scores in the streaming context

Model	Balancing Technique					
	SMOTE	SMOTE-Tomek	SMOTE-ENN	ADASYN	V-GAN	W-GAN
LR	0.811	0.811	0.811	0.691	0.842	0.875
KNN	0.854	0.855	0.855	0.774	0.870	0.896
DT	0.889	0.889	0.889	0.810	0.908	0.909
RF	0.896	0.898	0.898	0.821	**0.910**	0.911

Table 8. Mean Sensitivity scores in the streaming context

Model	Balancing Technique					
	SMOTE	SMOTE-Tomek	SMOTE-ENN	ADASYN	V-GAN	W-GAN
LR	0.806	0.806	0.806	0.667	0.848	0.914
KNN	0.860	0.861	0.861	0.746	0.828	0.859
DT	0.895	0.895	0.895	0.779	0.865	0.867
RF	0.899	0.895	0.895	0.793	**0.871**	0.873

Table 9. Mean Specificity scores obtained by various ML techniques over sliding windows

Model	Balancing Technique					
	SMOTE	SMOTE-Tomek	SMOTE-ENN	ADASYN	V-GAN	W-GAN
LR	0.815	0.816	0.816	0.715	0.836	0.836
KNN	0.849	0.850	0.850	0.801	0.913	0.933
DT	0.882	0.882	0.882	0.842	**0.950**	0.952
RF	0.893	0.892	0.892	0.810	0.948	0.949

Table 10. Rules obtained by best DT model in the streaming context

Rule No.	RULE
1	IF $X_9 \leq 0.0$ AND $X_2 \leq 0.02$ AND $X_8 \leq 0.50$ AND $X_3 \leq 0.02$ THEN **Negative Class**
2	IF $X_9 \leq 0.0$ AND $0.02 \leq X_2 \leq 0.47$ AND $X_8 \leq 0.50$ AND $X_3 \leq 0.02$ THEN **Positive Class**
3	IF $X_9 \leq 0.0$ AND $X_2 \leq 0.31$ AND $X_8 \leq 0.50$ AND $X_3 > 0.02$ THEN **Negative Class**
4	IF $X_9 \leq 0.0$ AND $0.31 < X_2 \leq 0.47$ AND $X_8 \leq 0.50$ AND $X_3 > 0.02$ THEN **Negative Class**
5	IF $X_9 \leq 0.0$ AND $X_2 > 0.47$ $X_8 > 0.50$ THEN **Negative Class**
6	IF $X_9 \leq 0.0$ AND $0.47 < X_2 \leq 0.50$ AND $X_8 \leq 0.50$ THEN **Positive Class**
7	IF $X_9 \leq 0.0$ AND $0.47 < X_2 \leq 0.50$ AND $X_8 > 0.50$ THEN **Negative Class**
8	IF $X_9 \leq 0.0$ AND $0.47 < X_2 \leq 0.50$ AND $0.41 \leq X_{11} \leq 0.81$ OR $X_{11} > 0.94$ THEN **Negative Class**
9	IF $X_9 > 0.0$ AND $X_3 \leq 0.00$ AND $X_8 \leq 0.50$ AND $X_9 \leq 0.83$ THEN **Negative Class**
10	IF $X_9 > 0.0$ AND $X_3 \leq 0.00$ AND $X_8 \leq 0.50$ AND $X_9 > 0.83$ AND $X_0 \leq 0.22$ THEN **Negative Class**
11	IF $X_9 > 0.0$ AND $X_3 \leq 0.00$ AND $X_8 \leq 0.50$ AND $X_9 > 0.83$ AND $X_0 > 0.22$ THEN **Positive Class**
12	IF $X_9 > 0.0$ AND $X_3 \leq 0.00$ AND $X_8 > 0.50$ AND $X_2 \leq 0.56$ THEN **Negative Class**
13	IF $X_9 > 0.0$ AND $X_3 \leq 0.00$ AND $X_8 > 0.50$ AND $X_2 \leq 0.56$ THEN **Positive Class**
14	IF $0.0 < X_9 \leq 0.94$ AND $X_3 > 0.00$ AND $X_8 \leq 0.38$ THEN **Negative Class**
15	IF $0.0 < X_9 \leq 0.94$ AND $X_3 > 0.00$ AND $X_8 > 0.38$ THEN **Positive Class**

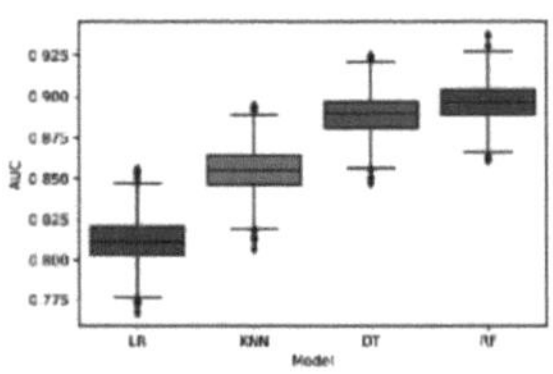

Fig. 3. AUC obtained when SMOTE is employed in the streaming context

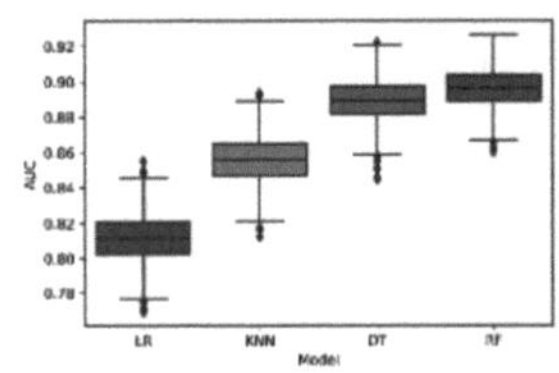

Fig. 4. AUC obtained when SMOTE-Tomek is employed in the streaming context

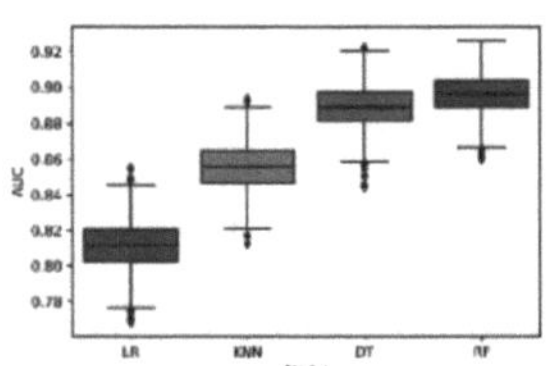

Fig. 5. AUC obtained when SMOTE-ENN is employed in the streaming context

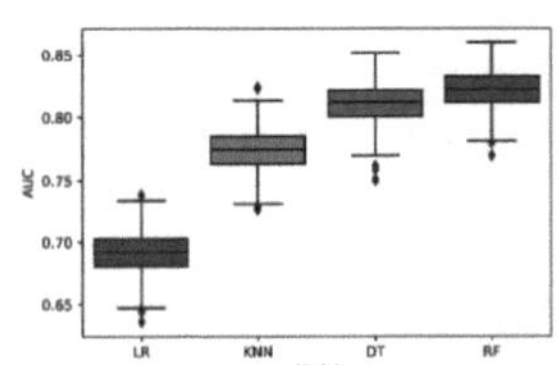

Fig. 6. AUC obtained when ADASYN is employed in the streaming context

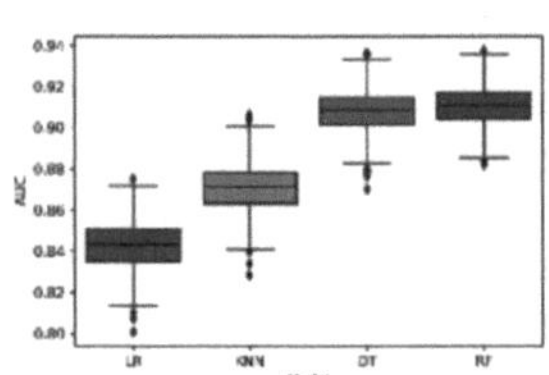

Fig. 7. AUC obtained when V-GAN is employed in the streaming context

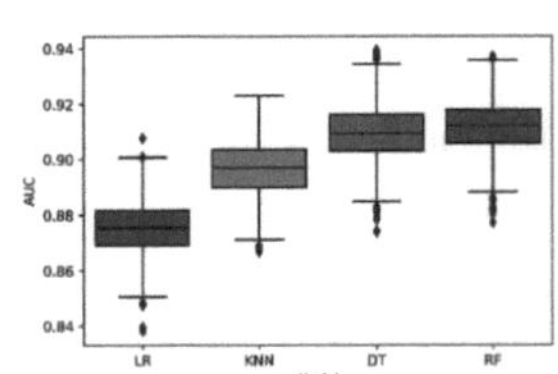

Fig. 8. AUC obtained when W-GAN is employed in the streaming context

5.4 Statistical Testing of the Results

The Wilcoxon test is a non-parametric t-test performed on top-2 models by considering the AUC obtained over several windows and results are presented in Table 11.

The null hypothesis is, H_0: *Both the algorithms are statistically equal.*

The alternate hypothesis is, H_1: *Both the algorithms are statistically not equal.*

Table 11. Paired Wilcoxon-test results

Model	Parameter	
	Wilcoxon test-statistic	p-value
DT vs RF	204017.0	$5.34\text{x}10^{-07}$

The p-value determines whether to accept or reject the null hypothesis. The significance level is chosen to be 5%. Hence, the p-value should be less than 5% to reject the null hypothesis. Table 11 infers that in the streaming context, RF is more statistically significant than the next-best performing model DT.

5.5 Practical Challenges

While deploying these fraud detection models using streaming analytics could face latency, resource and integration issues which are handled in this study as follows:

- ***Latency:*** Real time fraud detection demands models to be less complex. Hence, we have chosen our models to be light weight.
- ***Resource utilization:*** Underlying framework should handle large transaction volumes effectively. This is well handled by Spark.
- ***Integration issues:*** While deploying a new framework, it is normal to arise **integration issues with legacy banking systems**. To meet this requirement, we have chosen Spark framework which have wider options and could be integrated easily.

6 Conclusions, Limitations & Future Scope

In the current study, we developed a scalable ML model for the identification of ATM fraudulent transactions in the ATM transaction dataset collected from India. We devised two different mechanisms suitable for both streaming and static environments respectively. The collected dataset is observed to be highly imbalanced. Therefore, we incorporated various sampling techniques viz., SMOTE, ADASYN, and GAN. Of all the sampling techniques we employed, it turned out that V-GAN significantly outperformed the rest of the models. Further, we employed various machine learning techniques in the static and streaming contexts. In the static context, RF outperformed the rest of the models by achieving an AUC of 0.973, followed by GBT with a similar AUC, and in the streaming context, RF obtained an AUC of 0.910. RF turned out to be statistically significant in both the static and streaming contexts.

In the future, the following work can be taken up: current proposed sliding window approach helps to mitigate the risk of correctly identifying fraudulent ATM transactions.

However, we observed that there is one certain limitation, which is as follows: every window assumes that it has received a certain amount of positive sample information. To mitigate this risk, (i) One Class Classification (OCC) model could be employed instead of binary classification models; (ii) Exploring deep learning methods such as auto encoder, variational auto encoder can also be taken up in the future.

References

1. Phua, C., Lee, V., Smith, K., & Gayler, R.: A comprehensive survey of data mining-based fraud detection research. Artif. Intell. Rev. **24**(1), 1–14. abs/1009.6119 (2005)
2. Wang, J., Liao, Y., Tsai, T., Hung, G.: Technology-based financial frauds in Taiwan: issue and approaches. In IEEE Conference on: Systems, Man and Cyberspace Oct, pp. 1120–1124 (2006)
3. Srivastava, A., Kundu, A., Majumdar, A.K.: Credit card fraud detection using hidden Markov model. IEEE Trans. Dependable Secure Comput. **5**(1), 37–48 (2008)
4. Zaharia, M., et al.: Apache Spark: A unified engine for big data processing. Commun. ACM **59**(11), 56–65 (2016)
5. Ngai, E.W.T., Hu, Y., Yong, Wong, Y.H., Chen, Y., Sun, X.: The application of data mining techniques in financial fraud detection: A classification framework and an academic review of literature. Decis. Support Syst. **50**, 559–569, (2011)
6. Edge, M.E., Sampaio, P.R.F.: The design of FFML: A rule-based policy modelling language for proactive fraud management in financial data streams. Expert Syst. Appl. **39**(8), 9966–9985 (2012)
7. Krishna, G.J., Ravi, V.: Evolutionary computing applied to customer relationship management: A survey. Eng. Appl. Artif. Intell. **56**, 30–59 (2016)
8. Kumar, B.S., Ravi, V.: A survey of the applications of text mining in financial domain. Knowl.-Based Syst. **114**, 128–147 (2015)
9. Wei, W., Li, L., Cao, Y., Chen, J.: Effective detection of sophisticated online banking fraud on extremely imbalanced data. World Wide Web (2012). https://doi.org/10.1007/s11280-012-0178-0
10. Chan, P.K., Fan, W., Prodromidis, A.L., Stolfo, S.J.: Distributed data mining in credit card fraud detection. IEEE Intell. Syst. **14**(6), 67–74 (1999)
11. Chen, R., Chen, T., Lin, C.: A new binary support vector system for increasing detection rate of credit card fraud. Int. J. Pattern Recognit Artif Intell. **20**(2), 227–239 (2006)
12. Dorronsoro, J.R., Ginel, F., Sánchez, C., Cruz, C.S.: Neural fraud detection in credit card operations. IEEE Trans. Neural Networks **8**(4), 827–834 (1997)
13. Quah, J.T.S., Sriganesh, M.: Real-time credit card fraud detection using computational intelligence. Expert Syst. Appl. **35**(4), 1721–1731 (2008)
14. Sánchez, D., Vila, M.A., Cerda, L., Serrano, J.M.: Association rules applied to credit card fraud detection. Expert Syst. Appl. **36**(2), 3630–3640 (2009)
15. Zaslavsky, V., Strizhak, A.: Credit card fraud detection using self-organizing maps. Inf. Secur. **18**, 48–63 (2006)
16. Syeda, M., Zhang, Y., & Pan, Y. (2002). Parallel granular neural networks for fast credit card fraud detection. In 2002 IEEE International Conference on Fuzzy Systems 1, pp. 572–577
17. Ayeni, R.K., Adebiyi, A.A., Okesola, J.O., Igbekele, E.: Phishing attacks and detection techniques: a systematic review. In: 2024 International Conference on Science, Engineering and Business for Driving Sustainable Development Goals (SEB4SDG), pp. 1–17 IEEE (2024)
18. Arya, A.S., Ravi, V., Tejasviram, V., Sengupta, N., Kasabov, N.: Cyber fraud detection using evolving spiking neural network. In: 2016 11th International Conference on Industrial and Information Systems (ICIIS), pp. 263–268 IEEE (2016)

19. Ravisankar, P., Ravi, V., Rao, G.R., Bose, I.: Detection of financial statement fraud and feature selection using data mining techniques. Decis. Support Sys. **50**(2), 491–500 (2011)
20. Farquad, M.A.H., Ravi, V., Raju, S.B.: Analytical CRM in banking and finance using SVM: a modified active learning-based rule extraction approach. Int. J. Electr. Customer Relationsh. Manag. **6**(1), 48–73 (2012)
21. Vasu, M., Ravi, V.: A hybrid under-sampling approach for mining unbalanced datasets: applications to banking and insurance. Int. J. Data Min. Model. Manag. **3**(1), 75–105 (2011)
22. Kumar, G.S., Ravi, V.: A new approach for credit card fraud detection using unsupervised learning. In: Proceedings of IEEE International Conference on Computational Intelligence and Computing Research, pp. 1–6 (2013)
23. Dhanalakshmi, R., Prabhu, C., Chellapan, C.: Detection of phishing websites and secure transactions. In: International Journal Communication & Network Security (IJCNS), pp. 15–21 (2011)
24. Lakshmia, V., Vijaya, M.: Efficient prediction of phishing websites using supervised learning algorithms. In Proceedings of International Conference on Communication Technology and System Design 2011. Procedia Engineering, 30, 798–805 (2012)
25. Tayal, K., Ravi, V.: Particle swarm optimization trained class association rule mining: application to phishing detection. In Proceedings of the International Conference on Informatics and Analytics (ICIA-16), pp. 1–8. Association for Computing Machinery (2016)
26. Tayal, K., Ravi, V.: Fuzzy association rule mining using binary particle swarm optimization: application to cyber fraud analytics. IEEE International Conference on Computational Intelligence and Computing Research (ICCIC2015), pp. 318–322 (2015)
27. Pradeep, G., Ravi, V., Kaushik, N., Deekshatulu, B. L., Bose, I., Aditya, A.: Fraud detection in financial statements using evolutionary computation based rule miners. In Panigrahi, B., Suganthan, P., Das, S. (Eds.), Swarm, Evolutionary, and Memetic Computing. SEMCCO 2014. Lecture Notes in Computer Science, 8947 (2014)
28. Sundarkumar, G. G., Ravi, V., Nwogu, I., Govindaraju, V.: Malware detection via API calls, topic models and machine learning. 2015 IEEE International Conference on Automation Science and Engineering (CASE), pp. 1212–1217 (2015)
29. Sheen, S., Anitha, R., Sirisha, P.: Malware detection by pruning of parallel ensembles using harmony search. Pattern Recogn. Lett. **34**(14), 1679–1686 (2013)
30. Kamaruddin, S., Ravi, V.: Credit card fraud detection using big data analytics: use of PSOAANN based one-class classification. In Proceedings of the International Conference on Informatics and Analytics (ICIA-16), pp. 1–8. Association for Computing Machinery (2016)
31. Sundarkumar, G.G., Ravi, V.: A novel hybrid undersampling method for mining unbalanced datasets in banking and insurance. Eng. Appl. Artif. Intell. **37**, 368–377 (2015)
32. Thennakoon, A., Bhagyani, C., Premadasa, S., Mihiranga, S., Kuruwitaarachchi, N.: Real-time credit card fraud detection using machine learning. In 2019 9th International Conference on Cloud Computing, Data Science & Engineering (Confluence), pp. 488–493. Noida, India (2019)
33. Arya, M., Sastry, H.: DEAL - Deep Ensemble algorithm framework for credit card fraud detection in real time analytics. J. Stat. Manag. Syst. **8**(2), 71–83 (2020)
34. Mittal, S., Tyagi, S.: Computational Techniques for Real-Time Credit Card Fraud Detection. In: Gupta, B., Perez, G., Agrawal, D., Gupta, D. (eds.) Handbook of Computer Networks and Cyber Security, pp. 473–489. Springer (2020)
35. Abakarim, Y., Lahby, M., Attioui, A.: An efficient real time model for credit card fraud detection based on deep learning. In Proceedings of the 12th International Conference on Intelligent Systems: Theories and Applications (SITA'18), pp. 1–7. Association for Computing Machinery (2018)

36. Rajeshwari, U., Babu, B.S.: Real-time credit card fraud detection using Streaming Analytics. In 2016 2nd International Conference on Applied and Theoretical Computing and Communication Technology (iCATccT), pp. 439–444 (2016)
37. Carcillo, F., Dal Pozzolo, A., Le Borgne, Y., Caelen, O., Mazzer, Y., Bontempi, G.: SCARFF: A scalable framework for streaming credit card fraud detection with spark. Inform. Fusion **41**, 182–194 (2018)
38. Zaharia, M., Chowdhury, M., Das, T., Dave, A., Ma, J., McCauley, M., ... & Stoica, I.: Resilient distributed datasets: A fault-tolerant abstraction for in-memory cluster computing. In 9th USENIX Conference on Networked Systems Design and Implementation (NSDI'12) (p. 2) (2012, April)
39. Zaharia, M., Das, T., Li, H., Shenker, S., Stoica, I.: Discretized streams: an efficient and fault-tolerant model for stream processing on large clusters. In Proceedings of the 4th USENIX conference on Hot topics in cloud computing, p. 10 (2012)
40. Wahid, A., Msahli, M., Bifet, A., Memmi, G.: NFA: A neural factorization autoencoder based online telephony fraud detection. Digit. Commun. Netw. **10**(1), 158–167 (2024)
41. Abbassi, H., Mendili, S.E., Gahi, Y.: Digital banking fortification: a real-time isolation forest architecture for detecting online transaction fraud. Eng. Res. Express **6**(2), 025214 (2024)
42. Khodabandehlou, S., Golpayegani, A.H.: FiFrauD: unsupervised financial fraud detection in dynamic graph streams. ACM Trans. Knowl. Discov. Data **18**(5), 1–29 (2024)
43. Nguyen, T.T., Phan, T.C., Pham, H.T., Nguyen, T.T., Jo, J., Nguyen, Q.V.H.: Example-based explanations for streaming fraud detection on graphs. Inf. Sci. **621**, 319–340 (2023)

FedELF: A Privacy-Preserving Federated Classification Using Iterative Extreme Learning Factory

Polaki Durga Prasad, Yelleti Vivek, and Vadlamani Ravi(✉)

Center for Artificial Intelligence and Machine Learning, Institute for Development and Research in Banking Technology, Hyderabad 500076, India
vravi@idrbt.ac.in

Abstract. Federated Learning (FL) has gained significant traction due to its decentralized training approach while preserving privacy. The Extreme Learning Machine (ELM) emerged as a competitive technique for diverse classification and regression tasks, owing to its advantages such as fewer trainable parameters and expedited training times. Nonetheless, it suffers from overfitting and suboptimal generalization. Motivated by these considerations, we introduce FedELM-SGD, an iterative variant of ELM which integrates mini-batch stochastic gradient descent (SGD) within privacy-preserving federated learning settings. We also propose an extreme learning factory under federated settings (FedELF) to counteract the impact of random parameter initialization in ELM. In this study, we outline a two-stage methodology: firstly, the generation of privacy-preserving datasets through perturbation of real datasets using Differential Privacy techniques; secondly, the invocation of either FedELM-SGD or FedELF to construct a globally shared classification model. We demonstrate the efficacy of our proposed methods across various finance and medical-related datasets.

Keywords: Federated Learning · ELM · Ensemble ELM · SGD

1 Introduction

In the realm of data privacy and security, the advent of Federated Learning (FL), as introduced by Konecny et al. [1], proffers a novel approach to train machine learning (ML) models without the necessity to transmit sensitive data to a central server. This method streamlines training across remote, siloed data centers that maintain private data, with a central server orchestrating the collaboration of participating clients to develop a higher-quality model. FL operates beneath the assumption that many clients possess limited computational resources, non-independent and identically distributed (non-i.i.d) data, and constrained communication network capabilities, rendering it a domain-agnostic methodology. Its applications are particularly predominant in the financial sector for detecting fraudulent activities, such as credit or debit card fraud [2, 3], which have been identified as significant contributors to financial losses [2].

S. Dhar et al. (Eds.): AGC 2025, CCIS 2686, pp. 223–254, 2026.
https://doi.org/10.1007/978-3-032-09153-6_14

Research by Yang et al. [3] showcased the efficacy of a federated fraud detection system based on artificial neural networks, outperforming traditional fraud detection systems trained solely on data from individual banks. Moreover, FL-based fraud detection systems offer enhanced privacy protection. Nonetheless, adopting FL techniques involves several challenges, including the statistical heterogeneity of client data, communication overhead, system heterogeneity, and privacy concerns.

Federated Learning (FL) encompasses distinct categories, as delineated by Yang et al. [4]: (i) Horizontal federated learning (HFL), wherein data points are partitioned while feature space overlaps; (ii) Vertical federated learning (VFL), where sample space aligns while feature space varies, and (iii) Federated Transfer learning (FTL), characterized by an overlap in small sample space but varied feature space. In the context of the present study, the methodology proposed is tailored for HFL to address real-world demands effectively. The list of more frequent abbreviations used in the current study is presented in Table A.1 in the Appendix.

In data management within FL frameworks, various privacy-preserving strategies are often integrated to mitigate the risk of information leakage. Notable techniques, as identified by Xuefei et al. [5], include

- Differential Privacy (DP), leverages randomized algorithms to discern patterns within dataset groups while preserving individual privacy.
- Secure multi-party communication protocols to ensure privacy, authenticity, anintegrity during network communication among multiple parties.
- Homomorphic encryption schemes enable mathematical computations directly on encrypted data, partially or fully.

Among these, DP was chosen, particularly due to its lower computational overhead and simplicity, which are achievable through data perturbation techniques employing Laplacian, Gaussian, or Exponential distributions. The Laplacian distribution, widely employed in anomaly detection [7] and intrusion detection [8], was chosen for its popularity and simplicity in the current study.

The parameterization of FL typically involves local models trained on data specific to each client, denoted as {1,..., k} (refer to Table 1), generated by k distinct nodes. Further, data at each node $t \in k$ is non-i.i.d across the network and is generated by a distinct distribution $Xt \sim \mathbb{P}t$, reflecting statistical heterogeneity. These FL configurations necessitate multiple communication rounds between clients and servers, resulting in communication overhead bottlenecks.

The Extreme Learning Machine (ELM), introduced by Huang et al. [9], is a single-layered neural network architecture. It serves as an efficient alternative to traditional neural networks and is capable of learning the non-linearity of complex datasets. ELM finds application in diverse ML tasks such as classification [10] and regression problems [11]. ELM learning occurs through two primary methods: (i) one-pass learning and (ii) iterative methods. Iterative methods, including standalone ELM model learning and ensemble ELM models—referred to as ELM Factory—are ranked higher due to their superior learning capabilities, handling issues like poorer convergence, inability to discern underlying patterns, and propensity for overfitting observed in one-pass learning.

This prompted an iterative ELM approach within federated learning settings, utilizing mini-batch stochastic gradient descent (SGD) over gradient methods for expedited convergence.

This study presents a two-stage methodology addressing privacy concerns and optimizing model performance within a federated learning (FL) framework. In the initial phase, we prioritize data privacy through Differential Privacy (DP) perturbation techniques, aligning with regulatory standards such as GDPR. Moving to the second stage, we propose FedELF, an iterative approach to Extreme Learning Machine (ELM) tailored for federated settings. FedELF incorporates a majority voting principle and leverages Federated Averaging (FedAvg) for aggregation. Additionally, we introduce FedELM-SGD, a standalone ELM variant featuring mini-batch Stochastic Gradient Descent (SGD) learning optimized for federated environments.

Our methodologies are validated using tabular datasets from finance and medical domains, with a focus on Horizontal Federated Learning (HFL), where the feature space is shared among clients while differing in the sample space. This comprehensive approach aims to reconcile privacy imperatives with model efficacy in FL paradigms, adhering to rigorous research standards.

The following are the significant contributions to the current study:

- Introduced a two-phase methodology tailored for training iterative learning techniques, explicitly addressing tabular datasets from the finance and medical sectors.
- Presented FedELM-SGD, a novel approach involving iterative training of ELM within FL framework, employing Stochastic Gradient Descent (SGD).
- Proposed FedELF, an ensemble methodology comprising several standalone FedELM-SGD models, enhancing model robustness and performance in federated settings.

The remainder of the paper is structured as follows: The literature review was presented in Sect. 2. Section 3 provides background, and Sect. 4 presents the proposed methodology. Section 5 describes datasets, while Sect. 6 provides an overview of the experimental setup. Section 7 presents experimental results, while then Sect. 8 concludes the paper.

2 Literature Survey

We Liang et al. [12] introduced the Online Sequential Extreme Learning Machine (OS-ELM), a Single Layer Feedforward Network (SLFN) incorporating both additive and radial basis function nodes within a unified framework. Their iterative ELM algorithm updates model parameters with newly acquired training data, demonstrating superior performance over other sequential learning algorithms within constrained timeframes. Qin et al. [13] proposed a model selection and modified aggregation approach for Federated Learning (FL) focused on anomaly detection. They utilized OS-ELM models trained at edge devices, with iterative parameter aggregation at the server based on local model performance metrics. Chen et al. [14] devised Fed-ELMS, an online federated ELM algorithm building upon OS-ELM principles. The global model parameters are initialized with in-house training data and updated via aggregated local model weights.

Xiao et al. [15] developed ISPW-ADMM, a decentralized FL framework integrating stochastic parallel random walk with alternating direction thod of multipliers algorithm. Their method updates weights between ELM model layers using stochastic gradients from local client models. Ito et al. [16] proposed an on-device FL approach combining OS-ELM with an autoencoder. Autoencoder filters input data with high loss, while Elastic ELM merges training and intermediate client results. Sun et al. [17] introduced FedIO, a one-pass Vertical FL method jointly training FL models with both inner- and outer-hospital data. Clients aggregate hidden layer activations using least square estimates to estimate ELM output weights. Qin et al. [18] presented MLMG, a multi-local and multi-global FL mechanism clustering non-iid user data for model training. Local OS-ELM models are trained on clustered data batches, with server clustering and updating parameters accordingly. Wang et al. [10] proposed FLTrELM, an FL framework leveraging federated transfer learning with ELM. Clients update local models from a global ELM model trained on a cloud server, utilizing the Lagrange method for model parameter updates. Yu [19] devised a fall detection algorithm in FL using ELM, updating local models with OS-ELM and aggregating at the server. Qiu [20] introduced NTFLELM, a network traffic classification method employing FL and ELM. Global ELM model parameters are initialized and updated using least square solutions and local feature extraction. Ravi and Singh [21] proposed AAELF, a one-class classification algorithm employing Auto Associative Extreme Learning Machines in an ensemble framework, utilizing Evolving Clustering for dataset preprocessing before ELM input.

Our current study presents the following distinctive features not addressed in existing literature:

- ***Privacy Preservation with Differential Privacy (DP):*** Unlike prior research, our study prioritizes data privacy and compliance with data protection regulations. We employ Differential Privacy (DP) techniques in the initial phase, ensuring the confidentiality of sensitive information. To the best of our knowledge, even though employing DP is popular, it is not explored in the context of tabular datasets in conjunction with ELM models.
- ***Ensemble ELM in Horizontal Federated Learning (HFL) settings:*** While earlier works primarily focused on utilizing OS-ELM as an iterative learning method within VFL settings, our study diverges by proposing an ensemble approach within HFL settings. Further, recognizing the advantages of ensemble models in mitigating overfitting, we introduce an ensemble ELM model optimized through Stochastic Gradient Descent (SGD) optimization. This novel application of ensemble ELM models in HFL settings represents a unique contribution to the field.

3 Preliminaries

This section discusses the employed architectures for classification in the current study.

3.1 Differential Privacy

Differential privacy (DP) [6] is a privacy-preserving technique to publicly share information about a database by describing data patterns of groups within the database without revealing uniquely identifiable information. A randomized mechanism is differentially private if an observer could not able to tell whether a particular sample was used in the computation. The formal definition of ϵ-DP (a variant of DP) is given below.

Definition of ϵ-DP: A randomized algorithm or mechanism (M) satisfies ϵ-DP if all the neighbouring datasets of $X, X\prime \in \mathrm{X}^{\mathrm{n}}$ which differ only by the single record and for all $S \subseteq Y$, where Y is the set of all possible outcomes of the mechanism M, satisfies the following condition mentioned in Eq. 1.

$$\Pr[M[X] \in S] \leq e^{\epsilon} Pr[M[X\prime] \in S] \tag{1}$$

where ϵ is the privacy budget, and the smaller the value, the greater the privacy.

Using DP, we can either add noise to statistical queries (Randomized mechanism) or directly to the original datasets (Privacy Preserving Datasets) [6]. Central differential privacy [6] deals with adding noise to the dataset before the training process and to the metadata of the trained model.

Laplacian mechanism [6]: For a given query function, f: $\mathrm{X}^{\mathrm{n}} \rightarrow \mathrm{Y}$ and $\epsilon > 0$. The Laplacian mechanism M is defined as given in Eq. 2.

$$M(X) = f(X) + \mathcal{L}(0, \frac{\Delta f}{\epsilon}) \tag{2}$$

where Δf is the magnitude of the maximum difference between output values of query function 'f'.

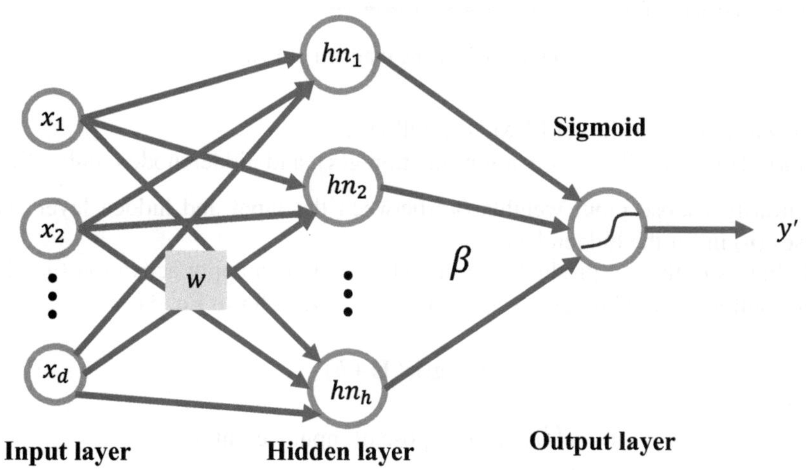

Fig. 1. Simple Architecture of ELM

3.2 Extreme Learning Machine

ELM [9] is a fast-learning algorithm for single hidden layer feed-forward networks (SLFNs). It mainly comprises three layers (i) Input layer, (ii) Hidden layer, and (iii)

Output layer. The architecture of ELM is depicted in Fig. 1. The weights between the input layer and hidden layer and biases in the hidden layer are initialized randomly, and weights between the hidden and output layers are estimated using the least square solution.

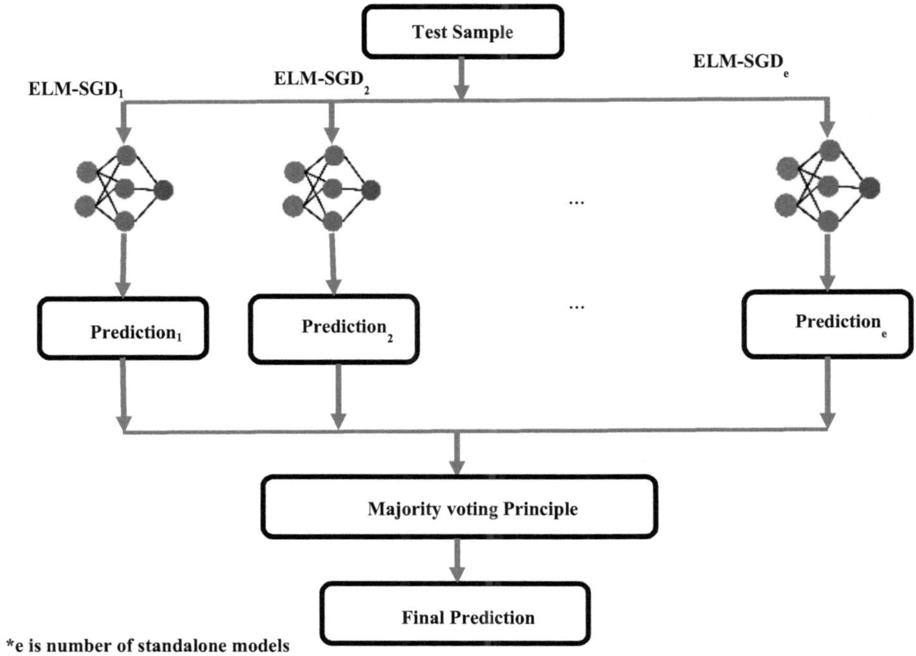

Fig. 2. Schematic diagram of ELF

The learning algorithm of ELM is as follows:
Input: Train data X_{train}, activation function g(x), and hidden node number 'h.'

1. Arbitrarily choose input weights ($\boldsymbol{w}$) between the input and hidden layer and the biases ($\boldsymbol{b}$) in for the hidden layer.
2. Add biases to the dot product of train data and $\boldsymbol{w}$ then apply activation function on this result to get the hidden layer output matrix **H**. (refer to Eq. 3)

$$\mathbf{H} = \mathbf{g}(\boldsymbol{w}^T\mathbf{H} + \boldsymbol{b}) \tag{3}$$

$(\because \boldsymbol{w}^T$ is the transpose of input weights)

3. Now, the weights $\widehat{\boldsymbol{\beta}}$ between the hidden and output layer are estimated using least-square solutions of a generalized linear system $\mathbf{T} = \mathbf{H}\,\beta$ i.e. solution that minimizes the loss function (refer to Eq. 4).

$$\min\left|H\widehat{\beta} - T\right| \tag{4}$$

4. Authors have shown that $\widehat{\beta} = \mathbf{H}^{+}\mathbf{T}$ is a unique minimum norm least-square solution of $\mathbf{H}\,\boldsymbol{\beta} = \mathbf{T}$ where $\mathbf{H}^{+}$ is the moose-Penrose generalized inverse [11, 34] of $\mathbf{H}$.
5. Using the estimated output weights, predicted values of input data can be calculated by (refer to Eq. 5)

$$y = \mathbf{H}'\widehat{\beta} \tag{5}$$

where $\mathbf{H}'$ is the output of the hidden layer when test data is applied to the input layer.

3.3 Extreme Learning Factory

It is an ensemble of multiple vanilla ELM models [21] (refer to Fig. 2). The ensemble is done based on an aggregation schema called majority voting. All standalone ELM models are trained individually on a different set of data batches. After these models are trained, a test data sample is given to each of these individual models for prediction. The majority voting is applied to these predictions to find the most probable prediction and assign this to the given input test sample.

4 Proposed Methodology

This section presents our two-stage methodology in detail, accompanied by a schematic diagram depicted in Fig. 3. In Phase I, we generate perturbed data using differential privacy with Laplacian and Gaussian distributions. This phase ensures data privacy and compliance with regulatory standards such as GDPR. Moving to Phase II, we commence FL training. We introduce two algorithms for this phase: FedELF and FedELM-SGD, outlined in Algorithms 1 and 2. It is crucial to note that Phase I, involving the generation of perturbed data, is a standard procedure regardless of the employed ELM technique.

Phase I: Generation of perturbed data for implementing DP
This phase consists of two steps and is part of the non-federated settings (refer to Fig. 4).

- For experimentation, the dataset is partitioned among K distant clients and server by employing stratified and simple random sampling techniques respectively.
- Generating perturbed datasets at each client and server using DP, independently.

Step a: Data Partitioning
For experimentation, the original dataset is partitioned among clients and the server in the following way:

- We used the stratified random sampling technique to adopt the *m:(1-m)* proportion between server and client nodes. Importantly, *m%* of the perturbed dataset is reserved for server-side evaluation. This aids in evaluating the updated global model.
- The rest of the *(1-m) %* of the dataset is shared among clients. We used a random sampling strategy to distribute the dataset among clients, which ensures the non-i.i.d nature of the dataset among clients. Now, client $\{c_1, c_2, \ldots, c_k\}$ receives dataset$\{p_1, p_2, \ldots, p_k\}$ respectively (Fig. 3). Further, each client receives a varied number of samples that meet real-world requirements.

- In our approach, we fixed m to be 10%, and hence the rest 90% of the dataset is partitioned, and shared among clients (Fig. 3). Suppose, there are 1000 total samples present in a dataset, then 100 ($m = 10\%$) is reserved for the server, and the remaining 900 samples (*1-m%*) are shared among clients by following a random distribution. For instance, there are two client nodes, then one may receive more samples than the other. Here, in our example, let us consider one client received 550 samples and the other received 350 samples.
- Once the dataset is shared among the clients, if any of the datasets is imbalanced in nature, then the oversampling technique (SMOTE [24] or ADASYN [31]) is employed on the training set of the minority class.

Step b: Perturbed dataset generation using DP

The privacy-preserving dataset is a way of preserving privacy for deep learning, as described in Differential privacy (DP) (refer to Sect. 3.1) is employed at each node individually. This ensures that by employing DP technique does not violate the federated learning principles. Sun et al. [17] proposed a way to generate perturbed data by normalizing data and iteratively calculating the amount of random noise that could be added to the dataset by training a Decision Tree on the generated perturbed dataset.

Inspired by Sun et al. [17], we also maintained the calibration of the noise for each attribute and handled it individually. Notably, the noise is taken from Laplacian distribution. Later, we applied a mean query individually on each attribute to find their own Laplacian distribution from where random noise can be sampled. Once the distribution is identified, it is used to sample noise and further added to the samples.

The dataset collected at each client is split into train and test datasets in a 4:1 ratio by adopting stratified random sampling. Then apply oversampling or under sampling techniques to balance the class imbalanced train dataset. Then after concatenate the train and test dataset and apply the below algorithm to generate a perturbed dataset. Finally, the, generated perturbed dataset is split into train and test datasets using indices from the previous train dataset (Fig. 3).

The algorithm for generating the perturbed dataset is given below:

- Pick an attribute in the dataset and find that feature's maximum and minimum values.
- Now, calculate the maximum magnitude of these values.
- Find the ratio of the maximum magnitude value to a total number of samples which acts as L_1 sensitivity (Δf) of the mean query for that feature (refer to Eq. 6).
- Choose ϵ, , the privacy budget, and then calculate the standard deviation of noise distribution as ($\frac{\Delta f}{\epsilon}$).
- Now each value in the feature is added by a noise value sampled from Laplacian distribution $\mathcal{L}(0, \frac{\Delta f}{\epsilon})$.
- Repeat the above procedure for all the features.

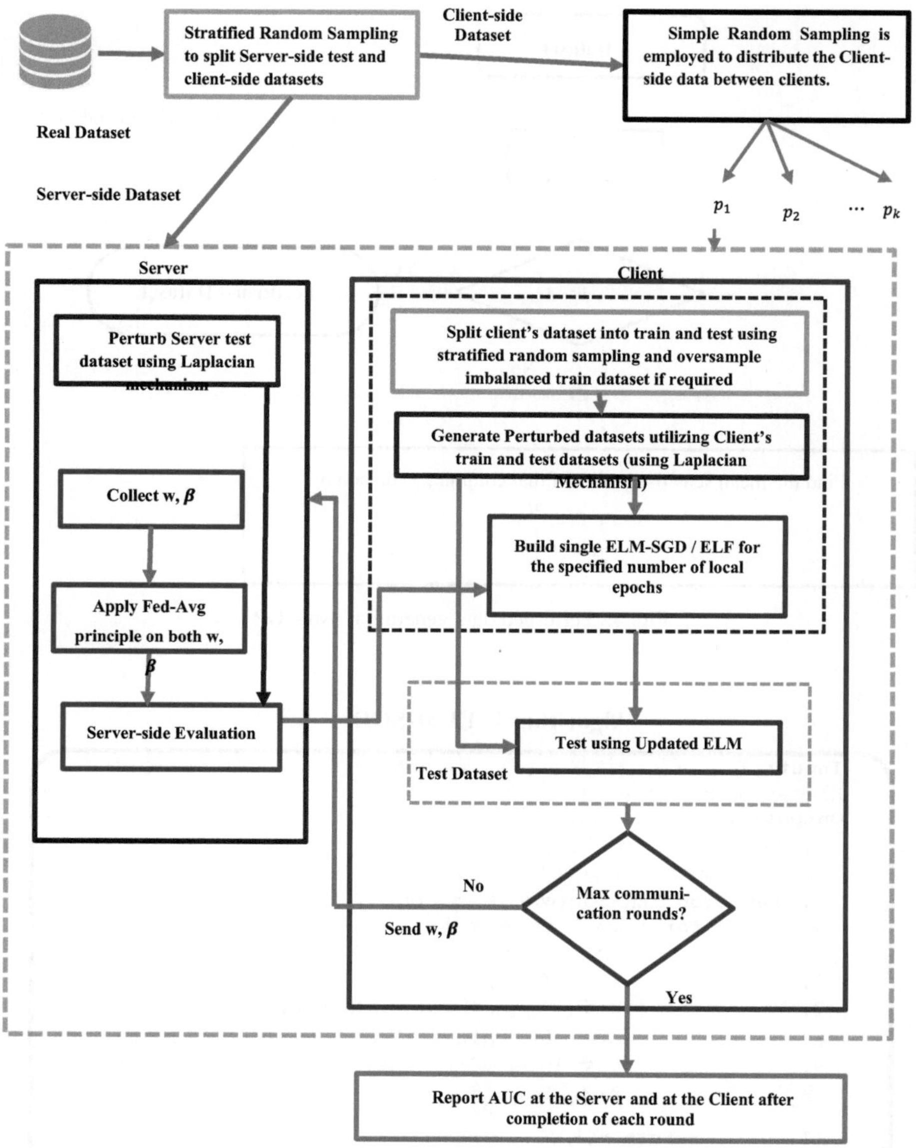

Fig. 3. Schematic diagram of the proposed methodology

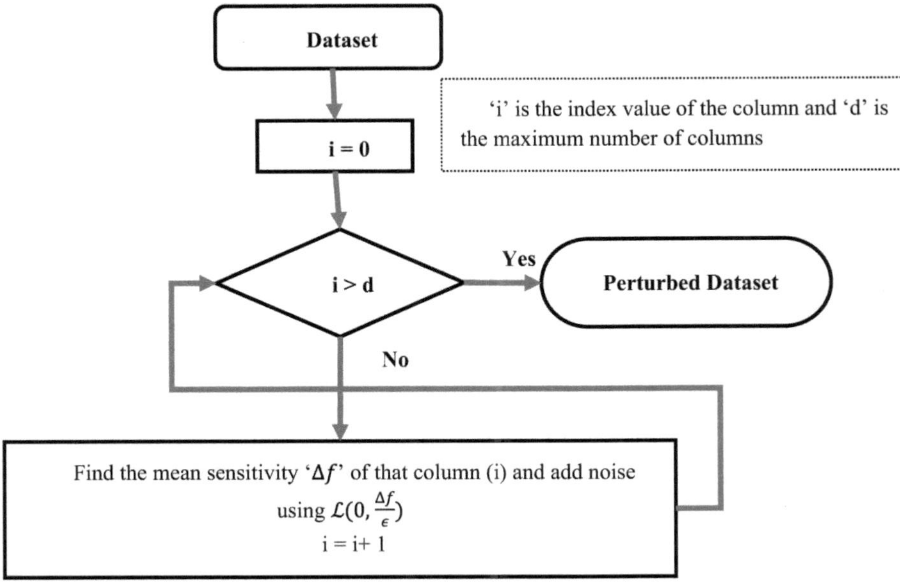

Fig. 4. Perturbed data generation using DP

Algorithm 1: ELM-SGD

```
Input: X_train, y_train, X_test, y_test, activation function (g(x),
Hidden node number (h)
Output: weights, score
X_train, y_train ← X_train
u, d ← dimentionOf(X_train)
function ELM-SGD(w,b, β̂, η, batch_size, nepochs)
        for i := 1 to nepochs
                X_batch ← sampleBatch(X_train, batch_size)
                Y_batch ← X_train[indexOf(X_batch)]
                H ←g(X_batch dot(w) + [b,b,...]_batch_size)
                pred ← Evaluate(X_batch , y_batch, w, b, β̂)
                L ← binaryCrossEntropyLoss(y_batch, pred)
                β̂ ←β̂ − η * ∂L/∂β̂
        return w,b, β̂
function Evaluate(X_test , y_test, w, b, β̂)
        H ← g(X_test dot(w) + b)
        y_pred ← H dot(β̂)
        return y_pred
```

Algorithm 2: FedELF

```
Server executes:
Input: K, X_test, number of rounds (nrounds), e number of models to ensemble
Output: AUC at the server and clients
    //initialize e number of global models in the first round of communi-
    cation using aggregated weights of clients.
    for i := 1 to e do
        $w_1^i, b_1^i, \beta_1^i \leftarrow \frac{\sum_{k=1}^{K} w_1^{i,k}}{K}, \frac{\sum_{k=1}^{K} b_1^{i,k}}{K}, \frac{\sum_{k=1}^{K} \beta_1^{i,k}}{K}$ (∵ $w_1^{i,k}$ weights of i^th model at k^th
    client)
    end for
    for j :=2 to nrounds do
        $\hat{\beta}_{j+1} \leftarrow$ clientUpdate($j, W_j, B_j, \hat{\beta}_j$)
        for i := 1 to e do //parllelly aggregate based on model index
            $\hat{\beta}^i_j \leftarrow \sum_{k=1}^{K} \hat{\beta}^{i,k}_j / K$
        end for
    end for

Client executes:
Input: D_k: private dataset at the client, e number of models to ensemble
Output: model weights, AUC
    train, test ← D_k
    X_train, y_train ← train
    X_test, y_test ← test
    // server invokes the below method to receive the result
    function clientUpdate(server_round, parameters):
        for i := 1 to e do
            if server_round equals 1:
                Randomly initialize the weights and biases of elm
    ($w_{dxh}^i, b_h^i, \hat{\beta}_h^i$)
            else
                Initialize ($w_{dxh}^i, b_h^i, \hat{\beta}_h^i$) using global parameters
    paremeters[i]
            end if
            $w_{dxh}^i, b_h^i, \hat{\beta}_h^i$ ← ELM-SGD($w_{dxh}^i, b_h^i, \hat{\beta}_h^i$, η, batch_size, nepochs)
            y_pred^i ← clientEvaluate(X_test , y_test, $w_{dxh}^i, b_h^i, \hat{\beta}_h^i$)
        end for
        y_pred = $\sum_{i=1}^{e} y_{pred}^i / e$ //ensemble the results
        for j := 0 to len(y_pred) do
            y_pred[j] ← 0 if y_pred[j] < 0.5 else 1 // this condition
    statement is used when activation function is sigmoid.
        end for
        score ← computeAUC(y_test, y_pred)
        if server_round equals 1
            return $W, B, \hat{\beta}$, score ∵ $W, B, \hat{\beta}$ are sets of meta data of lo-
    cal models.
        else
            return $\hat{\beta}$, score
```

This algorithm is also depicted in Fig. 4.

In this study, we experimented with different ϵ values and fixed $\epsilon = 1$

$$L_1 sensitivity = \frac{maximum(|\min|, |max|)}{n} \tag{6}$$

(|.| is the magnitude of the given value, 'min' and 'max' is the minimum and maximum value of that attribute, and 'n' is the number of samples).

Suppose if a node received 550 samples and 5 features; then the L_1 sensitivity is called for each feature individually, and then the laplacian noise is added to that particular feature. The same is repeated for all the features. Similarly, it is repeated for the other client node and the server node individually.

For, the Server-side test dataset, simply uses the above algorithm to generate a perturbed dataset, which can then be used for testing the aggregated model at the server after each communication round (Fig. 3).

4.1 Phase II: Iterative FedELF Training

In this phase the actual federated training occurs where each client trains their own ELF model on their perturbed train dataset and send updated model parameters for global aggregation at the server. In the section, we will discuss two algorithms i.e., ELM-SGD and ELF and their implementations in federated setup. ELF is ensemble implementation of ELM-SGD. The federated implementation of ELF is termed as FedELF similarly for ELM-SGD as FedELM-SGD.

The algorithm for ELM-SGD and FedELF (refer to Fig. 5) is given in Algorithms 1 and 2. The following are the steps for training the FedELF model:

1. In the initial round of FL training, information like the number of local models at a client, sampling distribution, and learning hyperparameters of ELM like the number of hidden nodes, learning rate, activation function, local batch size, number of local epochs, etc., will be sent to all the available clients for model initializations.
2. In the first round of FL training, each client initializes a given number of local ELM models based on parameters sent by the server. Thus, initialized models are indexed so that models with the same index will be aggregated at the server. This initialization step is done to make the clients train their personalized models, i.e., different clients will have different random initial weights and biases.
3. Now, each client will iteratively update the output weights of the standalone local ELM-SGD models by training on local train data.
4. Each ELM-SGD model in ELM Factory (refer to Algorithm 1) is trained as follows:
   a. For each local epoch, sample a batch from the local train dataset
   b. Predict the target value and then calculate the loss function
   c. Find the gradient of the loss function with respect to output weights
   d. Now, update output weights (step 10, Algorithm 1)
5. The above steps 2–4 are executed as same for each client, which is performed parallel to each other. Once the above steps are completed, weights are sent to the server, and then by utilizing the indexing information, the FedAvg is applied to weights and biases and sent back the updated ELF to each client. Then, the client-side evaluation

is performed, thereby assessing the quality of the updated global ELF model. For the subsequent rounds, clients will not re-initialize the input weights and biases; instead uses the metadata of global models, where the weights and biases are taken from the server after employing the FedAvg princip
6. By executing steps 1–5 completes one communication round. The above process is repeated until the specified number of communication rounds are completed, completing the training phase.
7. During the test phase, the test samples are given to thus trained ELF, and compute the predictions based on the majority voting principle. This is performed at each client.

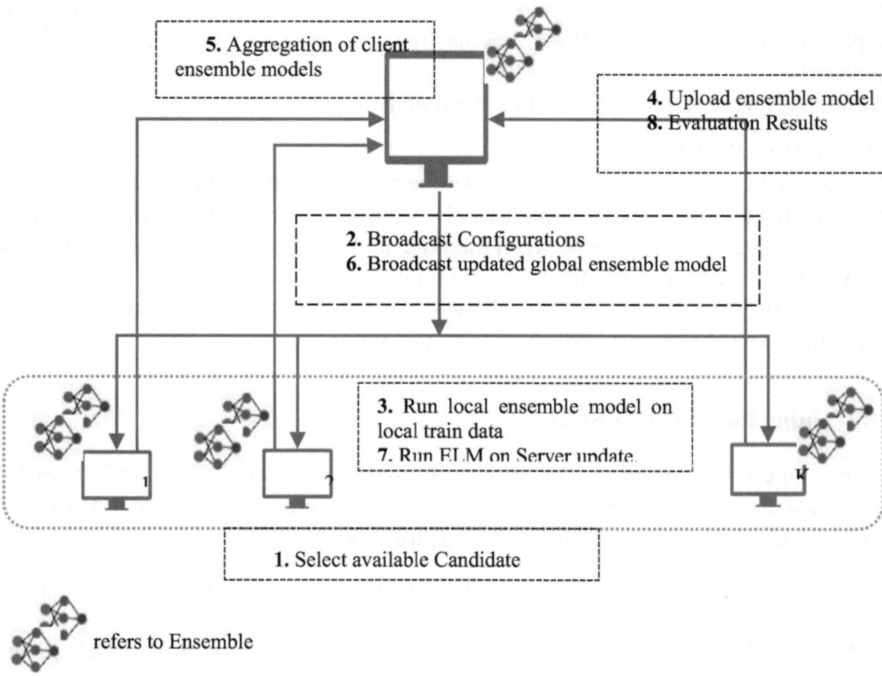

Fig. 5. Architecture of FedELF

Let us understand the proposed methodology using a dummy dataset and through the phase wise analysis, which is described as follows:

Phase-I Generation of perturbed dataset:

- *Step a: Data partitioning:* Suppose we have a dataset with 1000 samples and 5 features. In this case, 100 samples (m = 10%) are reserved for the server, while the remaining 900 samples (*1-m%*) are shared among clients.
- The distribution is random, ensuring each class receives a varied number of samples with different class proportions. For instance, in this example, one client receives 550 samples and the other client receives 350 samples.
- If the dataset turns out to be an imbalanced dataset, then one of the balancing techniques is employed on the training set of the majority class accordingly.
- *Step b Perturbation using DP:* Since one of the client nodes has received 550 samples and 5 features, the L_1 sensitivity is called for each feature individually, and then the laplacian noise is added to that particular feature. The same is repeated for all the features.
- Similarly, it is repeated for the other client node and the server node individually.

Phase-II FedELF training:

Based on the specified number of standalone models, all these models are initialized and trained for the specified number of local epochs. After each communication round, each model shares the weights between the node and to server. Suppose the number of hidden nodes in each model is 10. Then, the following is the length of the vector shared per model to the server is 10* (5 + 1), where 5 is the number of features and 1 represents the weights connected between the hidden to the output layer.

4.2 Training for FedELM-SGD

The first stage (local model initialization) is common for both FedELF and FedELM-SGD; hence discussion is obviated. Further, only the differences pertained to FedELM-SGD, a standalone model is presented here as follows:

1. In Step 2 of Phase II, there is no need for indexing the model parameters since there is only a single stand-alone model.
2. In Step 7 of Phase II, the predictions are only based on a trained local model, and no majority voting principle is employed.

The rest of the steps are the same, so discussion is obviated.

5 Dataset Description

This study evaluated five benchmark datasets spanning the banking and medical domains, as detailed in Table 1. Notably, except for the Wilcoxon Breast Cancer dataset, four of the five datasets exhibit severe class imbalance. For the OVA Omentum dataset, originally comprising 10,936 features, we employed t-statistic-based feature selection to identify the top 500 characteristics for our analysis. Given the highly imbalanced nature of the four datasets, we employed the Synthetic Minority Sampling Technique (SMOTE) [24] and Adaptive Synthetic Sampling (ADASYN) [31] to oversample the minority class, maintaining a balanced ratio of 3:2. We performed the proposed method with the FedMLP-SGD and FedLR-SGD as they are state-of-the-art methods and follows FedAvg as the central aggregation principle.

During Phase I, we allocated privacy budgets listed as [0.67, 1, 1.5, 2.25, 3.375], incrementing each value in the list by a factor of '1.5'. We applied Laplacian and Gaussian distribution variants separately to perturb the dataset. The description of the generated perturbed datasets is provided in Table 2.

Table 1. Description of Benchmark datasets.

Name of the Dataset	Field of Study	Samples	Features	Class Distribution ratio	
				Negative Class (%)	Positive Class (%)
Breast Cancer [25, 32]	Health care	699	11	65.50	34.50
AP Colon Lung [26, 33]	Health care	412	10936	69.42	30.58
Ova Omentum [26, 33]	Bio-informatics	1545	10936	95.00	5.00
Ova Uterus [26, 33]	Bio-informatics	1545	10936	91.72	8.28
Credit Card Fraud [27]	Banking	2,84,807	32	99.83	0.17
Santander Customer Transaction Prediction [28]	Banking	2,00,000	202	89.99	10.11
Polish Bankruptcy [28]	Banking	10,172	65	96.10	3.90

Table 2. Description of Benchmark datasets after minority oversampling

Name of the Dataset	Samples	Features	Class Distribution ratio	
			Negative Class (%)	Positive Class (%)
Breast Cancer	699	10	65.50	34.50
AP Colon Lung	412	100	69.42	30.58
Ova Omentum	2,318	100	70.09	29.9
Ova Uterus	2,318	100	70.09	29.9
Credit Card Fraud	4,27,211	30	69.9	30.1
Santander Customer Transaction Prediction	3,00,000	201	69.97	30.02
Polish Bankruptcy	15,258	65	69.98	30.01

6 Environmental Setup

In our experimental setup, we utilized one server and two clients, each equipped with Intel i7 6th generation processors and 32 GB of RAM. To implement Federated Learning (FL) algorithms, we leveraged the Flower framework [23], which is developed in Python programming language. This framework supports horizontal scaling and offers flexibility for seamlessly integrating newly proposed approaches with minimal engineering overhead.

6.1 AUC

AUC, i.e., an average of specificity and sensitivity (Eq. (8)), is chosen since it is proven to be a robust measure while handling imbalanced datasets.

$$AUC = \frac{(Sensitivity + Specificity)}{2} \tag{8}$$

where

$$Sensitivity = \frac{TP}{TP + FN} \tag{9}$$

and

$$Specificity = \frac{TN}{TN + FP} \tag{10}$$

where TP is a true positive, FN is a false negative, TN is a true negative, and FP is a false positive.

7 Results and Discussions

All hyperparameters used for experiments on FedELF, FedELM-SGD, and federated-based MLP are summarized in Table 3. For SGD-based Logistic Regression (LR), all hyperparameters are applicable except for hidden layer nodes. We observed that the convergence is reached after 3 epochs across all datasets. Therefore, we conducted experiments up to 10 communication rounds and maintained this number consistently for all datasets. Tables 4, 5, 6, 7, 8, 9, 10 and 11 provide a comparative analysis of the Area Under the Curve (AUC) obtained by each model after each round for Breast Cancer, Credit Card Fraud, OVA Omentum, Santander Customer Transaction Prediction, Polish Bankruptcy, AP Colon Lung and Ova Uterus datasets, respectively. Furthermore, it is worth mentioning that the number of base models (e) in FedELF is arbitrarily set to ten.

Table 3. Hyperparameters for Federated training of Iterative ELM

Dataset	Hyperparameters				
	Local epochs	Batch Size	Learning rate	Hidden Nodes	Communication rounds
Breast cancer	5	10	0.5	10	10
AP Colon Lung	5	6	0.3	30	10
Ova Omentum	8	10	0.1	65	10
Ova Uterus	5	10	0.2	45	10
Credit Card Fraud	10	32	0.05	16	10
Santander Customer Transaction Prediction	300	32	0.005	200	10
Polish Bankruptcy	50	32	0.0005	32	10

Mean AUC results obtained during the communication rounds are presented in Tables 6, 8, 10, 12 and 14 for the ADASYN oversampled dataset and in Tables 7, 9, 11, 13 and 15 for the SMOTE oversampled dataset and the same is depicted in Figs. 6, 7 and 8. Importantly, Breast cancer and AP Colon lung cancer datasets being balanced datasets, are not oversampled. Instead, the performance of various algorithms on these datasets is presented in Tables 4 and 5, respectively. In these tables, we also studied the performance of techniques when injected with Gaussian and Laplacian noise by varying five different thresholds.

While ranking the techniques, we conducted a cost-benefit analysis, by giving higher preference to the server mean AUC. When the difference in mean AUC between two different distributions is minimal (0.01–0.02), then the preference is given to the algorithms having equally well-performing client models.

- For instance, the mean AUC at the server node obtained by FedELF when Laplacian distribution is employed (refer to Table 6) is 0.979, which is identical when Gaussian distribution is employed. However, the similarity, the mean AUC at the server node in FedLR-SGD is 0.97, at the client nodes is better in the case of Laplacian distribution, which ranks it better and turned to be obtaining better generalization.
- A similar observation is noticed with the Santander transaction dataset is oversampled with SMOTE (refer to Table 13). The generalization ability of the FedELF with Gaussian is better than Laplacian distribution at the client nodes.

After following the above cost-benefit analysis, the best combination is presented in Table 16. The important observations noticed are as follows:

- FedELF and FedLR-SGD are the top two performing algorithms are where the former turned out to be the best in five out of seven datasets except Breast cancer and Polish bankruptcy datasets.
- With ADASYN data balancing technique, FedELF turned out to be the best in 4 out of 5 datasets except for the Polish Bankruptcy (Table 14) dataset.
- Similar observation is noticed when the other balancing technique, SMOTE was employed. FedLR-SGD turned out to be the best in only one case i.e., with polish bankruptcy dataset. In the rest of the cases, FedELF is turned out to be the best and makes it a clear winner.
- Among balanced datasets, there is no clear winner since FedLR-SGD turned out to be the best in the Breast cancer dataset, and FedELF secured the first position by obtaining mean AUC in the case of AP Colon cancer dataset.
- Additionally, FedELF outperformed the rest of the algorithms with substantial mean AUC differences in all the other datasets (5 out of 7 datasets). This ranks FedELF superior irrespective of the oversampling technique.

Further, the following advantages with FedELF makes it even more robust which are as follows:

- FedELF is an ensemble model which makes it less susceptible to overfitting.
- The generalization ability of FedELF is better than the standalone algorithms. It is well noticed on several occasions, where it is observed a large deviation between the mean AUC obtained at the server node and client nodes.
- However, on the downside, the computational complexity of FedELF is higher than that of the remaining standalone algorithms.

Among Gaussian and Laplacian-based noise, superior results are obtained by Gaussian (in four out of five datasets) irrespective of the type of the oversampling technique due to the following reasons:

- Laplacian distribution follows heavy-tailed distributions which makes it more sensitive to outliers compared to Gaussian distribution. This could degrade the performance of the employed federated technique (FedELF, FedLR-SGD, etc.).
- Most of these datasets follow Gaussian distribution which makes the Gaussian distribution injected-noise DP technique more amenable to these datasets rather than Laplacian distribution injected-noise.
- Gaussian distribution introduces more smoothness so that the perturbed dataset is less deviated concerning the original data.

Further, we presented the trend analysis observed through different epsilon values across different nodes in Figs. 6, 7, 8, 9, 10 and 11. The following are the important observations:

- The impact of varying epsilon remains almost stable in the majority of the datasets. This observation is noticed across all the algorithms.
- However, a decrease in mean AUC trend is noticed in the case of client nodes. This decline is more prevalent in the case of FedMLP as depicted in Figs. 8, 9, 10 and 11.

7.1 Computational Time Analysis

We performed computational time analysis of all the employed algorithms and presented the training time (see Figs. 12, 13, 14, 15, 16, 17 and 18) for the best combination as presented in Table 16. While our proposed model, FedELF, owing to its ensemble nature, has higher computational complexity compared to standalone models, as far as implementation is concerned, the caching mechanism inherent to FL helps reduce the computational time. We noticed that when the number of features are small in number, then the standalone model is taking less time than the ensemble model. However, when the features are large, then the caching mechanism has helped the ensemble model to complete quicker than the standalone model. This efficiency is primarily attributed to the model's caching capabilities. For example, with a batch size of 80, a FedELM-SGD model can cache up to 80 samples per round. In contrast, FedELF, assuming it uses 10 standalone models, can cache up to 800 samples per round. This larger cache size reduces the need for repetitive data processing and minimizes data access times, leading to significantly reduced processing times from the second communication round onward.

Table 4. Comparative analysis of Mean AUC results on the Breast Cancer dataset

Algo-rithm	Mean of AUCs obtained after global training over ten communication rounds														
	ε = 0.67 (G/L)			ε = 1 (G/L)			ε = 1.5 (G/L)			ε = 2.25 (G/L)			ε = 3.375 (G/L)		
	S	C_1	C_2	S	C_1	C_2	S	C_1	C_2	S	C_1	C_2	S	C_1	C_2
FedELF	0.925/ 0.94	0.984/ 0.964	0.92/ 0.924	0.974/ 0.972	0.973/ 0.98	0.936/ 0.953	0.951/ 0.93	0.986/ 0.975	0.957/ 0.957	0.957/ 0.968	0.985/ 0.987	0.957/ 0.952	0.944/ 0.971	0.981/ 0.983	0.968/ 0.968
FedELM-SGD	0.88/ 0.946	0.955/ 0.972	0.948/ 0.955	0.954/ 0.953	0.959/ 0.957	0.955/ 0.955	0.949/ 0.954	0.959/ 0.957	0.955/ 0.955	0.954/ 0.954	0.957/ 0.959	0.955/ 0.955	0.954/ 0.954	0.959/ 0.959	0.955/ 0.955
FedMLP-SGD	0.845/ 0.598	0.953/ 0.596	0.865/ 0.561	0.817/ 0.892	0.938/ 0.687	0.905/ 0.675	0.916/ 0.683	0.831/ 0.893	0.809/ 0.873	0.812/ 0.844	0.823/ 0.809	0.8/ 0.794	0.887/ 0.861	0.816/ 0.896	0.865/ 0.893
FedLR-SGD	0.929/ 0.971	0.988/ 0.988	0.915/ 0.965	0.974/ 0.966	0.957/ 0.986	0.977/ 0.97	0.957/ 0.976	0.988/ 0.988	0.965/ 0.971	0.976/ 0.957	0.992/ 0.974	0.945/ 0.977	**0.979/ 0.979**	**0.984/ 0.986**	**0.961/ 0.967**

* where G is Gaussian and L is logistic distribution respectively; S: server node; C_1 : Client node$_1$; C_2: Client node$_2$

Table 5. Comparative analysis of Mean AUC results on the AP Colon Lung dataset

Algo-rithm	Mean of AUCs obtained after global training over ten communication rounds														
	ε = 0.67 (G/L)			ε = 1 (G/L)			ε = 1.5 (G/L)			ε = 2.25 (G/L)			ε = 3.375 (G/L)		
	S	C_1	C_2	S	C_1	C_2	S	C_1	C_2	S	C_1	C_2	S	C_1	C_2
FedELF	0.554/ 0.694	0.882/ 0.954	0.907/ 0.974	**0.912/ 0.955**	**0.905/ 0.885**	**0.93/ 0.926**	0.905/ 0.91	0.913/ 0.937	0.924/ 0.962	0.924/ 0.95	0.926/ 0.93	0.93/ 0.968	0.922/ 0.942	0.931/ 0.913	0.951/ 0.939
FedELM-SGD	0.635/ 0.688	0.851/ 0.865	0.893/ 0.901	0.894/ 0.892	0.929/ 0.917	0.951/ 0.947	0.908/ 0.904	0.91/ 0.924	0.947/ 0.943	0.915/ 0.908	0.915/ 0.917	0.947/ 0.947	0.919/ 0.916	0.917/ 0.913	0.947/ 0.943
FedMLP-SGD	0.745/ 0.688	0.832/ 0.796	0.872/ 0.81	0.873/ 0.929	0.87/ 0.919	0.97/ 0.946	0.945/ 0.86	0.883/ 0.954	0.886/ 0.929	0.907/ 0.889	0.858/ 0.881	0.943/ 0.931	0.941/ 0.846	0.951/ 0.854	0.944/ 0.902
FedLR-SGD	0.819/ 0.846	0.89/ 0.887	0.938/ 0.912	0.86/ 0.8	0.818/ 0.817	0.834/ 0.908	0.852/ 0.833	0.9/ 0.868	0.911/ 0.851	0.826/ 0.819	0.833/ 0.914	0.9/ 0.882	0.907/ 0.86	0.894/ 0.822	0.948/ 0.846

* where G is Gaussian and L is logistic distribution respectively; S: server node; C_1 : Client node$_1$; C_2: Client node$_2$

Table 6. Comparative analysis of Mean AUC results on the Ova Omentum dataset using ADASYN

Algo-rithm	Mean of AUCs obtained after global training over ten communication rounds														
	ε = 0.67 (G/L)			ε = 1 (G/L)			ε = 1.5 (G/L)			ε = 2.25 (G/L)			ε = 3.375 (G/L)		
	S	C_1	C_2	S	C_1	C_2	S	C_1	C_2	S	C_1	C_2	S	C_1	C_2
FedELF	0.587/ 0.796	0.868/ 0.873	0.709/ 0.726	0.819/ 0.834	0.854/ 0.84	0.708/ 0.71	0.826/ 0.821	0.847/ 0.854	0.707/ 0.711	**0.822/ 0.861**	**0.855/ 0.85**	**0.712/ 0.702**	0.842/ 0.838	0.843/ 0.847	0.712/ 0.708
FedELM-SGD	0.605/ 0.662	0.786/ 0.814	0.676/ 0.629	0.827/ 0.834	0.804/ 0.804	0.655/ 0.655	0.839/ 0.834	0.804/ 0.804	0.655/ 0.656	0.827/ 0.821	0.803/ 0.804	0.654/ 0.655	0.833/ 0.821	0.804/ 0.804	0.655/ 0.655
FedMLP-SGD	0.643/ 0.616	0.534/ 0.721	0.493/ 0.679	0.744/ 0.652	0.733/ 0.819	0.66/ 0.742	0.697/ 0.779	0.798/ 0.649	0.619/ 0.62	0.623/ 0.75	0.662/ 0.759	0.606/ 0.611	0.616/ 0.717	0.61/ 0.775	0.55/ 0.643
FedLR-SGD	0.716/ 0.739	0.71/ 0.767	0.617/ 0.726	0.712/ 0.725	0.794/ 0.825	0.619/ 0.675	0.763/ 0.767	0.784/ 0.761	0.61/ 0.669	0.77/ 0.738	0.768/ 0.76	0.65/ 0.621	0.759/ 0.79	0.763/ 0.808	0.688/ 0.741

* where G is Gaussian and L is logistic distribution respectively; S: server node; C_1 : Client node$_1$; C_2: Client node$_2$

Table 7. Comparative analysis of Mean AUC results on the Ova Omentum dataset using SMOTE

Algo-rithm	Mean of AUCs obtained after global training over ten communication rounds														
	ε = 0.67 (G/L)			ε = 1 (G/L)			ε = 1.5 (G/L)			ε = 2.25 (G/L)			ε = 3.375 (G/L)		
	S	C_1	C_2	S	C_1	C_2	S	C_1	C_2	S	C_1	C_2	S	C_1	C_2
FedELF	0.598/ 0.722	0.758/ 0.805	0.859/ 0.705	0.85/ 0.827	0.857/ 0.847	0.708/ 0.709	0.822/ 0.832	0.844/ 0.843	0.708/ 0.689	0.82/ 0.825	0.853/ 0.84	0.708/ 0.718	0.838/ 0.852	0.847/ 0.858	0.707/ 0.71
FedELM-SGD	0.605/ 0.662	0.786/ 0.814	0.676/ 0.629	0.827/ 0.834	0.804/ 0.804	0.655/ 0.655	0.839/ 0.834	0.804/ 0.804	0.655/ 0.656	0.827/ 0.821	0.803/ 0.804	0.654/ 0.655	0.833/ 0.821	0.804/ 0.804	0.655/ 0.655
FedMLP-SGD	0.634/ 0.758	0.636/ 0.668	0.731/ 0.577	0.735/ 0.659	0.695/ 0.733	0.724/ 0.617	0.729/ 0.722	0.754/ 0.686	0.677/ 0.569	0.665/ 0.8	0.685/ 0.774	0.648/ 0.661	0.699/ 0.499	0.598/ 0.497	0.58/ 0.494
FedLR-SGD	0.786/ 0.764	0.814/ 0.749	0.732/ 0.659	0.739/ 0.721	0.801/ 0.749	0.576/ 0.766	0.688/ 0.759	0.748/ 0.727	0.715/ 0.716	0.73/ 0.707	0.79/ 0.727	0.613/ 0.745	0.667/ 0.729	0.734/ 0.747	0.73/ 0.707

* where G is Gaussian and L is logistic distribution respectively; S: server node; C_1 : Client node$_1$; C_2: Client node$_2$

Table 8. Comparative analysis of Mean AUC results on the Ova Uterus dataset using ADASYN

Algorithm	Mean of AUCs obtained after global training over ten communication rounds														
	ε = 0.67 (G/L)			ε = 1 (G/L)			ε = 1.5 (G/L)			ε = 2.25 (G/L)			ε = 3.375 (G/L)		
	S	C_1	C_2	S	C_1	C_2	S	C_1	C_2	S	C_1	C_2	S	C_1	C_2
FedELF	0.712/ 0.8	0.76/ 0.757	0.887/ 0.794	0.818/ 0.84	0.785/ 0.805	0.792/ 0.793	0.832/ 0.83	0.696/ 0.797	0.787/ 0.832	0.842/ 0.799	0.715/ 0.805	0.785/ 0.781	0.82/ 0.81	0.732/ 0.694	0.789/ 0.77
FedELM-SGD	0.607/ 0.645	0.755/ 0.732	0.738/ 0.746	0.776/ 0.776	0.738/ 0.739	0.765/ 0.764	0.775/ 0.78	0.739/ 0.743	0.769/ 0.764	0.781/ 0.78	0.739/ 0.738	0.765/ 0.765	0.782/ 0.785	0.739/ 0.739	0.764/ 0.765
FedMLP-SGD	0.504/ 0.622	0.553/ 0.814	0.539/ 0.794	0.746/ 0.831	0.835/ 0.551	0.81/ 0.619	0.833/ 0.553	0.668/ 0.827	0.704/ 0.891	0.644/ 0.776	0.551/ 0.857	0.619/ 0.817	0.756/ 0.571	0.763/ 0.576	0.725/ 0.564
FedLR-SGD	0.747/ 0.697	0.744/ 0.741	0.774/ 0.717	0.775/ 0.814	0.776/ 0.832	0.779/ 0.773	0.76/ 0.786	0.838/ 0.819	0.795/ 0.722	0.716/ 0.791	0.75/ 0.83	0.744/ 0.787	0.806/ 0.78	0.843/ 0.812	0.801/ 0.793

* where G is Gaussian and L is logistic distribution respectively; S: server node; C_1 : Client node$_1$; C_2: Client node$_2$

Table 9. Comparative analysis of Mean AUC results on the Ova Uterus dataset using SMOTE

Algorithm	Mean of AUCs obtained after global training over ten communication rounds														
	ε = 0.67 (G/L)			ε = 1 (G/L)			ε = 1.5 (G/L)			ε = 2.25 (G/L)			ε = 3.375 (G/L)		
	S	C_1	C_2	S	C_1	C_2	S	C_1	C_2	S	C_1	C_2	S	C_1	C_2
FedELF	0.656/ 0.771	0.742/ 0.719	0.778/ 0.777	**0.821/ 0.843**	**0.777/ 0.766**	**0.804/ 0.772**	0.812/ 0.822	0.745/ 0.763	0.781/ 0.808	0.811/ 0.817	0.732/ 0.755	0.789/ 0.784	0.83/ 0.797	0.767/ 0.752	0.803/ 0.756
FedELM-SGD	0.607/ 0.645	0.755/ 0.732	0.738/ 0.746	0.776/ 0.776	0.738/ 0.739	0.765/ 0.764	0.775/ 0.78	0.739/ 0.743	0.769/ 0.764	0.781/ 0.78	0.739/ 0.738	0.765/ 0.765	0.782/ 0.785	0.739/ 0.739	0.764/ 0.765
FedMLP-SGD	0.566/ 0.714	0.604/ 0.692	0.655/ 0.694	0.675/ 0.791	0.586/ 0.575	0.627/ 0.698	0.749/ 0.769	0.761/ 0.629	0.674/ 0.772	0.808/ 0.828	0.714/ 0.692	0.828/ 0.854	0.73/ 0.729	0.611/ 0.73	0.738/ 0.676
FedLR-SGD	0.654/ 0.767	0.685/ 0.739	0.756/ 0.794	0.788/ 0.793	0.84/ 0.828	0.819/ 0.762	0.785/ 0.808	0.807/ 0.817	0.733/ 0.778	0.799/ 0.791	0.843/ 0.798	0.8/ 0.768	0.799/ 0.777	0.813/ 0.797	0.766/ 0.725

* where G is Gaussian and L is logistic distribution respectively; S: server node; C_1 : Client node$_1$; C_2: Client node$_2$

Table 10. Comparative analysis of Mean AUC results on the Credit card fraud dataset using ADASYN

Algo-rithm	Mean of AUCs obtained after global training over ten communication rounds														
	ε = 0.67 (G/L)			ε = 1 (G/L)			ε = 1.5 (G/L)			ε = 2.25 (G/L)			ε = 3.375 (G/L)		
	S	C_1	C_2	S	C_1	C_2	S	C_1	C_2	S	C_1	C_2	S	C_1	C_2
FedELF	0.86/ 0.895	0.942/ 0.936	0.89/ 0.872	0.889/ 0.876	0.933/ 0.933	0.883/ 0.887	0.893/ 0.883	0.939/ 0.94	0.885/ 0.89	**0.889/ 0.91**	**0.934/ 0.924**	**0.886/ 0.882**	0.878/ 0.873	0.93/ 0.937	0.873/ 0.872
FedELM-SGD	0.775/ 0.776	0.921/ 0.921	0.903/ 0.903	0.776/ 0.776	0.921/ 0.921	0.903/ 0.903	0.776/ 0.776	0.921/ 0.921	0.903/ 0.903	0.776/ 0.776	0.921/ 0.921	0.903/ 0.903	0.776/ 0.776	0.921/ 0.921	0.903/ 0.903
FedMLP-SGD	0.692/ 0.75	0.585/ 0.788	0.569/ 0.779	0.711/ 0.558	0.5/ 0.767	0.5/ 0.822	0.5/ 0.709	0.768/ 0.619	0.665/ 0.646	0.772/ 0.383	0.882/ 0.409	0.863/ 0.386	0.495/ 0.41	0.519/ 0.408	0.499/ 0.407
FedLR-SGD	0.733/ 0.77	0.844/ 0.882	0.857/ 0.869	0.754/ 0.715	0.922/ 0.881	0.898/ 0.845	0.778/ 0.757	0.911/ 0.853	0.91/ 0.852	0.75/ 0.752	0.896/ 0.915	0.856/ 0.912	0.773/ 0.783	0.917/ 0.917	0.917/ 0.898

* where G is Gaussian and L is logistic distribution respectively; S: server node; C_1 : Client node$_1$; C_2: Client node$_2$

Table 11. Comparative analysis of Mean AUC results on the Credit card fraud dataset using SMOTE

Algo-rithm	Mean of AUCs obtained after global training over ten communication rounds														
	ε = 0.67 (G/L)			ε = 1 (G/L)			ε = 1.5 (G/L)			ε = 2.25 (G/L)			ε = 3.375 (G/L)		
	S	C_1	C_2	S	C_1	C_2	S	C_1	C_2	S	C_1	C_2	S	C_1	C_2
FedELF	0.876/ 0.861	0.893/ 0.9	0.887/ 0.897	0.858/ 0.869	0.888/ 0.891	0.887/ 0.894	0.832/ 0.861	0.889/ 0.89	0.892/ 0.902	0.84/ 0.854	0.895/ 0.89	0.896/ 0.886	0.868/ 0.846	0.888/ 0.882	0.882/ 0.901
FedELM-SGD	0.735/ 0.736	0.895/ 0.895	0.895/ 0.895	0.736/ 0.736	0.894/ 0.895	0.895/ 0.895	0.736/ 0.736	0.894/ 0.894	0.895/ 0.895	0.736/ 0.736	0.894/ 0.894	0.895/ 0.895	0.736/ 0.736	0.894/ 0.894	0.895/ 0.895
FedMLP-SGD	0.491/ 0.771	0.551/ 0.795	0.59/ 0.788	0.496/ 0.75	0.648/ 0.439	0.619/ 0.428	0.614/ 0.36	0.496/ 0.8	0.496/ 0.83	0.67/ 0.61	0.73/ 0.68	0.738/ 0.655	0.499/ 0.5	0.509/ 0.5	0.507/ 0.5
FedLR-SGD	0.819/ 0.744	0.891/ 0.919	0.908/ 0.917	0.808/ 0.785	0.913/ 0.885	0.92/ 0.874	0.767/ 0.751	0.91/ 0.931	0.907/ 0.907	0.734/ 0.756	0.909/ 0.926	0.898/ 0.921	0.777/ 0.767	0.903/ 0.898	0.898/ 0.885

* where G is Gaussian and L is logistic distribution respectively; S: server node; C_1 : Client node$_1$; C_2: Client node$_2$

Table 12. Comparative analysis of Mean AUC results on the Santander customer transaction prediction dataset using ADASYN

Algo-rithm	Mean of AUCs obtained after global training over ten communication rounds														
	ε = 0.67 (G/L)			ε = 1 (G/L)			ε = 1.5 (G/L)			ε = 2.25 (G/L)			ε = 3.375 (G/L)		
	S	C_1	C_2	S	C_1	C_2	S	C_1	C_2	S	C_1	C_2	S	C_1	C_2
FedELF	0.727/ 0.723	0.705/ 0.705	0.703/ 0.701	0.724/ 0.726	0.695/ 0.69	0.695/ 0.686	0.723/ 0.718	0.698/ 0.7	0.702/ 0.696	0.728/ 0.724	0.705/ 0.7	0.701/ 0.698	0.731/ 0.719	0.708/ 0.691	0.703/ 0.691
FedELM-SGD	0.671/ 0.671	0.665/ 0.665	0.666/ 0.666	0.67/ 0.67	0.665/ 0.665	0.666/ 0.666	0.67/ 0.67	0.665/ 0.665	0.666/ 0.666	0.67/ 0.67	0.665/ 0.665	0.666/ 0.666	0.67/ 0.67	0.665/ 0.665	0.666/ 0.666
FedMLP-SGD	0.501/ 0.537	0.503/ 0.553	0.5/ 0.546	0.521/ 0.562	0.536/ 0.491	0.521/ 0.488	0.529/ 0.472	0.52/ 0.553	0.523/ 0.549	0.502/ 0.516	0.501/ 0.517	0.503/ 0.52	0.508/ 0.532	0.509/ 0.527	0.51/ 0.53
FedLR-SGD	0.643/ 0.647	0.638/ 0.649	0.645/ 0.648	0.619/ 0.643	0.63/ 0.653	0.632/ 0.652	0.622/ 0.65	0.619/ 0.647	0.62/ 0.64	0.613/ 0.652	0.618/ 0.662	0.613/ 0.655	0.616/ 0.607	0.625/ 0.611	0.611/ 0.611

* where G is Gaussian and L is logistic distribution respectively; S: server node; C_1 : Client node$_1$; C_2: Client node$_2$

Table 13. Comparative analysis of Mean AUC results on the Santander customer transaction prediction dataset using SMOTE

Algo-rithm	Mean of AUCs obtained after global training over ten communication rounds														
	ε = 0.67 (G/L)			ε = 1 (G/L)			ε = 1.5 (G/L)			ε = 2.25 (G/L)			ε = 3.375 (G/L)		
	S	C_1	C_2	S	C_1	C_2	S	C_1	C_2	S	C_1	C_2	S	C_1	C_2
FedELF	**0.728/ 0.722**	**0.704/ 0.688**	**0.701/ 0.689**	0.716/ 0.708	0.7/ 0.682	0.699/ 0.678	0.725/ 0.71	0.683/ 0.68	0.683/ 0.677	0.724/ 0.718	0.696/ 0.696	0.699/ 0.691	0.721/ 0.726	0.695/ 0.705	0.697/ 0.705
FedELM-SGD	0.676/ 0.676	0.681/ 0.681	0.681/ 0.681	0.676/ 0.676	0.681/ 0.681	0.681/ 0.681	0.676/ 0.676	0.681/ 0.681	0.681/ 0.681	0.676/ 0.676	0.681/ 0.681	0.681/ 0.681	0.676/ 0.676	0.681/ 0.681	0.681/ 0.681
FedMLP-SGD	0.498/ 0.503	0.503/ 0.505	0.498/ 0.504	0.509/ 0.505	0.497/ 0.507	0.495/ 0.495	0.506/ 0.505	0.51/ 0.504	0.512/ 0.499	0.534/ 0.533	0.552/ 0.528	0.544/ 0.537	0.496/ 0.495	0.498/ 0.497	0.491/ 0.493
FedLR-SGD	0.618/ 0.614	0.624/ 0.62	0.616/ 0.605	0.601/ 0.628	0.622/ 0.614	0.624/ 0.621	0.614/ 0.613	0.61/ 0.631	0.61/ 0.62	0.613/ 0.625	0.619/ 0.636	0.618/ 0.636	0.612/ 0.609	0.619/ 0.61	0.618/ 0.608

* where G is Gaussian and L is logistic distribution respectively; S: server node; C_1 : Client node$_1$; C_2: Client node$_2$

Table 14. Comparative analysis of Mean AUC results on Polish Bankruptcy dataset using ADASYN

Algo-rithm	Mean of AUCs obtained after global training over ten communication rounds														
	ε = 0.67 (G/L)			ε = 1 (G/L)			ε = 1.5 (G/L)			ε = 2.25 (G/L)			ε = 3.375 (G/L)		
	S	C_1	C_2	S	C_1	C_2	S	C_1	C_2	S	C_1	C_2	S	C_1	C_2
FedELF	0.499/ 0.504	0.51/ 0.499	0.5/ 0.503	0.498/ 0.502	0.508/ 0.5	0.5/ 0.506	0.513/ 0.5	0.506/ 0.5	0.512/ 0.514	0.511/ 0.499	0.5/ 0.5	0.518/ 0.501	0.5/ 0.499	0.5/ 0.515	0.501/ 0.514
FedELM-SGD	0.475/ 0.518	0.544/ 0.548	0.597/ 0.54	0.53/ 0.529	0.507/ 0.503	0.561/ 0.554	0.515/ 0.518	0.504/ 0.505	0.552/ 0.555	0.547/ 0.521	0.505/ 0.505	0.556/ 0.555	0.526/ 0.53	0.504/ 0.504	0.554/ 0.554
FedMLP-SGD	0.503/ 0.464	0.495/ 0.526	0.479/ 0.527	0.504/ 0.496	0.542/ 0.39	0.55/ 0.404	0.552/ 0.431	0.511/ 0.519	0.514/ 0.491	0.483/ 0.536	0.542/ 0.519	0.508/ 0.509	0.5/ 0.486	0.5/ 0.483	0.5/ 0.512
FedLR-SGD	0.563/ 0.467	0.531/ 0.528	0.571/ 0.603	0.471/ 0.464	0.477/ 0.488	0.487/ 0.499	0.464/ 0.47	0.494/ 0.5	0.51/ 0.477	0.432/ 0.486	0.471/ 0.506	0.444/ 0.567	0.642/ 0.484	0.453/ 0.497	0.528/ 0.639

* where G is Gaussian and L is logistic distribution respectively; S: server node; C_1 : Client node$_1$; C_2: Client node$_2$

Table 15. Comparative analysis of Mean AUC results on Polish Bankruptcy dataset using SMOTE

Algo-rithm	Mean of AUCs obtained after global training over ten communication rounds														
	ε = 0.67 (G/L)			ε = 1 (G/L)			ε = 1.5 (G/L)			ε = 2.25 (G/L)			ε = 3.375 (G/L)		
	S	C_1	C_2	S	C_1	C_2	S	C_1	C_2	S	C_1	C_2	S	C_1	C_2
FedELF	0.52/ 0.507	0.5/ 0.505	0.508/ 0.502	0.5/ 0.5	0.504/ 0.506	0.52/ 0.502	0.498/ 0.5	0.511/ 0.499	0.516/ 0.499	0.5/ 0.498	0.499/ 0.499	0.5/ 0.512	0.499/ 0.5	0.5/ 0.504	0.5/ 0.548
FedELM-SGD	0.564/ 0.554	0.494/ 0.541	0.58/ 0.544	0.51/ 0.514	0.502/ 0.51	0.556/ 0.561	0.532/ 0.55	0.509/ 0.505	0.563/ 0.563	0.542/ 0.53	0.5/ 0.501	0.554/ 0.554	0.526/ 0.522	0.501/ 0.503	0.554/ 0.557
FedMLP-SGD	0.486/ 0.436	0.604/ 0.424	0.546/ 0.498	0.5/ 0.36	0.473/ 0.487	0.486/ 0.566	0.53/ 0.52	0.485/ 0.417	0.499/ 0.525	0.505/ 0.5	0.5/ 0.5	0.502/ 0.5	0.499/ 0.52	0.5/ 0.501	0.5/ 0.386
FedLR-SGD	**0.641/ 0.515**	**0.588/ 0.509**	**0.528/ 0.493**	0.544/ 0.613	0.502/ 0.516	0.607/ 0.647	0.482/ 0.505	0.54/ 0.558	0.609/ 0.363	0.457/ 0.439	0.479/ 0.465	0.5/ 0.479	0.506/ 0.487	0.501/ 0.554	0.568/ 0.662

* where G is Gaussian and L is logistic distribution respectively; S: server node; C_1 : Client node$_1$; C_2: Client node$_2$

Table 16. Best combinations of sampling technique, noise distribution type and epsilon

Dataset	Sampling Technique	Noise Distribution	Epsilon
Breast Cancer	–	Laplacian	3.375
Credit Card Fraud	ADASYN	Gaussian	2.25
Ova Omentum	ADASYN	Gaussian	2.25
Santander Customer Transaction Prediction	SMOTE	Laplacian	0.67
Polish Bankruptcy	SMOTE	Laplacian	0.67
AP Colon Lung	–	Gaussian	1
Ova Uterus	SMOTE	Gaussian	1

Table 17. t-test results on AUC obtained by FedELF versus FedLR-SGD

Datasets	Server	
	t-statistic	p-value
Breast Cancer	5.31	0.48×10^{-4}
AP Colon Lung	11.004	0.201×10^{-8}
Ova Omentum	9.834	0.115×10^{-7}
Ova Uterus	2.9	0.00952
Credit Card Fraud	18.693	0.308×10^{-12}
Polish Bankruptcy	42.455	0.169×10^{-18}
Santander Customer Transaction Prediction	5.688	0.214×10^{-4}

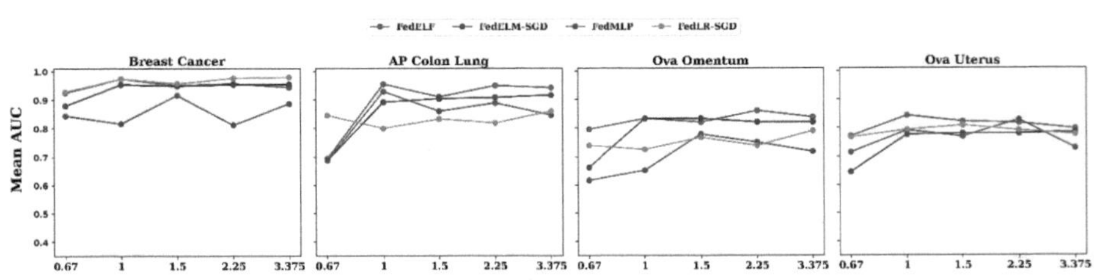

Fig. 6. Behaviour of the Server across various epsilons

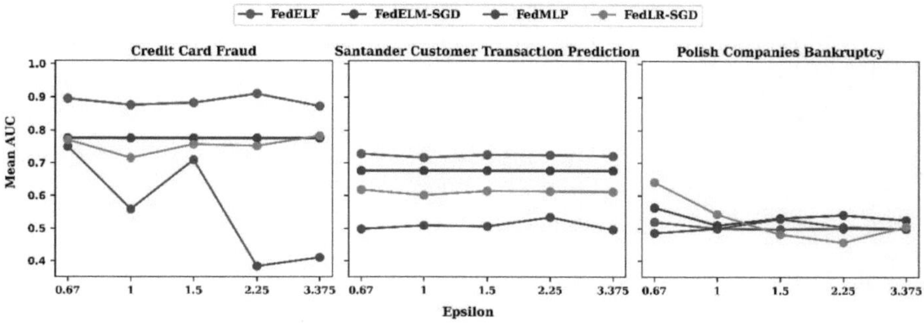

Fig. 7. Behaviour of the Server across various epsilons (contd.)

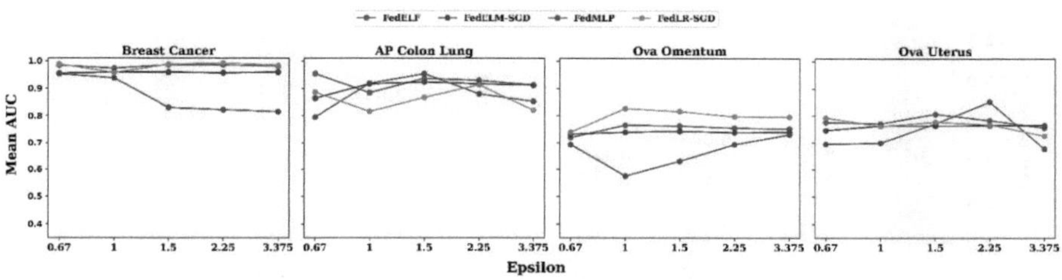

Fig. 8. Behaviour of $Client_1$ across various epsilons

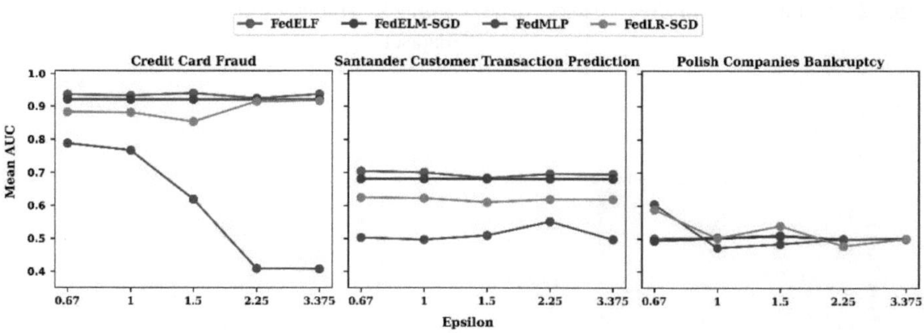

Fig. 9. Behaviour of $Client_1$ across various epsilons (contd.)

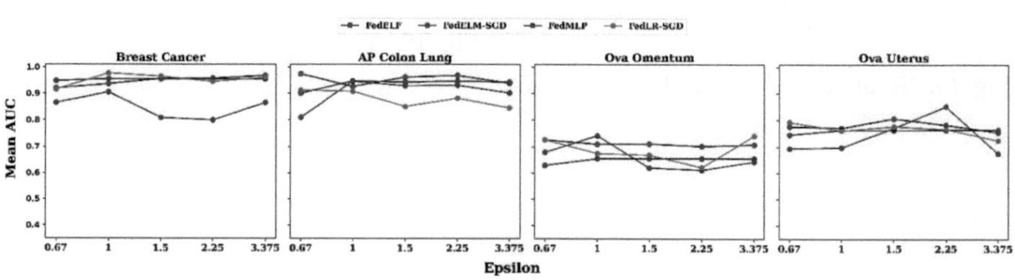

Fig. 10. Behaviour of $Client_2$ across various epsilons

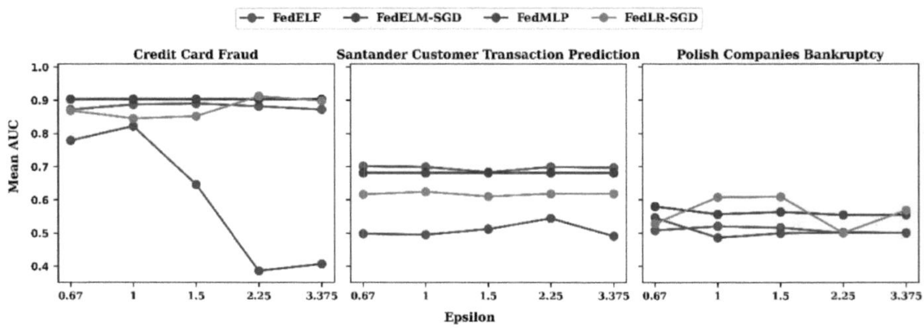

Fig. 11. Behaviour of $Client_2$ across various epsilons (contd.)

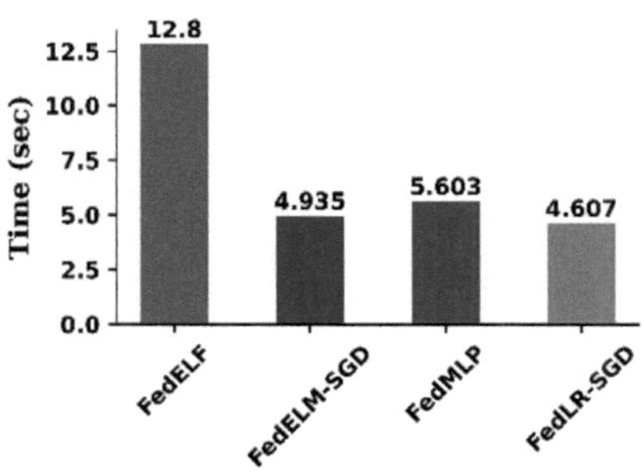

Fig. 12. Training time in Breast Cancer dataset

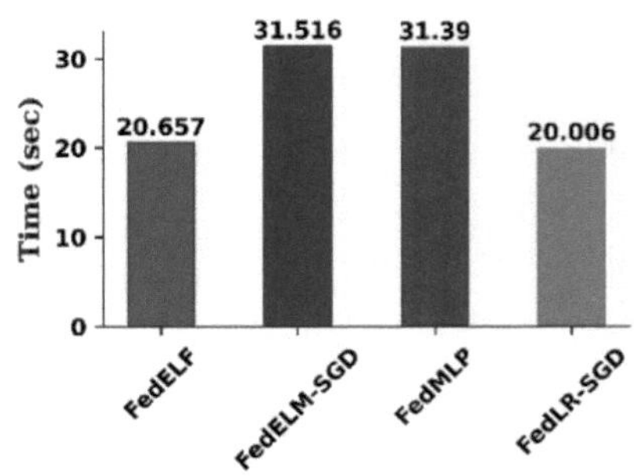

Fig. 13. Training time in AP Colon Lung dataset

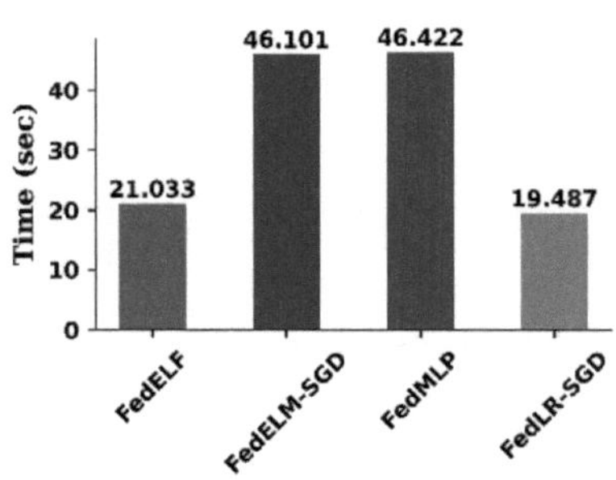

Fig. 14. Training time in Ova Omentum dataset

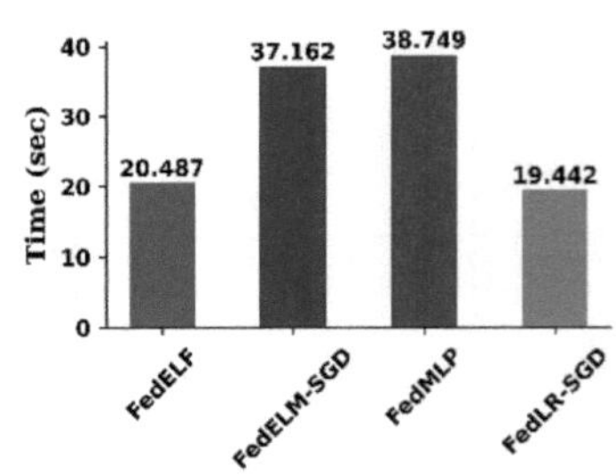

Fig. 15. Training time in Ova Uterus dataset

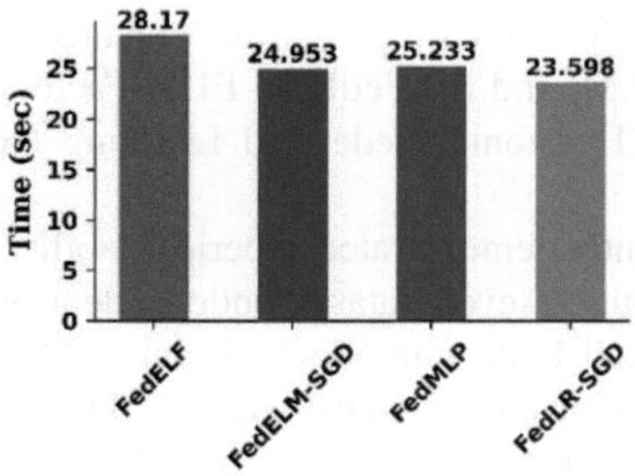

Fig. 16. Training time in Credit card Fraud dataset

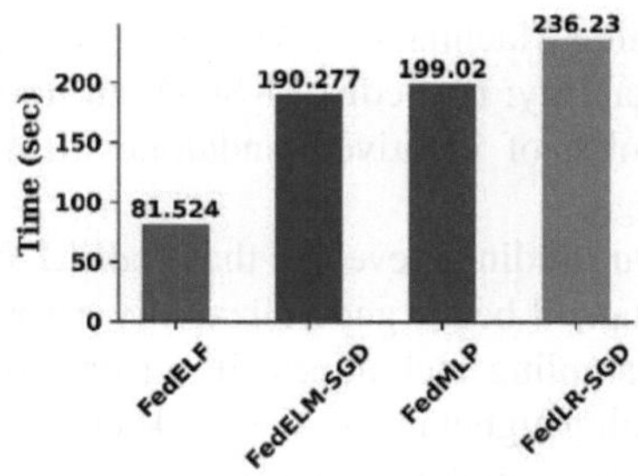

Fig. 17. Training time in Santander Consumer Transaction Prediction dataset

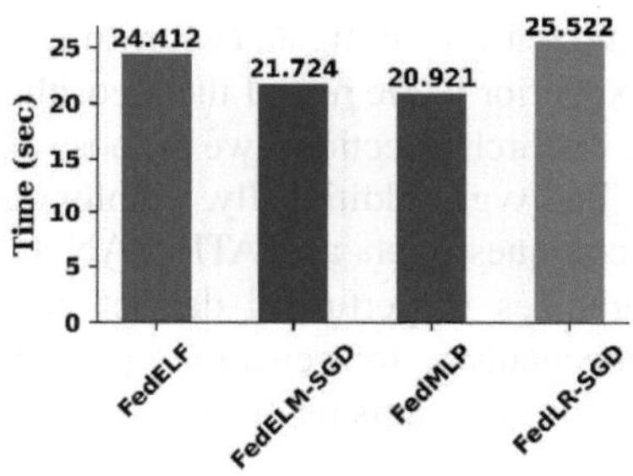

Fig. 18. Training time in Polish bankruptcy dataset

7.2 Statistical Testing of the Results

The following pairwise t-test analysis is conducted on the AUC (see Table 17) with 5% significance level and the degrees of the freedom are taken as 18 (10 + 10–2). This analysis is conducted for all the nodes over the ten communication rounds across the top 2 best-performing algorithms, i.e., FedELF and FedLR-SGD. The statistical test emphasizes the statistically valid statements about the performance across various datasets.

The null hypothesis and the alternate hypothesis are taken as follows:

H_0: *Both the algorithms are statistically equal*
H_1: *Both the algorithms are statistically not equal*

The results show that in the server node, FedELF turned out to be statistically significant than the FedELM-SGD in all the datasets. This makes FedELF to be superior both in terms of mean AUC and computational time makes it a clear winner.

8 Conclusions and Future Directions

In this study, we introduced a two-phase Federated Learning (FL) methodology. In the initial stage, we implemented Differential Privacy (DP) to generate privacy-preserved data in compliance with data privacy regulations. We proposed two iterative Extreme

Learning Machine (ELM) techniques renowned for their simplicity and enhanced generalizability: (i) FedELM-SGD, an iterative ELM, and (ii) FedELF, ELM Factory, an ensemble of iterative standalone ELM under Horizontal Federated Learning (HFL) settings.

Our findings revealed that FedELF consistently demonstrated superior stability and maintained better generalizability across five out of seven datasets independent of the oversampling techniques. It performed very well both with SMOTE and ADASYN, obtaining higher mean AUC and better generalization compared to FedLR-SGD. Further, the time complexity is also well analyzed through training time, where the FedELF turned out to be the winner owing to its faster convergence. Among the distributions, Gaussian distribution yielded superior results in four out of five datasets, making it a better choice than the Laplacian distribution for these domains. While t-statistic results suggested superior performance of FedELF in all the datasets it a clear winner and primer choice owing to the following reasons: mitigated overfitting and enhanced stability and generalization, rendered it superior to the rest of the algorithms.

Looking towards future research directions, we propose exploring alternative aggregation algorithms beyond FedAvg. Additionally, employing Generative Adversarial Network (GAN)-based techniques such as PATE-GAN [29] and DP-WGAN [30] for enhanced privacy guarantees in perturbed dataset generation could be investigated. Developing robust mechanisms for generating perturbed datasets with minimal computational constraints is also a promising avenue for future exploration.

Appendix

Table A.1. List of more frequent abbreviations in the current study

Abbreviations	Full forms and meaning
ELM	Extreme Learning Machine
SGD	Stochastic Gradient Descent
ELM-SGD	Iterative version of ELM based on mini-batch SGD
FedELM-SGD	Federated Learning using an iterative Extreme Learning Machine based on SGD.
ELF	Extreme Learning Factory
FedELF	Federated Extreme Learning Factory
Non-i.i.d	Non independent and identically distributed
DP	Differential Privacy
AUC	Area under curve
FedMLP-SGD	Iterative version of MLP based on mini-batch SGD
FedLR-SGD	Iterative version of LR based on mini-batch SGD
SMOTE	Synthetic Minority Oversampling Techniques
ADASYN	Adaptive Synthetic Sampling approach

References

1. Konecny, J., McMahan, H. B., Felix, X.Y., Richtarik, P., Suresh, A.T., Bacon, D.: Federated Learning: Strategies for Improving Communication Efficiency. (2016). https://doi.org/10.48550/arXiv.1610.05492
2. Xuan, S., Liu, G., Li, Z., Zheng, L., Wang, S., Jiang, C.: Random forest for credit card fraud detection. In 2018 IEEE 15th international conference on networking, sensing, and control (ICNSC), pp. 1–6 (2018)
3. Yang, W., Zhang, Y., Ye, K., Li, L., Xu, C.Z.: Ffd: A federated learning based method for credit card fraud detection. In Big Data–BigData 2019: 8th International Congress, Held as Part of the Services Conference Federation, SCF 2019, San Diego, CA, USA, June 25–30, 2019, Proceedings 2019; 8, 18–32
4. Yang, Q., Liu, Y., Chen, T., Tong, Y.: Federated machine learning: Concept and applications. ACM Trans. Intell. Syst. Technol. (TIST) **10**(2), 1–19 (2019)
5. Yin, X., Zhu, Y., Hu, J.: A comprehensive survey of privacy-preserving federated learning: A taxonomy, review, and future directions. ACM Comput. Surv. (CSUR) **54**(6), 1–36 (2021)
6. Dwork, C., McSherry, F., Nissim, K., Smith, A.: Calibrating noise to sensitivity in private data analysis. In Theory of Cryptography: Third Theory of Cryptography Conference, TCC 2006, New York, NY, USA, March 4–7, 2006. Proceedings 3 pp. 265–284
7. Erlingsson, Ú., Pihur, V., Korolova, A.: Rappor: Randomized aggregatable privacy-preserving ordinal response. In Proceedings of the 2014 ACM SIGSAC conference on computer and communications security. pp. 1054–1067 (2014)
8. Zhao, Y., Li, M., Lai, L., Suda, N., Civin, D., Chandra, V.: Federated learning with non-iid data. arXiv preprint arXiv:1806.00582 (2018)
9. Huang, G.B., Zhu, Q.Y., Siew, C.K.: Extreme learning machine: Theory and applications. Neurocomputing **70**(1–3), 489–501 (2006)
10. Wang, K., Li, J., Wu, W.: An efficient intrusion detection method based on federated transfer learning and an extreme learning machine with privacy preservation. Secur. Commun. Netw. **1**, 2913293 (2022)
11. Huang, G.B., Zhou, H., Ding, X., Zhang, R.: Extreme learning machine for regression and multiclass classification. IEEE Trans. Syst. Man, Cybern. Part B (Cybernetics), **42**(2), 513–529 (2011)
12. Liang, N.Y., Huang, G.B., Saratchandran, P., Sundararajan, N.: A fast and accurate online sequential learning algorithm for feedforward networks. IEEE Trans. Neural Netw. **17**(6), 1411–1423 (2006)
13. Qin, Y., Matsutani, H., Kondo, M. A selective model aggregation approach in federated learning for online anomaly detection. In 2020 International Conferences on Internet of Things (iThings) and IEEE Green Computing and Communications (GreenCom) and IEEE Cyber, Physical and Social Computing (CPSCom) and IEEE Smart Data (SmartData) and IEEE Congress on Cybermatics (Cybermatics), pp. 684–691 (2020)
14. Chen, Y.T., Chuang, Y.C., Wu, A.Y.: Online extreme learning machine design for the application of federated learning. In 2020 2nd IEEE International Conference on Artificial Intelligence Circuits and Systems (AICAS), pp. 188–192 (2020)
15. Xiao, Y., et al.: Fully decentralized federated learning-based on-board mission for UAV swarm system. IEEE Commun. Lett. **25**(10), 3296–3300 (2021)
16. Ito, R., Tsukada, M., Matsutani, H.: An on-device federated learning approach for cooperative model update between edge devices. IEEE Access **9**, 92986–92998 (2021)
17. Sun, W., Chen, Y., Yang, X., Cao, J., Song, Y.: FedIO: Bridge inner-and outer-hospital information for perioperative complications prognostic prediction via federated learning. In 2021 IEEE International Conference on Bioinformatics and Biomedicine (BIBM), pp. 3215–3221 (2021)

18. Qin, Y., Kondo, M.: Mlmg: Multi-local and multi-global model aggregation for federated learning. In 2021 IEEE international conference on pervasive computing and communications workshops and other affiliated events (PerCom Workshops), pp. 565–571 (2021)
19. Yu, Z., Liu, J., Yang, M., Cheng, Y., Hu, J., Li, X.: An elderly fall detection method based on federated learning and extreme learning machine (fed-elm). IEEE Access **10**, 130816–130824 (2022)
20. Qiu, C.: A network traffic classification method based on federated learning and extreme learning machine. In 2023 IEEE International Conference on Control, Electronics and Computer Technology (ICCECT), pp. 284–289 (2023)
21. Ravi, V., Singh, P.: Auto-associative extreme learning factory as a single class classifier. In 2014 IEEE International Conference on Computational Intelligence and Computing Research, pp. 1–6 (2014)
22. Sun, Z., Wang, Y., Shu, M., Liu, R., Zhao, H.: Differential privacy for data and model publishing of medical data. IEEE Access **7**, 152103–152114 (2019)
23. Beutel, D.J., Topal, T., Mathur, A., Qiu, X., Fernandez-Marques, J., Gao, Y., Lorenzo, S., Kwing, H.L., Titouan, P., Pedro, P.B.de G., Lane, N.D. Flower: A friendly federated learning research framework. arXiv preprint arXiv:2007.14390 (2020)
24. Chawla, N.V., Bowyer, K.W., Hall, L.O., Kegelmeyer, W.P.: SMOTE: synthetic minority over-sampling technique. J. Artif. Intell. Res. **16**, 321–357 (2002)
25. Wolberg, W.H., Mangasarian, O.L.: Multisurface method of pattern separation for medical diagnosis applied to breast cytology. Proc. Natl. Acad. Sci. **87**(23), 9193–9196 (1990)
26. Stiglic, G., Kokol, P.: Stability of ranked gene lists in large microarray analysis studies. Biomed. Res. Int. **1**, 616358 (2010)
27. Zięba, M., Tomczak, S. K., Tomczak, J. M.: Ensemble boosted trees with synthetic features generation in application to bankruptcy prediction. Expert Syst. Appl., **58**, 93–101. (2016) The dataset can be downloaded from: https://doi.org/10.1016/j.eswa.2016.04.001
28. Kaggle repository link: https://www.kaggle.com/
29. Jinsung, Y., James, J., Mihaela van der S.: PATE-GAN: Generating synthetic data with differential privacy guarantees. In Proceedings of the International Conference on Learning Representations. IEEE, Los Alamitos, CA, 2019, pp. 536–545
30. Xie, L., Lin, K., Wang, S., Wang, F., Zhou, J.: Differentially private generative adversarial network. arXiv preprint arXiv:1802.06739, (2018)
31. He, H., Bai, Y., Garcia, E. A., Li, S.: ADASYN: Adaptive synthetic sampling approach for imbalanced learning. In 2008 IEEE international joint conference on neural networks (IEEE world congress on computational intelligence), 1322–1328 (2008)
32. UCI repository: https://archive.ics.uci.edu/
33. Openml: https://www.openml.org/search?type=data&sort=runs&status=active
34. Penrose, R.:A generalized inverse for matrices. In Mathematical proceedings of the Cambridge philosophical society. Cambridge University Press. vol. 51, 3, (1955), pp. 406–413

Gen AI-Assisted Copilot for Risk Analysts

Pallabi Banerjee(✉)

R.S. Software, Kolkata 700091, India
pallabib@rssoftware.co.in

Abstract. As digital payments surge, the complexity of fraud continues to escalate, demanding more sophisticated tools to support risk analysts. This paper introduces the development of a Gen AI-assisted copilot integrated within a fraud detection solution, aimed at optimizing fraud alert investigation. The copilot leverages a Retrieval-Augmented Generation (RAG) model that combines a Random Forest (RF) machine learning model as the retriever and a fine-tuned Large Language Model (LLM) for generating contextual responses. By classifying fraud alerts, measuring similarity with historical data, and predicting the next action, the system automates decision-making while generating natural language recommendations for analysts. This approach significantly reduces manual workload, increases operational efficiency, and allows risk analysts to focus on critical, high-priority cases. The result is a robust system that enhances the speed and accuracy of fraud detection, investigation, and resolution, ultimately improving risk analyst productivity by almost 2.5 times.

Keywords: Digital Payments · Fraud Detection · Gen AI-Assisted Copilot · Risk Analysts · Operational Efficiency

1 Introduction

The rise of fraud in digital payments [1, 2] necessitates the development of efficient and intelligent tools to aid risk analysts in fraud detection. As fraud becomes more complex [3, 4], the need for sophisticated tools that can predict fraud, suggest appropriate actions, and provide intelligent recommendations has grown.

This paper introduces a Gen AI-assisted [4, 5] copilot for risk analysts integrated into fraud detection solutions. The copilot employs a Retrieval-Augmented Generation (RAG) [6, 7] model combining a Random Forest machine learning model as the retriever and a fine-tuned Large Language Model (LLM) [8, 9] model. By leveraging this two-step approach, the copilot retrieves relevant historical data using the RF model [10, 11] and generates contextual responses with a fine-tuned LLM. This integration enables the system to predict the next best action [12, 13] based on past fraud patterns while providing well-structured, natural language recommendations to guide risk analysts in managing cases efficiently.

S. Dhar et al. (Eds.): AGC 2025, CCIS 2686, pp. 255–266, 2026.
https://doi.org/10.1007/978-3-032-09153-6_15

2 Literature Review

The growing complexity of fraud in digital payments highlights the need for intelligent systems to aid risk analysts. Insights from research on generative AI configurations, AI copilots, and operational implications validate the potential of tools like the Gen AI-assisted copilot to enhance fraud detection through predictive analytics and actionable recommendations.

The classification framework proposed in [14] explores the interaction between human expertise and generative AI systems. It highlights three configurations: Auto-Pilot, where AI takes the lead with minimal human intervention; Co-Pilot, which emphasizes collaboration between human and AI; and Pilot, where human users maintain full control while AI provides support. This framework underscores the adaptability of AI systems like the Gen AI-assisted copilot, which operate as a Co-Pilot to assist risk analysts in fraud detection by offering insights without fully automating critical decision-making.

The study [15] emphasizes the significance of usability and outcome-driven metrics in evaluating AI-assisted systems. It provides insights into the iterative design process for copilots, focusing on their ability to enhance user performance while maintaining trust and minimizing cognitive overload. These principles align with the development of the Gen AI-assisted copilot, where the system leverages Retrieval-Augmented Generation (RAG) models to ensure both accuracy and explainability in fraud detection tasks.

The paper [16] examines the transformative effects of AI tools on organizational efficiency and user interaction. It discusses how AI systems can streamline workflows, reduce manual effort, and enhance decision-making speed, which are directly relevant to the Gen AI-assisted copilot goals of improving fraud alert resolution times and enabling analysts to focus on high-priority cases. This integration of AI enhances operational productivity while mitigating risks in digital payment systems.

In [17], the risks of over-reliance on AI systems are explored, particularly when transitioning from a Co-Pilot to an Auto-Pilot configuration. The paper highlights the importance of safeguards, such as transparency and user oversight, to ensure AI systems support rather than replace human judgment. These considerations are critical to the Gen AI-assisted copilot design, where AI-generated recommendations are explicitly contextualized and structured to assist analysts without removing their autonomy.

3 Challenge

Fraud detection solutions continuously monitor transactions and generate alerts when potential fraud is detected. However, determining the appropriate next steps requires significant human effort, as risk analysts must evaluate the alert to take the next action.

Risk analysts investigate alerts by systematically assessing suspicious activities flagged by automated systems to determine whether they represent a genuine risk, fraud, or a false positive. The investigation process involves reviewing alert details, analyzing transaction patterns, verifying customer or entity information, checking for related alerts, and gathering additional information if needed. Based on the findings, the next steps include escalating the case if a significant risk is detected, dismissing it if deemed a false positive, or flagging the account for continued monitoring in cases of moderate suspicion. The steps taken depend on the severity of the alert, the risk involved, and the findings during the investigation. Fraud analysts spend significant time sifting through data, cross-referencing information, and generating reports, limiting their ability to focus on high-priority cases.

The lack of automation and intelligent assistance in the current system makes it difficult for analysts to handle high volumes of cases [18] efficiently, potentially leading to missed or delayed responses to genuine fraud incidents. Moreover, the repetitive nature of tasks like data entry, alert triaging, and report generation contributes to operational inefficiencies.

4 Method

To address these challenges, a Co-Pilot has been developed to aid Risk analysts with the alert investigation in the case management system workflow. This AI-driven co-pilot will assist fraud analysts by providing real-time insights and offering intelligent recommendations for alert investigation. The co-pilot will streamline the case management process, enabling faster detection, investigation, and resolution of fraudulent activities while reducing manual effort and improving overall efficiency.

In addition to AI-driven co-pilots assisting fraud analysts, other methods exist to enhance fraud detection and management. One approach involves integrating feedback loops into fraud detection models, allowing for continuous learning and reducing the need for extensive human intervention in decision-making. By leveraging historical fraud patterns and analyst feedback, these systems refine their accuracy over time 19]. Another widely used method focuses on automating transaction fraud detection using artificial intelligence and data mining techniques. These techniques help reduce the manual effort required for fraud investigation by identifying anomalies and potential fraudulent transactions more efficiently [20]. Such automation streamlines the detection and response process, improving overall fraud management effectiveness.

We are utilizing a variation of the Retrieval-Augmented Generation (RAG) [6, 7] model in conjunction with a fine-tuned Large Language Model (LLM) [8, 9] to deliver contextual and accurate responses. In this setup, the retrieval process is handled by a Random Forest (RF) [10, 11] model, which serves as the retriever within the RAG framework. The RF model is refreshed daily with data from the past X days to ensure that it accurately reflects recent patterns and violations. The fine-tuned LLM generates responses by leveraging the augmented prompt, which combines historical data retrieved by the RF model with real-time insights. Fine-tuning of the LLM is only required when new violation types or action steps are introduced, ensuring that the system remains up-to-date and relevant while minimizing unnecessary fine-tuning operations.

To streamline this process, we introduce a Gen AI-driven copilot that automates the decision-making process by:

1. Classifying the violation to determine whether it was model-based or rule-based, and if rule-based, determining the specific rule type.
2. Calculating the percentage of similarity with past fraud alerts.

3. Using this information to predict the optimal 'Next-Action' using a Random Forest model.
4. Providing a natural language recommendation to assist the risk analyst, leveraging a fine-tuned Large Language Model (DistilGPT-2).

5 The Workflow: Automating Risk Analyst Assistance

This process flow diagram outlines the core components of the Gen AI system, divided into two key workflows represented by blue and green boxes:

1. Blue Boxes: RF Model Refresh (Retriever)

   The blue boxes depict the daily process of refreshing the Random Forest (RF) model, which acts as the system's retriever. To maintain accuracy and relevance, the RF model is updated with data from the past X days. This ensures that the retriever has the most recent information available, enhancing the retrieval process and keeping the system aligned with any new trends or patterns in violation types and actions.
2. Green Boxes: Online Flow with Fine-Tuned LLM (Augmented Prompt)

   The green boxes represent the online flow, which utilizes a fine-tuned Large Language Model (LLM). The LLM is provided with an augmented prompt generated from the RF model's retrieved data to offer more contextual and insightful responses. By combining the retriever's latest data with the LLM's advanced language generation capabilities, the system delivers highly relevant and precise results in real-time.

This combination of daily RF model refreshes and the augmented prompts used by the fine-tuned LLM ensures that the system remains both data-driven and responsive, providing timely and context-rich insights (Fig. 1).

5.1 Step 1: Identifying Violation Type

When a new alert is generated, the first task is to identify the 'Violation_Type' as depicted in ① of the above flow. In this step, the co-pilot determines the highest priority rule or model that was involved in the generation of the alert. The violation type can have 4 values:

1. **List-rules** which are hotlist rules, flagging transactions involving known fraudulent entities or behaviors.
2. **Transaction-rules** which are filters based on transaction thresholds, involving analysis of specific threshold values and transaction patterns.
3. **Models** that are driven by machine learning models, where ML scores guide decisions.
4. **Profile** which involves the utilization of profile data to evaluate transaction risk.

5.2 Step 2: Calculating Similarity

Next, the system calculates the 'Similarity_Percentage' as depicted in task ②, which quantifies how similar the new alert is to previous fraud alerts generated within the past X days. To calculate this similarity score, the past X days of fraud alerts are accessed to see what percentage of them can be identified as a significant (greater than 50%) match as depicted in task ③.

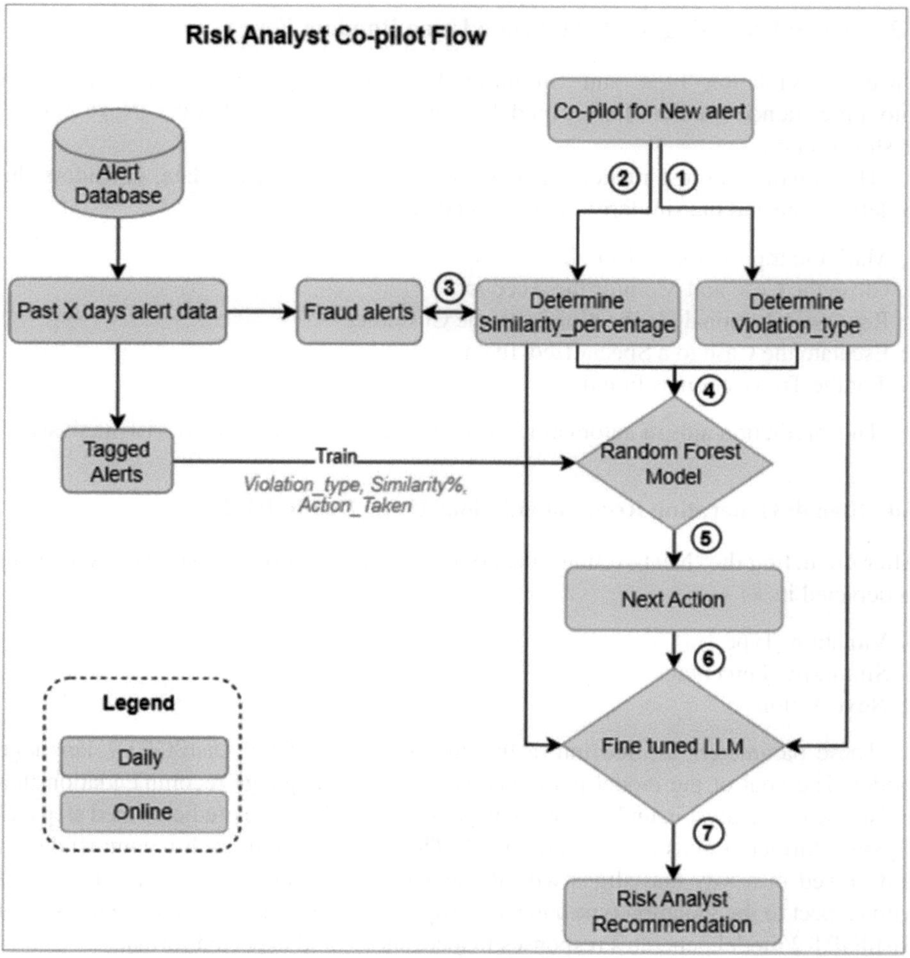

Fig. 1. Risk analyst Co-pilot process Flow

We have used cosine similarity with Natural Language Processing (NLP) in Python to achieve a percentage match with past fraud data. This method involved representing the description of the violation type as numerical vectors and measuring its similarity with past fraud alerts using cosine similarity. By applying this approach, we have been able to efficiently compare the current alert data with historical fraud alerts, indicating the similarity between the current case summary and known fraud scenarios.

A high match percentage suggests a strong correlation with historical fraud patterns, signaling a potentially fraudulent transaction that requires immediate attention. Conversely, a low match percentage may indicate a lower likelihood of fraud, allowing the case to be processed more efficiently with minimal manual intervention. This similarity score helps the Random Forest model predict the optimal next step.

5.3 Step 3: Predicting the Next-Action Using Random Forest

Once the 'Violation_Type' and 'Similarity_Percentage' are determined, they are fed into a pre-trained Random Forest model as depicted in ④ to predict the 'Next-Action' as shown in ⑤.

The Random Forest model, trained on historical fraud case data, considers the violation type and the similarity score to predict actions such as:

1. Mark the transaction as Genuine
2. Close the Case and Monitor the Account
3. Request Additional Verification from the Customer
4. Escalate the Case to a Specialized Team
5. Tag the Transaction as Fraud

This prediction aids in automating the decision-making process for risk analysts.

5.4 Step 4: Generating Recommendations Using DistilGPT-2

After predicting the 'Next-Action', the copilot takes the following parameters as input as depicted in ⑥:

1. Violation_Type
2. Similarity_Percentage
3. Next-Action

These parameters are fed into a fine-tuned version of the DistilGPT-2 language model. The goal of the model is to generate a natural language recommendation that assists the risk analyst in understanding the reasoning behind the predicted next step and suggests further actions as illustrated in ⑦. The first prompt in each recommendation is designed in a way that aligns with the analyst's expectations, by being hardcoded with respect to the violation type and similarity percentage, ensuring that the fine-tuned DistilGPT-2 model generates responses in a consistent and expected format.

DistilGPT-2 was chosen for its balance of performance, efficiency, and flexibility. It enables fast, accurate, and cost-effective recommendations tailored to the specific needs of risk analysts in fraud solutions. The model's smaller size, high-speed inference, and strong language generation capabilities make it an excellent fit for generating fraud case recommendations, driving operational efficiency, and assisting in the resolution of fraud cases.

6 Results

6.1 Fine-Tuning DistilGPT-2

Fine-tuning DistilGPT-2 involved taking the pre-trained DistilGPT-2 model and training it on a specific dataset to adapt it for a particular task, such as generating recommendations in a fraud detection system. The process began by preparing a custom dataset that included the input parameters and desired output (in this case, 'Violation_Type', 'Similarity_Percentage', 'Next-Action', and the corresponding recommendation). The dataset was tokenized using the same tokenizer as the pre-trained model to ensure compatibility and padding/truncation was applied to ensure consistent sequence lengths. During fine-tuning, the model's weights were adjusted based on the new data using supervised learning, through gradient descent. This allowed the model to learn the nuances of the new task while retaining the general language understanding from its pre-training.

The fine-tuning process leveraged a small learning rate and a relatively short number of training epochs to avoid overfitting while maintaining high-quality output generation. Once the model was fine-tuned, it was validated on unseen data to ensure it could generate accurate and contextually appropriate recommendations. The result was a specialized version of DistilGPT-2, fine-tuned to provide actionable recommendations based on the specific input parameters from the previous steps.

The model went through the data three times (epochs) to improve accuracy, and a weight decay of 0.01 was applied to prevent overfitting, helping the model generalize well to new data. The model was fine-tuned by evaluating performance at the end of each training round (epoch). A small learning rate of 2e-5 was used to ensure gradual adjustments, with two data samples processed at a time during both training and evaluation.

The results from the training process are as follows:

Epoch	Training Loss	Validation Loss
1	No log	0.005962
2	0.865600	0.003407
3	0.005400	0.002772

Training Loss: The substantial drop from 0.865600 to 0.005400 indicated that the model learned efficiently across the epochs and could perform with very low error on the training data by the third epoch.

Validation Loss: The consistent decrease from 0.005962 to 0.002772 showed that the model had improved its generalization capabilities and not overfitting the training data.

Based on the above, we inferred that the model's performance had improved significantly across the epochs. The large drop in training loss and the continued decrease in validation loss suggested the model had learned effectively and generalizing well to new data. This indicated a well-performing model by the end of the third epoch, with no signs of overfitting.

6.2 Creating the Data for Fine-Tuning DistilGPT-2

To fine-tune DistilGPT-2, we used a dataset with the following columns:

1. Violation_Type
2. Similarity_Percentage
3. Next-Action
4. Recommendation: A 3-5 sentence recommendation tailored to the input parameters, guiding the risk analyst on the appropriate next steps.

The fine-tuning process was done by tokenizing the ‘Recommendation’ field and training the model using the inputs (Violation_Type, Similarity_Percentage, Next-Action) as context.

For example, a sample recommendation that was used to train the LLM for a Violation_Type of ‘Model’ and Similarity_Percentage of 85% for which the Random Forest model predicted the next action as ‘Tag the Transaction as Fraud’ was:

> “A bank risk analyst is investigating a case with an 85% match to cases marked as fraud in the past 30 days. The transaction was flagged by a model rule, suggesting a high likelihood of fraud based on historical data. The analyst should tag the transaction as fraud and initiate immediate response actions to mitigate potential losses.”

Samples corresponding to each violation type were created with varying percentages of similarity to get the corresponding values of ‘Next action’ from the Random Forest model which were then fed to the fine-tuned LLM to generate recommendations.

6.2.1 Test Data Creation for LLM Fine-Tuning

The test dataset was created by sampling historical fraud cases and generating additional variations to capture a wide range of fraud scenarios and corresponding recommendations. Each test entry in the dataset was based on combinations of:

1. Violation_Type: selected randomly from ‘List-rule’, ‘Transaction-rule’, ‘Model’, ‘Profile’.
2. Similarity_Percentage: A range from 0% to 100%, divided into categories as described above.
3. Next-Action: Predicted by the Random Forest model.

The goal of the test dataset was to simulate a variety of real-world scenarios that a risk analyst would encounter, ensuring the language model could handle diverse situations and recommend appropriate actions.

6.2.2 Model Testing

The fine-tuned DistilGPT-2 model was tested with unseen data to evaluate its ability to:

- Generate accurate and relevant recommendations.
- Adhere to the structure and style of the training data.
- Provide actionable insights that risk analysts can easily understand and implement.

We evaluated the model based on its ability to produce clear and coherent recommendations aligned with the predicted ‘Next-Action’ (Table 1).

Table 1. Test results of Copilot recommendations

Violation_Type	Similarity_Percentage	Copilot recommendations
List-rule	25	The case shows a 25% similarity to previous fraud cases, triggered by a hotlist rule indicating the involvement of a previously compromised entity. Close the Case and Monitor the Account for any unusual patterns or suspicious activities.
List-rule	85	The case shows an 85% similarity to previous fraud cases, triggered by a hotlist rule indicating the involvement of a previously compromised entity. Tag the Transaction as Fraud and take immediate action to prevent further suspicious activities.
Transaction-rule	15	The case has a 15% match with historical fraud cases and was flagged by a single transaction rule due to abnormal transaction parameters. Escalate the Case to a Specialized Team
Transaction-rule	50	The case has a 50% match with historical fraud cases and was flagged by a single transaction rule due to abnormal transaction parameters. Request Additional Verification from the Customer Relations team for additional verification.
Model	15	A bank risk analyst is investigating a case with a 15% match to cases marked as fraud in the past 30 days. The transaction was flagged by a model rule, suggesting a high likelihood of fraud based on historical data. Close the Case and Monitor the Account for any unusual patterns or suspicious activities.
Model	75	A bank risk analyst is investigating a case with a 75% match to cases marked as fraud in the past 30 days. The transaction was flagged by a model rule, suggesting a high likelihood of fraud based on historical data. Tag the Transaction as Fraud and take immediate action to prevent further suspicious activities.
Profile	15	A recent transaction matches 15% of past fraud cases, with the alert triggered by unusual activity patterns detected over the last 24 h. Escalate the Case to a Specialized Team
Profile	85	A recent transaction matches 85% of past fraud cases, with the alert triggered by unusual activity patterns detected over the last 24 h. Escalate the Case to a Specialized Team for thorough review and investigation.

6.3 Metrics for Measuring Effectiveness

The integration of the Gen AI-assisted copilot within fraud solution provides a framework to evaluate its impact on speed, accuracy, and overall effectiveness in fraud detection, investigation, and resolution. To quantify these enhancements, several key metrics can be monitored:

1. Speed Metrics
   - Average Case Investigation Time: Measures the reduction in time taken by analysts to investigate fraud alerts from initiation to closure.
2. Accuracy Metrics
   - Prediction Accuracy: Assesses the system's ability to correctly predict the next best action using the Random Forest model.
3. Resolution Metrics
   - Time-to-Resolution (TTR): Captures the duration from the moment a fraud alert is generated to its successful resolution, including escalations or customer verifications.
   - Escalation Rate: Tracks the percentage of cases requiring manual escalation to specialized teams, showcasing how effectively the copilot resolves cases autonomously.
4. Risk Analyst Effectiveness Metrics
   - High-Priority Case Focus: Monitors the percentage of time analysts spend on high-priority cases, enabled by automation of repetitive tasks.
   - Analyst Productivity: Measures the number of cases resolved per analyst within a given timeframe, reflecting improved operational efficiency.

By tracking these metrics over time, organizations can validate the tangible benefits of the Gen AI-assisted copilot in enhancing fraud management capabilities and delivering superior outcomes for risk analysts.

The below table provides an illustrative view of the improvements in speed, accuracy, and overall operational efficiency (Table 2).

Table 2. Overview of increase in overall operational efficiency

Metric	Before Copilot	After Copilot	Improvement
Average Case Investigation Time	45 min per case	15 min per case	67% reduction
Prediction Accuracy	80%	95%	15% improvement
Time-to-Resolution (TTR)	2 days per case	8 h per case	83% reduction
Escalation Rate	30%	10%	67% reduction
High-Priority Case Focus	40% of analyst time	70% of analyst time	75% increase
Analyst Productivity	10 cases resolved/day	25 cases resolved/day	2.5x increase

7 Conclusion

The AI-assisted copilot for fraud detection solutions is a groundbreaking innovation that combines the strengths of traditional machine learning models with modern language models to revolutionize the decision-making process for risk analysts. By leveraging a Random Forest model for next-action predictions and a fine-tuned DistilGPT-2 model for generating actionable recommendations, the copilot has demonstrated measurable improvements in speed, accuracy, and overall operational efficiency. With reductions in investigation time, false positive rates, and escalation rates, coupled with increased throughput and analyst productivity, the copilot ensures timely and precise handling of fraud alerts. This transformative tool empowers risk analysts to focus on high-priority cases, enhancing the overall effectiveness of fraud detection, investigation, and resolution processes.

Acknowledgements. This study was supported by the valuable contributions of Dr. Sumit Misra, whose encouragement, meticulous review, and validation of the results were instrumental in its success. Sincere gratitude is also extended to Soma Das for her invaluable assistance with the citations and for formatting the paper to meet all the required criteria.

Disclosure of Interests. The author has no competing interests to declare that are relevant to the content of this article.

References

1. Putrevu, J., Mertzanis, C.: The adoption of digital payments in emerging economies: challenges and policy responses. Digit. Policy Regul. Gov. **26**(5), 476–500 (2024). https://doi.org/10.1108/DPRG-06-2023-0077
2. Pizzan-Tomanguillo, N. del P., Pereyra-Gonzales, T.V., León-Ramírez, S.V., Bautista-Fasabi, J., Rosales-Bardalez, C.D., Gómez-Apaza, R.D.: Evolution and Trends in Digital Wallet Research: A Bibliometric Analysis in Scopus and Web of Science. Publications, **12**(4), 34 (2024). https://doi.org/10.3390/publications12040034
3. Aivaz, K.-A., Florea, I.O., Munteanu, I.: Economic fraud and associated risks: An integrated bibliometric analysis approach. Risks **12**(5), 74 (2024). https://doi.org/10.3390/risks1205 0074
4. Ozili, P.K.: Advances and Issues in Fraud Research: A Commentary. J. Financ. Crime, **27**(1) (2020). https://doi.org/10.2139/ssrn.3132210
5. Yetiştiren, B., Özsoy, I., Ayerdem, M., Tüzün, E.: Evaluating the Code Quality of AI-Assisted Code Generation Tools: An Empirical Study on GitHub Copilot, Amazon CodeWhisperer, and ChatGPT. arXiv (2023). https://doi.org/10.48550/arXiv.2304.10778
6. Gao, Y., Xiong, Y., Gao, X., Jia, K., Pan, J., Bi, Y., Dai, Y., Sun, J., Wang, M., Wang, H.: Retrieval-Augmented Generation for Large Language Models: A Survey. arXiv (2024). https://doi.org/10.48550/arXiv.2312.10997
7. Zhao, S., Yang, Y., Wang, Z., He, Z., Qiu, L.K., Qiu, L.: Retrieval Augmented Generation (RAG) and Beyond: A Comprehensive Survey on How to Make your LLMs use External Data More Wisely. arXiv (2024). https://doi.org/10.48550/arXiv.2409.14924

8. Shui, Z., Karypis, P., Karls, D.S., Wen, M., Manchanda, S., Tadmor, E.B., Karypis, G.: Fine-Tuning Language Models on Multiple Datasets for Citation Intention Classification. Findings of the Association for Computational Linguistics: EMNLP 2024, 16718–16732 (2024). https://doi.org/10.18653/v1/2024.findings-emnlp.974
9. Chow, Y., Tennenholtz, G., Gur, I., Zhuang, V., Dai, B., Thiagarajan, S., Boutilier, C., Agarwal, R., Kumar, A.: Inference-Aware Fine-Tuning for Best-of-N Sampling in Large Language Models. arXiv (2024). https://doi.org/10.48550/arXiv.2412.15287
10. Breiman, L.: Random forests. Mach. Learn. **45**, 5–32 (2001). https://doi.org/10.1023/A:1010933404324
11. Mentch, L., Zhou, S.: Randomization as regularization: A degrees of freedom explanation for random forest success. J. Mach. Learn. Res., 21(171), 1–36 (2020). https://jmlr.org/papers/v21/19-905.html
12. Cao, L., Zhu, C.: Personalized next-best action recommendation with multi-party interaction learning for automated decision-making. PLoS ONE **17**(1), e0263010 (2022). https://doi.org/10.1371/journal.pone.0263010
13. Agarwal, P., Gupta, A., Sindhgatta, R., Dechu, S.: Goal-Oriented Next Best Activity Recommendation using Reinforcement Learning. In: arXiv preprint arXiv:2205.03219 (2022). https://arxiv.org/abs/2205.03219
14. Safari, N., Techatassanasoontorn, A.A., & Diaz Andrade, A.: Auto-Pilot, Co-Pilot, and Pilot: Human and Generative AI Configurations in Software Development. In: ICIS 2024 Proceedings, p. 7. AIS eLibrary, New York (2024). https://aisel.aisnet.org/icis2024/isdesign/isdesign/7/
15. Furmakiewicz, M., Liu, C., Taylor, A., Venger, I: Design and Evaluation of AI Copilots. In: ACIS 2024 Proceedings, p. 69. AIS eLibrary, New York (2024). https://arxiv.org/abs/2407.09512
16. Durach, C.F., Gutierrez, L.: Operational Implications of Artificial Intelligence Chatbots. In: *International Journal of Physical Distribution & Logistics Management*, vol. 54, no. 1, pp. 10–25. Emerald Publishing, New York (2024). https://www.emerald.com/insight/content/doi/https://doi.org/10.1108/ijpdlm-01-2024-0031/full/html
17. Sarkar, A., Xu, X. (Tone), Toronto, N., Drosos, I., Poelitz, C.: When Copilot Becomes Autopilot: Generative AI's Critical Risk to Knowledge Work and a Critical Solution. In: AIS Transactions on Human-Computer Interaction, vol. 15(3), pp. 247–267. AIS eLibrary, New York (2023). https://arxiv.org/abs/2412.15030
18. Shen, H., & Kurshan, E. (2020, October 21). *Deep Q-network-based adaptive alert threshold selection policy for payment fraud systems in retail banking*. arXiv: Learning. https://arxiv.org/abs/2010.10120
19. Sohel, A., Alam, M. A., Waliullah, Md., Siddiki, A., & Shahab Uddin, M. K.: Fraud detection in financial transactions through data science for real-time monitoring and prevention. 1(01), 91–107 (2024). https://doi.org/10.69593/ajieet.v1i01.132
20. Arpitha, C. N., Pratheek, H. N., Shreyas, D. S., Sujith, D., & Kumar B S, V.: Transaction Fraud Detection Using Data Mining Techniques. 1–5 (2024). https://doi.org/10.1109/icait61638.2024.10690363

Nowcasting the Financial Time Series with Streaming Data Analytics Under Apache Spark

Mohammad Arafat Ali Khan, Chandra Bhushan, Vadlamani Ravi(✉), Vangala Sarveswara Rao, and Shiva Shankar Orsu

Centre of Excellence in Analytics, Institute for Development and Research in Banking Technology, Castle Hills Road No. 1,Masab Tank, Hyderabad 500057, India
vravi@idrbt.ac.in, rav_padma@yahoo.com

Abstract. This paper proposes nowcasting of high-frequency financial datasets in real-time with a 5-min interval using the streaming analytics feature of Apache Spark. The proposed 2-stage method consists of modelling chaos in the first stage followed by using a sliding window approach for separately training Lasso Regression (LR), Ridge Regression (RR), Generalized Linear Model (GLM), Gradient Boosting Tree (GBT) and Random Forest (RF) available in the MLLib library of Apache Spark in the second stage. The effectiveness of the proposed methodology is demonstrated on National Stock Exchange (NSE) dataset, Bombay Stock Exchange (BSE) dataset, and Bitcoin-INR conversion dataset. For evaluating the proposed methodology, we used measures such as Symmetric Mean Absolute Percentage Error (SMAPE), Directional Symmetry (DS) and Theil's U Coefficient. We tested the statistical significance of the performance of the models using the Diebold Mariano (DM) test.

Keywords: Nowcasting · Apache Spark · Financial Time Series Forecasting · Machine Learning

1 Introduction

As we move rapidly towards digital technology in service industries, the amount and the speed with which the data is generated in day-to-day operations is truly spectacular, signalling the presence of the volume and the velocity dimensions of big data. To make meaningful decisions from this kind of data, we need to make use of tools which can effectively handle huge amounts of data coming at a fast pace. Both batch processing the historical data and real-time stream processing of data are the hallmarks of the big data technologies. If the data is generated in real-time, then we can make use of streaming data analytics to derive the insights and take more informed and quick decisions based on it. Spark Streaming is capable of effectively handling data generated in real-time. It is an extension of the Core Spark API which enables high throughput, fault-tolerant and highly scalable processing. Spark streaming receives live data streams, divides it into batches and then processes them using the Spark engine to generate the stream of results in batches.

S. Dhar et al. (Eds.): AGC 2025, CCIS 2686, pp. 267–283, 2026.
https://doi.org/10.1007/978-3-032-09153-6_16

Nowcasting using financial data has become a recent trend due to the phenomenal advances made in machine learning & big data and the advent of new distributed and in-memory computing platforms. The term nowcasting is a portmanteau word for *now* and *forecasting* and has been used for a long time in meteorology and recently also in economics and finance. Nowcasting refers to the prediction of the present, the very near future and the very recent past. The principle behind nowcasting is making use of real-time data streams at high frequencies to make quick predictions immediately thereof.

This paper proposes nowcasting of stock prices and bitcoin-INR exchange rates under big data analytics paradigm. The main focus is on using the real-time data generated from these financial time series to forecast the very near future values using the machine learning algorithms. To the best of our knowledge, no work is proposed to use big data analytics for nowcasting yet.

The rest of the paper is organized as follows: Sect. 2 presents the motivation to the work; Sect. 3 presents the contribution made in this work. Section 4 presents the state-of-the-art in financial forecasting. The proposed methodology and experimental setup are described in Sect. 5. Section 6 presents the information about the dataset and evaluation measures. The results and discussion are presented in Sect. 7. The paper is concluded with some future directions of the work in Sect. 8.

2 Motivation

In the literature, to the best of our knowledge, not a single study was reported to real-time forecasting (a.k.a nowcasting) of financial time series under big data paradigm. No study dealt with two distinct financial time series data, namely, stock market price, and cryptocurrency price in a single paper. Nowcasting, unlike forecasting, helps decision makers arrive at decision quite quickly in near-real-time or real-time. Unlike traditional forecasting, nowcasting enables decision-makers to make timely, data-driven decisions in near-real-time or real-time, which is crucial for high frequency trading environments.

3 Contribution

We streamed real-time high frequency data at the intervals of 5 min from 3 different data sources and performed nowcasting experiments in a parallel processing environment in a YARN cluster. Real-time sliding window approach was implemented in our experiment to enable nowcasting. We adopted 5 distinct techniques available in machine learning library of Spark for the purpose of nowcasting. The same study can be implemented in nowcasting the macroeconomics time series data like consumer price index (CPI), if such a data is available and can be streamed at a high frequency.

4 Literature Survey

Bańbura et al. [1] and Bańbura et al. [2] defined the nowcasting as predicting the present, a very near future, where they nowcast the GDP of fourth-quarter using the multivariate dynamic model in the European region and the USA. Giannone et al. [3] illustrated

USA GDP current quarter nowcast using large factor models (FM) i.e. a large number of indicators around 200 which lead to the "curse of dimensionality" and later sorted by constructing principal components of the indicators. The parameters for the forecasting model are estimated using Ordinary Least Squares (OLS) regression. But due to staggered-data, this nowcast is "pseudo-real-time". Richardson et al. [4] performed New Zealand GDP quarterly nowcast using popular machine learning techniques namely Lasso, Ridge, Elastic Regression, and Least-squares boosting on historical data provided by Reserve Bank of New Zealand. Each model is evaluated using Symmetric Mean Absolute Percentage Error (SMAPE) and Mean Absolute Deviation (MAD). They reported that the ML techniques used by them outperformed the traditional statistical methods.

Ilamchezhian et al. [5] published a review on Spark framework in 2017, where they discussed the features of Spark, and its advantages and disadvantages. They also brought up the problem of using the Hadoop Distributed File System (HDFS) i.e. speed of transferring data from HDFS for processing is higher, which is solved by the Spark RDD (Resilient Distributed Dataset). The RDD is an immutable tuple of partitioned data, distributed across the cluster. Spark framework provides the Streaming and MLLib libraries. As Spark processing overcomes the speed issue of HDFS, we setup spark cluster over HDFS in Yarn mode. This framework affords us the freedom to stream and analyze as many datasets as we require.

Modugno [6] used the Dynamic Factor Model (DFM) to nowcast inflation by including the high-frequency data and raw material prices in the US. Bok et al. [7] also illustrated the DFM is parsimonious for high dimensional data. In another nowcasting experiment, Loermann et al. [8] in 2019, performed nowcasting of US GDP using artificial neural network (ANN) with DFM. They collected data from FRED-MD from 1999 to 2018 to predict in quasi-real-time. In 2016, Jayanthi et al. [9] described different tools to perform big data analytics in real-time. They mentioned many tools, cloud platforms like Hadoop, Apache-Spark, Flume, Sqoop, AWS, Amazon Kinesis. They also illustrated the comparative study of various resources and provided architecture to perform big data analytics.

Ranjitha [10] in a survey paper explained how real-time analytics help in monitoring data generated from sensors. The paper describes different sources of real-time data generated from different sensors and how to handle a large amount of data by using big data framework. The author mentioned different processing techniques, complex event processing (CEP) for the streaming data and how they monitor with the help of visualization. Girija et al. [11] performed stock price prediction based on two types of prediction analysis: technical and fundamental. Along with historical stock market data, the social media data was used here for the stock price prediction. They utilized news articles, tweets of the company's and historical stock data from Yahoo finance, whenever real-time data is generated it is streamed using Apache Flume and stored in the HDFS. Logistic regression is employed to predict the next day trend. In [12], the movement of the opening of the stock price of a company is predicted using sentiment of the user; for data ingestion they used Twitter streaming API. They adopted spark streaming and dumped the data into HDFS using Apache Flume. In the paper, lambda architecture was used for entire processing. Word embeddings were extracted from the sentiments. Recurrent neural network (RNN) was employed for analysis on Spark's architecture. The experiment is executed on the data of Google, Apple and Microsoft. In 2016, Menon et al. [13] performed a similar experiment where they used spark streaming using Scala language in big data paradigm to complete this task. They used multiple indices from the NSE stock market to predict the minute-wise indices' price. They took 50 min window size, 25 min sliding interval and run 20 experiments on Spark. They applied ARMA to predict the prices of the indices. Umadevi et al. [14] similar to [12] analyzed the datasets of Google, Apple and Microsoft. They trained the ARIMA model on historical data of closing value of stocks of all firms and 25 values/scores are nowcasted.

In another work, Wu et al. [15] adopted the chaotic modelling using lag and embedding dimension where they calculated the lag and embedding dimension using fast orthogonal programming. Then, they applied the wavelet neural network for forecasting the spot market price of electricity. In 2017, Selvin et al. [16] worked on the dataset of NSE listed companies Infosys, TCS and CIPLA with minute wise stock price with a sliding window approach of size 100 min with 90 min as training and 10 min for prediction. They implemented deep learning architectures, namely, convolutional neural network (CNN), RNN and long short term memory (LSTM) network for these experiments and compared with linear model ARIMA in terms of RMSE score. Raju and Tarif [17] modelled using LSTM and ARIMA in predicting real-time Bitcoin price using the merged dataset of Bitcoin historical and real-time data collected from different public API. The other dataset is the set of tweets related to the Bitcoin collected using Twitter API on which sentiment analysis is performed.

In this work, in a first-of-its-kind study, we proposed a novel nowcasting methodology based on Apache Spark's streaming analytics feature preceded by chaotic modelling and demonstrated its effectiveness on three datasets namely stock market and cryptocurrency to INR exchange series. We setup a Yarn cluster to perform parallel operations on all the datasets at the same time. Our work is different from others as follows: Wu et al. [15] applied chaotic modelling before applying any forecasting algorithm, but in this work, we used a different procedure for chaotic modelling. Menon et al. [13] and Selvin et al. [16] applied sliding window approach using Scala on Spark cluster but we applied the sliding window approach using Pyspark on Spark cluster in Yarn mode with different window size and sliding interval. Besides this, the main focus of this study lies in training the models in real-time for nowcasting. Kohns and Bhattarcharjee [29] proposed nowcasting model to forecast GDP growth based on google trends by employing Bayesian structural time series model. Another study also report to nowcast GDP [31]. Heikkinen and Heimonen [30] proposed nowcasting-forecasting model for ISM business survey by utilizing google weekly trends.

5 Proposed Methodology

We collected the historical stock market data (300 data points with the frequency of 5 min) from Yahoo finance API and the data is stored on HDFS in a multi-node YARN cluster. Our proposed methodology has two stages: (i) phase space reconstruction and (ii) nowcasting using machine learning models.

In stage 1, by using Lyapunov exponent we found chaos in our datasets. This step is important since it helps to capture nonlinear dynamics and deterministic chaos in complex systems, where traditional linear models fail. We performed chaotic modelling by estimating the lag/time delay using PACF function available in python package and embedding dimension by Cao's algorithm available in CRAN package of R. This process needs to be repeated on a daily basis before starting the experiment as lag and embedding dimension are observed to change with time (Fig. 1).

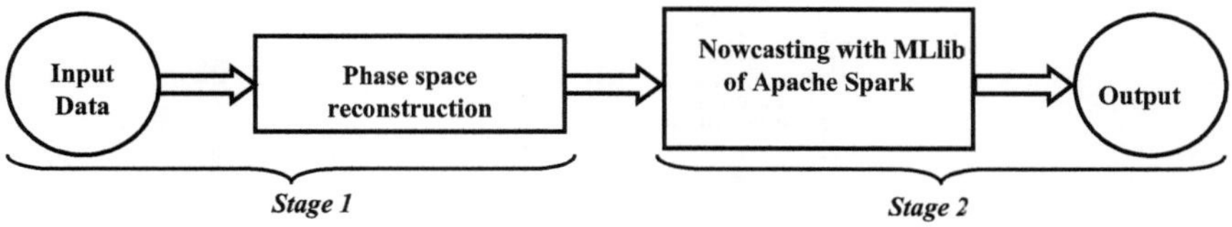

Fig. 1. Proposed Methodology

After estimating the time lag and embedding dimension, we applied Takens embedding to reconstruct datasets from time series to multi-input-single-output dataset, on which any forecasting technique can be trained. In stage 2, for nowcasting, we implemented 5 different machine learning models, namely, Lasso Regression, Ridge Regression, Random Forest, Gradient Boosting Tree (GBT), and Generalised Linear Model (GLM) available in Spark Ml library. For nowcasting, we made use of the Eq. (1).

$$\hat{y}_{t+1} = f\left(y_{t-(m-2)*\tau}, y_{t-(m-3)*\tau}, y_{t-(m-4)*\tau}, \ldots y_t\right)(1)$$

where $\hat{y}_{t+1}$ is forecasted value, f is the function takes actual values to calculate $\hat{y}_{t+1}$, y_t is the actual value at time t, m is the embedding dimension, and τ is the time delay.

In other words, let's say we have a lag of 1 and embedding dimension of 5 and have 300 data to train the model. So after phase space reconstruction we got a matrix of 60 rows and 5 columns shape. Then we consider 1st to 4^{th}columns as Independent features and 5^{th}column as target. Consider forecasting the 301st data point and for that, we will take the previousy_{297}, y_{298}, y_{299}, y_{300} values using the Eq. (1). As $\hat{y}_{t+1}$ is forecasted, we will compare the forecast with the actual value and calculate the squared error. The algorithms is retrained to best fit the available data, if the test MSE varies by more than 5% compared to training MSE. Otherwise, there is no need of retraining and the model will continue to forecast next value in the series.

After forecasting, we consecutively stream the actual value from the source API using Spark Streaming framework. In this case, the training data consists of previous training data + newly appended data points. We used a sliding window of size 300, which slides over 300 data points and forecasts the 301st value, in the next step it leaves the first point behind and appends the new data point to the training set for every training period and retraining the model. For example, assume if we a training set of 10 data points.

Using the trained model we forecast the $\hat{y_t}$, we also have y_t (actual value at that time stamp streamed using API). Here we check the squared error of actual and forecasted value and compare with the training MSE and if it varies more than 5%, then that denotes our model had forecasted a bad value and we need to tune the hyperparameters and retrain the model. Here we need to retrain the model by adding the y_t to the training set leaving the first data point in the training set nothing but a sliding window of 10 in this example.

As we are nowcasting at high frequency of time intervals we are not making the bad value forecasted ($\widehat{y_t}$) correct/more accurate, rather we are re-training our model to improve the forecasting in the next time stamp i.e., $y_{(t+1)}$ (Fig. 2)

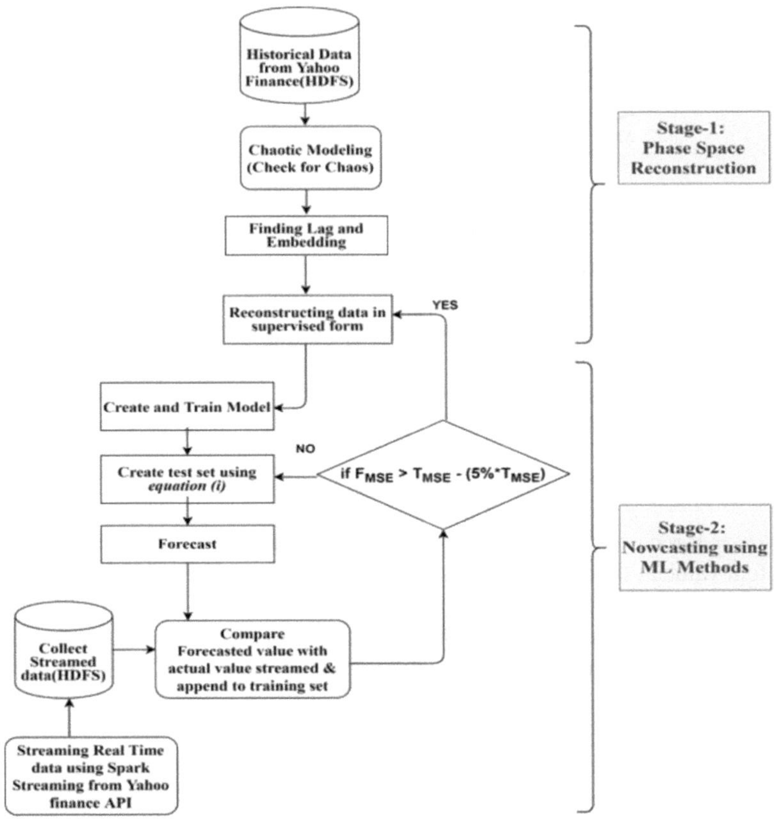

Fig. 2. The architecture of proposed methodology.

The forecasting is done for the entire day (i.e. till the market is closed) with the same frequency of 5 min. For the next day, we take the previous 4 days' historical data and find the lag and embedding of the data to convert the series into multi-input single-output (MISO) problem followed by nowcasting using algorithms. In this way, we conducted experiments consequently for 5 days.

Since we applied 5 models on the same dataset, we need to decide the best one out of them. To evaluate the performance of the models, we used three measures, namely, symmetric mean absolute percentage error (SMAPE), Directional Symmetry, and Theil's U Statistics. We also applied the Diebold-Mariano test to check the statistical significance between top two models.

5.1 Introduction to Chaos Theory

The theory of Chaos was proposed by Poincare in the late 1800s and later in 1963, the work was extended by Lorenz [18] to deal with complex nonlinear systems [19]. A chaotic system is deterministic, dynamic and evolves from the initial conditions and can be described in state space by the trajectories. The state space is represented by phase space, as the governing equations for the chaotic system are not known. The phase space can be rebuild using the original series using delay time and embedding dimension [20] as shown in Eq. (2).

$$Y_j = (Y_j, Y_{j+\tau}, Y_{(j+2\tau)}, \ldots\ldots, Y_{j+(m-1)\tau} \tag{2}$$

where Y_t is the target variable, t is independent feature representing time, τ is the time delay and m is the embedding dimension. After rebuilding the phase space, the univariate time series problem gets converted into a MISO problem.

5.1.1 Rosenstein's Method

Rosenstein's algorithm [21] estimates the largest Lyapunov exponent (λ) [22] from the given time series. If $\lambda \geq 0$ then it contains chaos; otherwise the chaos is not present in the given time series.

5.1.2 Cao's Method

To find the embedding dimension for a given time series we employed the method proposed by Cao[23] called false nearest neighbourhood.

5.2 Experimental Setup

A Spark Cluster is configured and setup for computations with 3 nodes as slaves and 1 master upon Hadoop cluster. Here, we configured Hadoop for using its Hadoop Distributed File System (HDFS) for storing the data in a distributed environment. To submit multiple job applications on a cluster we configured YARN upon Hadoop which shares the available resources to the jobs submitted. For these experiments, a total of 6 different clusters are used with the same configuration. Each machine has 32 GB RAM with 224 GB SSD to process fast data transfer between disk and RAM. The application can be allocated a minimum of 1 core and a maximum of 6 cores in a container. This sums up to have 90 GB of memory in a single cluster.

5.3 Streaming Methodology

Spark Streaming is a core Spark API that allows stream processing of live data streams. The data can be ingested from multiple sources like Kafka, Kinesis or TCP sockets. The processed data can be dumped into a local file system or distributed file system.

Spark Streaming provides high level of abstraction called DStreams (*discretized stream*) which is a collection of sequence of RDDs to stream the continuous data streams from any source. In fact, we can apply Spark's machine learning algorithms on the collection of DStreams.

In our experiments, we used the Streaming framework to stream the continuous data from Yahoo finance API and dumped the data into Hadoop Distributed File System (HDFS) and then we streamed the data for different pyspark applications from HDFS.

We streamed the data from the source with the help of readStream, which is the function available in Spark Streaming which supports Structured Streaming with the help of DataStreamReader Interface. DataStreamReader is an Interface that reads the source Data and converts it into Streaming Data. This reads the data and generates continuous DStreams, which are then captured and analyzed.

6 Dataset Description and Evaluation Measures

We analyzed three different datasets namely Bombay Stock Exchange (BSE), National Stock Exchange (NSE), and Bitcoin to Indian Rupee (BTC-INR) conversion. We continuously streamed the real-time data with a frequency of 5 min from these markets using an API provided by Yahoo finance. We used the Closed value from all of the markets. As for BSE, NSE and BTC INR operate from 9:30 a.m. to 3:30 p.m. five days in a week (Indian Market time), we conducted experiments with their respective operating time on 28th, 29th 30th of September and 1st and 5th of October in IST Zone. Whereas the FOREX exchanges operate round the clock, but we conducted the experiments on these datasets as per the Indian market time.

6.1 Evaluation Measures

We tested the models using three evaluation measures such as Symmetric Mean Absolute Percentage Error (SMAPE), Directional Statistics (DS) and Theil's U Coefficient.

6.1.1 Symmetric Mean Absolute Percentage Error (SMAPE)

This measure is not biased towards the scale of the datasets like Root Mean Squared Error. SMAPE gives us the error percentage as shown below.

$$SMAPE = \frac{100}{n}\sum_{i=1}^{n}\frac{|\hat{y}_i - y_i|}{(|\hat{y}_i| + |y_i|)/2}$$

Where $\hat{y_i}$ is the forecasted value, y_i is the actual value at that time stamp and n is the number of data points forecasted.

6.1.2 Directional Symmetry

Directional Symmetry evaluates the performance of the model by predicting the direction of change, whether positive/negative. Its equation form is shown below:

$$DS(t, \hat{t}) = \frac{100}{n-1}\sum_{i=2}^{n} d_i$$

$$where d_i = \begin{cases} 1, if\,(y_i - y_{i-1})(\hat{y_i} - \hat{y}_{i-1}) > 0 \\ 0, Otherwise \end{cases}$$

Where $\hat{y_i}$ is the forecasted value y_i is the actual value at that time stamp and n is the number of data points forecasted.

6.1.3 Theil's U Coefficient

Theil's U coefficient is a relative accuracy measure that gives more weightage to the large errors. If the value of U is greater than 1 then the naïve guess is better than forecast and if U value is close to zero, the forecast method is better.

$$TheilsUCoefficient = \sqrt{\frac{\sum_{t=1}^{n-1} \left(\hat{y}_{t+1} - y_{t+1}\right)^2}{\sum_{t=1}^{n-1} \left(\frac{y_{t+1} - y_t}{y_t}\right)^2}}$$

Where $\hat{y}_{t+1}$ is the forecasted value at $(t + 1)^{st}$ time stamp, y_{t+1} is the actual value at $(t + 1)^{st}$ time stamp, and n is the number of data points forecasted.

6.1.4 Diebold-Mariano Test

DM test is proposed by Diebold and Mariano [28]. It is a statistical test to check the equality of two forecast models. It determines whether forecasts of two or more models are significantly different or not.

7 Results and Discussion

Table 1 presents the combined 5 days forecast results of Lasso, Ridge, Random Forest, Gradient Boosting tree and Generalized Linear Regression across all three datasets in terms of SMAPE. These algorithms are specifically chosen owing to their less time complexity, less parameters to tune. In terms of SMAPE, GLM performed the best followed by ridge regression across all the datasets. The ensemble learning capability of random forest did not yield better results than less complex ridge or lasso regression in terms of SMAPE and Theil's U Coefficient.

GBT yielded the best Directional Symmetry (see Table 2.) over all algorithms across all datasets.

In terms of Theil's U coefficient values (see Table 3), both GLM and ridge outperformed the rest of the models but they yielded similar Theil's U Coefficient values because of regularization constants of GLM and Ridge.

Table 1. SMAPE on Combined 5 Days

Dataset / Model	Lasso	Ridge	RF	GBT	GLM
BSE	0.07077	0.06758	0.12980	0.11409	**0.06754**
NSE	0.06642	**0.06497**	0.11913	0.11025	**0.06497**
BTC-INR	0.05240	**0.04915**	0.07918	0.06836	0.04921

Table 2. Directional Symmetry on Combined 5 days

Dataset / Model	Lasso	Ridge	RF	GBT	GLM
BSE	44.87	44.59	50.41	**51.24**	44.59
NSE	45.55	48.51	**52.56**	49.32	48.51
BTC-INR	53.16	53.71	47.93	51.51	**53.99**

Table 3. Theil's U Coefficient for Combined 5 days

Dataset / Model	Lasso	Ridge	RF	GBT	GLM
BSE	0.700	**0.693**	1.261	1.319	**0.693**
NSE	**0.703**	0.705	1.280	1.298	0.705
BTC-INR	0.428	**0.419**	0.631	0.563	**0.419**

In terms of BSE data, with respect to all 5 days forecast the GLM turned out to be the best in terms of SMAPE followed by ridge regression. GBT achieved the best directional statistics. Ridge and GLM turned out to be the same but better than others in terms of Theil's U coefficient.

In NSE data, the combined SMAPE of all the 5 days forecast of ridge regression and GLM produced almost the same results and outperformed others. In terms of Directional Symmetry, the Random Forest was good in showing the directional correlation. Lasso Regression Theil's U coefficient is closer to zero compared to rest algorithms.

For BTC-INR, the same Ridge Regression and GLM having low percent of error in terms of SMAPE. GLM proved to be better in both measures Directional Symmetry and Theil's U coefficient.

In Diebold Mariano Test, we compared two models at alpha = 0.05 on SMAPE measures where if the DM test is greater than the p-value then the model is significantly different from each other and the models with less SMAPE value is better one. And if the DM test is lesser than p-value then the models are not significantly different.

In Table 4, we compared all models with each other and found some are significantly different and some are not on the BSE dataset. The models which are significantly different is highlighted with light orange and in that model which is highlighted in BOLD in the below tables had better forecast based on SMAPE value and another insignificant models are highlighted with light green. We found that Ridge-Lasso, GBT-Random Forest and GLM-Ridge forecast are almost the same. Likewise, within Table 4 the DM test performed on the NSE and BTCINR datasets respectively.

Table 4. Diebold Mariano for BSE, NSE and BTC-INR Results.

Model_1_Forecasts vs Model_2_Forecasts	**BSE**		**NSE**		**BTC_INR**	
	DM	**P-Value**	**DM**	**P-Value**	**DM**	**P-Value**
Ridge vs Lasso	0.3499	0.7265509	**0.1516**	**0.8795701**	0.6860	0.4931104
Random Forest vs **Lasso**	4.1153	4.79E-05	**3.5442**	**0.0004440**	5.0178	0.0000008
Random Forest vs **Ridge**	4.2901	2.30E-05	**3.5842**	**0.0003832**	4.7444	3.01E-06
GBT vs **Lasso**	1.5652	0.1184116	**2.4581**	**0.0144199**	3.5216	0.0004835
GBT vs **Ridge**	1.5761	0.1158637	**2.4679**	**0.0140409**	3.4033	0.0007399
GBT vs Random Forest	0.2109	0.8330729	**0.2655**	**0.7907101**	4.3945	1.46E-05
GLM vs Lasso	0.3679	0.7131292	**0.1512**	**0.8798286**	0.6535	0.5138093
GLM vs Ridge	1.6080	0.1086908	**1.0004**	**0.3177584**	1.0616	0.2890815
GLM vs Random Forest	4.2920	2.28E-05	**3.5842**	**0.0003832**	4.7343	0.0000031
GLM vs GBT	1.5766	0.1157492	2.4679	0.0140406	3.3944	0.0007634

From the Figs. 3,4 and 5 we can see that again Ridge and GLM are giving closer forecasts to the actual series values compared to other models experimented in this study. They are continuously performing better even with the fewer data in Day_1 and Day_2. Another detail from the figures is that the random forest and GBT are generating the worst forecasts compared to the remaining models even though both of them are ensemble models.

Figures 6, 7 and 8 depict the Box-plots for all the algorithms on a daily basis using SMAPE values. Using box plots, we can easily identify the model with high variability in results. i.e. model with less Inter Quartile Range (IQR) and less number of outliers is the best model. First we BSE and then followed by NSE and BTC-INR. Each chart contains 5 different boxplots where each boxplot shows models vs their corresponding

SMAPE day wise. We can conclude the lesser the IQR of the boxplot and presence of less no. of outlier will be the better of all. We can see few models having less IQR and no outlier. In most of the dataset Lasso, Ridge, and GLM having less IQR and no outliers whereas RF and GBT have high IQR and lots of outliers. We can observe from the boxplot of 4th day forecast of all the algorithms, the IQR is comparatively lesser than the rest of the days. Presence of high number of outlier represents that the model is not able to forecast the price efficiently. So we can conclude that Lasso, Ridge and GLM performed better than other models.

Boxplots depicted in Figs. 6, 7 and 8, shows the comparative result of Actual price vs forecasted price from all 5 models on all 3 datasets. These box plots also show how similar boxplot are with actual. In our experiment, Lasso, Ridge and GLM are proven to be most similar to actual price but RF and GBT fail to forecast accurately. In BTC-INR we can see boxplot having few outliers where it shows how many times the actual price varies from its own behaviour i.e. sudden rise or sudden fall. But an important thing to notice here is the models are able to forecast the sudden rise and sudden fall in the price. In this part again Lasso, Ridge, and GLM proved better than RF and GBT. In some boxplots like NSE and BSE there is no outlier point. Then in these cases we can find similarity between Actual price and forecasted price using median and IQR of each boxplot.

Table 5. Hyper parameters for Grid Search

Model	Hyper parameters
Lasso	regParam = [0.01,0.02,0.03,0.04,0.05,0.1,0.2,0.3,0.4] elasticNetParam = 0.0
Ridge	regParam = [0.01,0.02,0.03,0.04,0.05,0.1,0.2,0.3,0.4] elasticNetParam = 1.0
Random Forest	maxDepth = [1,2,3]; numTrees = [2,3,4] minInstancesPerNode = [1,2]
Gradient Boosting Tree	maxDepth = [1,2,3,4]; maxBins = [4,8,16] minInstancesPerNode = [1,2]
Generalised Linear Model	regParam = [0.01,0.02,0.03,0.04,0.05,0.1,0.2,0.3,0.4]

The hyperparameters combinations used for tuning all the models are presented in Table 5.

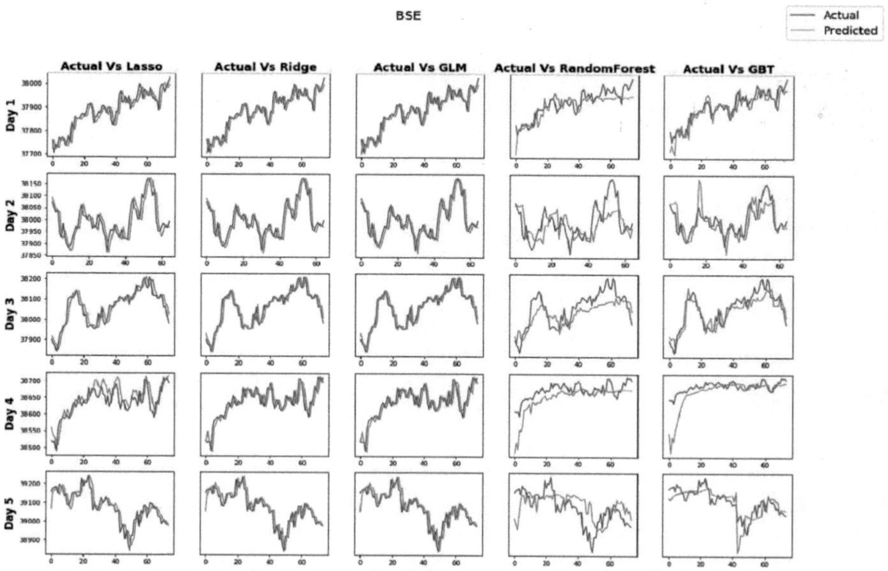

Fig. 3. Bombay Stock Exchange (BSE) Day-Wise forecasts plots.

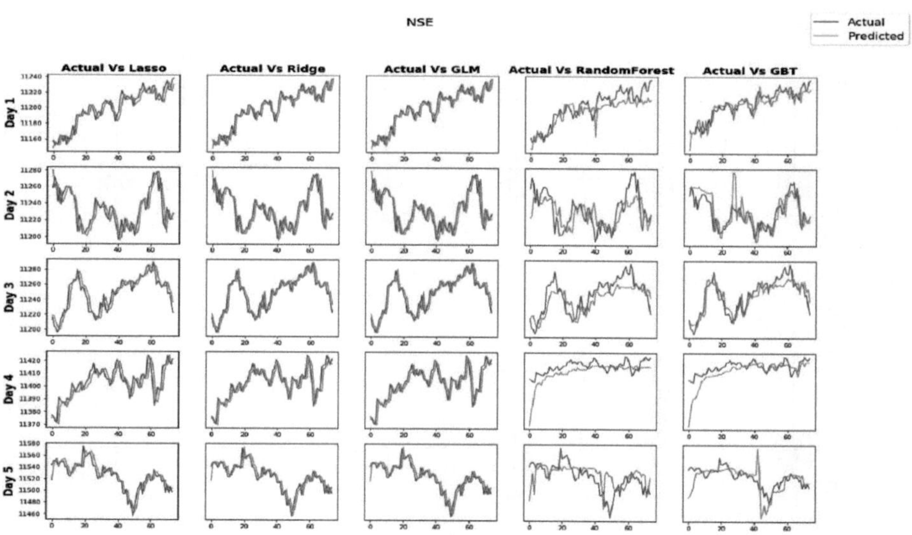

Fig. 4. National Stock Exchange (NSE) Day-Wise forecasts plots.

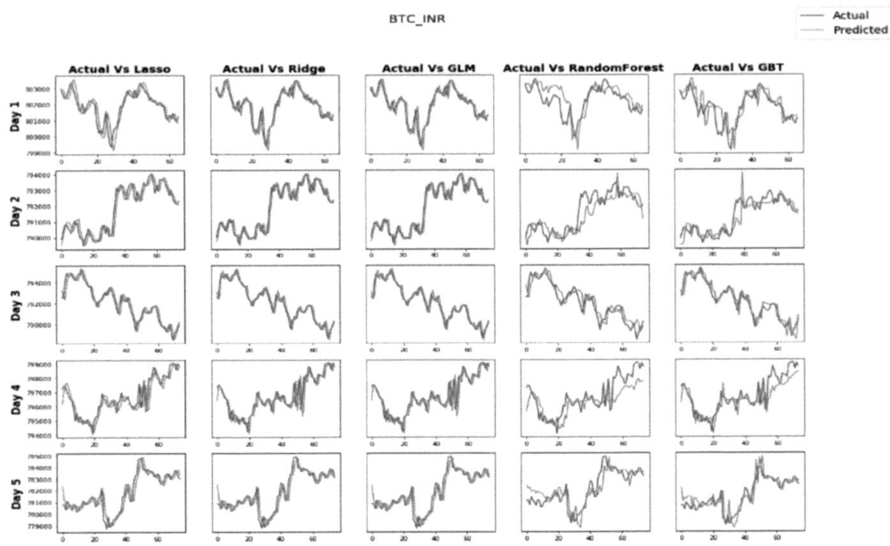

Fig. 5. BTC-INR Day-Wise forecasts plots.

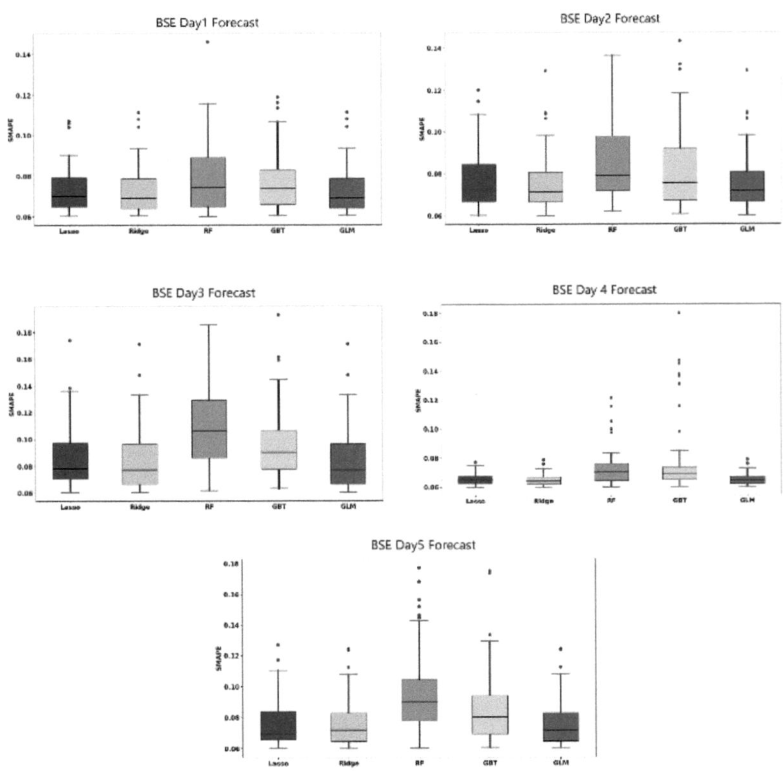

Fig. 6. Boxplot for BSE

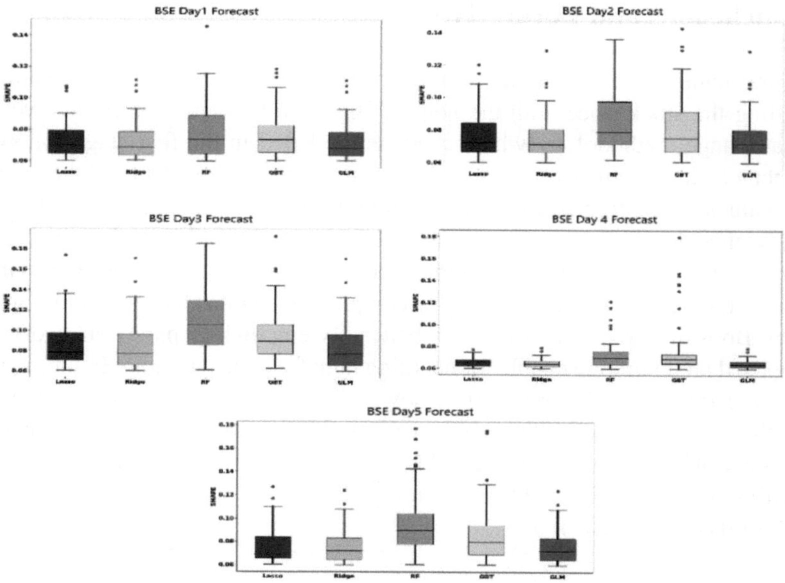

Fig. 7. Boxplot for NSE

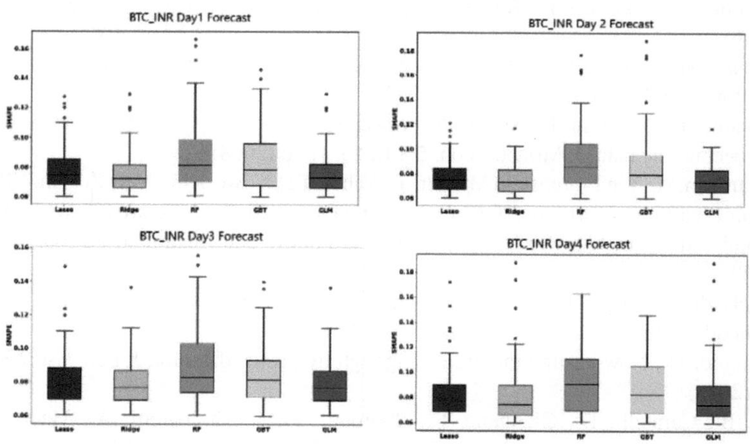

Fig. 8. Boxplot for BTC-INR

8 Conclusions and Future Work

This work mainly focussed on developing an architecture for Real-time streaming and nowcasting the stock price with the help of Big Data Frameworks. In this work proposed a 2-stage methodology where chaos is modelled in the first stage followed by implemented sliding window approach for training on real-time data and then nowcasting the values with a frequency of every 5 min. From the results, it is clear that simple models such as Ridge Regression, Lasso Regression and GLM is performing better on small volume of data better than ensemble methods such as Random Forest and Gradient Boosting Trees. But if the data grows the complex algorithms like Random Forest and Gradient Boosting Trees might perform better. By employing spark framework makes the proposed methodology enable for handling high frequency trading. In future, we can use Deep learning models for real-time nowcasting and also include other features for nowcasting like newsfeeds of the respective companies or markets, traded volumes. The same study can be also implemented on Macroeconomics time series data like Consumer price index (CPI). Further, handling volatility shocks or low liquidity periods is out of scope for this current research.

References

1. Banbura, M., Giannone, D., Reichlin, L.: Now-Casting. ECB Working Paper Series, (1275) (2010)
2. Bańbura, M., Giannone, D., Modugno, M., & Reichlin, L.: Now-casting and the real-time data flow. Handb. Econ. Forecast., **2**, (2013)
3. Giannone, D., Reichlin, L., Small, D.: Nowcasting: The real-time informational content of macroeconomic data. J. Monet. Econ. **55**(4), 665–676 (2008)
4. Richardson, A., van Florenstein Mulder, T., Vehbi, T.: "Nowcasting New Zealand GDP using machine learning algorithms," CAMA Working Papers 2018-47, Centre for Applied Macroeconomic Analysis, Crawford School of Public Policy, The Australian National University (2018)
5. Ilamchezhian, J., Cyril Raj, V.: Spark framework: A review. International Journal of Pure and Applied Mathematics, **117**(20 Special Issue), 861–866 (2017)
6. Modugno, M.: Now-casting inflation using high frequency data. Int. J. Forecast. **29**(4), 664–675 (2013)
7. Bok, B., Caratelli, D., Giannone, D., Sbordone, A.M., Tambalotti, A.: Macroeconomic nowcasting and forecasting with big data. Annu. Rev. Econ. **10**, 615–643 (2018)
8. Loermann, J., Maas, B.: Nowcasting US GDP with artificial neural networks," MPRA Paper 95459. University Library of Munich, Germany (2019)
9. Jayanthi, M.D., Sumathi, G., Sriperumbudur, S.: A Framework for Real-time Streaming Analytics using Machine Learning Approach. CompuSoft, 85–92 ISSN:2320-0790 (2016)
10. Ranjitha, P.: streaming analytics over real-time big data. Glob. J. Comput. Sci. Technol. [S.l.]. ISSN 0975-4172 (July 2015)
11. Attigeri, G.V. Manohara Pai, M.M, Pai, R.M., Nayak, A.: Stock market prediction: A big data approach, TENCON 2015 - 2015 IEEE Region 10 Conference, Macao, pp. 1–5 (2015)
12. Das, S., Behera, R.K., Kumar, M., Rath, S.K. Real-Time Sentiment Analysis of Twitter Streaming data for Stock Prediction. Proc. Comput. Sci., **132**(Iccids), 956–964. 7 (2018).

13. Menon V.K., Chekravarthi Vasireddy N., Jami S.A., Pedamallu V.T.N., Sureshkumar V., Soman K.P.: Bulk Price Forecasting Using Spark over NSE Data Set. In: Tan Y., Shi Y. (eds) Data Mining and Big Data. DMBD 2016. Lecture Notes in Computer Science, vol 9714. Springer, Cham (2016)
14. Umadevi, K.S., Gaonka, A., Kulkarni, R., Kannan, R.J.: Analysis of Stock Market using Streaming Data Framework. 2018 International Conference on Advances in Computing, Communications and Informatics, ICACCI 2018, 1388–1390 (2018)
15. Wu, W., Zhou, J. Z., Yu, J., Zhu, C. J., & Yang, J. J.: Prediction of spot market prices of electricity using chaotic time series. Proceedings of 2004 International Conference on Machine Learning and Cybernetics, 2(August), 888–893poin (2004)
16. Selvin, S., Vinayakumar, R., Gopalakrishnan, E.A., Menon, V.K., Soman, K.P.: Stock price prediction using LSTM, RNN and CNN-sliding window model. 2017 International Conference on Advances in Computing, Communications and Informatics, ICACCI 2017, 2017-Janua, 1643–1647 (2017)
17. Raju, S.M., Tarif, A.M.: Real-Time Prediction of BITCOIN Price using Machine Learning Techniques and Public Sentiment Analysis (2020)
18. Lorenz, E.N.: Deterministic nonperiodic flow. J. Atmos. Sci. **20**, 130–141 (1963)
19. Dhanya, C.T., Kumar, D.N.: Advances in Water Resources Nonlinear ensemble prediction of chaotic daily rainfall. Adv. Water Resour. **33**(3), 327–347 (2010)
20. Packard, N.H., Crutchfield, J.P., Farmer, J.D., Shaw, R.S.: Geometry from a time series. Phys. Rev. Lett. **45**, 712–716 (1980)
21. Rosenstein, M.T., Collins, J.J., De Luca, C.J.: A practical method for calculating largest Lyapunov from small data sets, **65**, 117–134 (1993)
22. Lyapunov, A., Problème général de la stabilité du mouvement, in: Ann. La Fac. Des Sci. Toulouse Mathématiques, pp. 203–474 (1907)
23. Cao, L.: Practical method for determining the minimum embedding dimension of a scalar time series **110**, 43–50 (1997)
24. Tishbirani, R.: Regression shrinkage and selection via the Lasso. J. R. Stat. Soc. Ser. B (Methodological) (1996)
25. Tin Kam Ho.: Random Decision Forests Tin Kam Ho Perceptron training. Proceedings of 3rd International Conference on Document Analysis and Recognition, 278–282 (1995)
26. Friedman, J.H.: Greedy function approximation: A gradient boosting machine. Ann. Stat. **29**(5), 1189–1232 (2001)
27. Nelder, J.A., Wedderburn, R.W.M.: Generalized Linear Models. J. R. Stat. Soc. Ser. A (General), **135**(3), 370–384 (1972)
28. Diebold, F.X., Mariano, R.S.: Comparing predictive accuracy. J. Bus. Econ. Stat. **13**(3), 253–263 (1995)
29. Kohns, D., Bhattacharjee, A.: Nowcasting growth using google trends data: A Bayesian structural time series model. Int. J. Forecast. **39**(3), 1384–1412 (2023)
30. Heikkinen, J., Heimonen, K.: Nowcasting the nowcasting-Forecasting ISM Business surveys (PMI and NSI) with weekly Google trends. Appl. Econ. **56**(51), 6300–6313 (2024)
31. Ferrara, L., Simoni, A.: When are Google data useful to nowcast GDP? An approach via preselection and shrinkage. J. Bus. Econ. Stat. **41**(4), 1188–1202 (2023)

Author Index

S. Dhar et al. (Eds.): AGC 2025, CCIS 2686, p. 285, 2026.
https://doi.org/10.1007/978-3-032-09153-6

MIX
Papier aus verantwortungsvollen Quellen
Paper from responsible sources
FSC® C105338

If you have any concerns about our products,
you can contact us on
ProductSafety@springernature.com

In case Publisher is established outside the EU,
the EU authorized representative is:
Springer Nature Customer Service Center GmbH
Europaplatz 3, 69115 Heidelberg, Germany

Printed by Libri Plureos GmbH
in Hamburg, Germany